STUDY GUIDE

FOR STARR'S
BIOLOGY
Concepts and Applications

JOHN D. JACKSON
North Hennepin Community College

JANE B. TAYLOR
Northern Virginia Community College

Wadsworth Publishing Company
Belmont, California
A Division of Wadsworth, Inc.

Biology Editor: Jack C. Carey
Editorial Assistant: Kathy Shea
Production Editor: Karen Garrison
Managing Designer: Stephen Rapley
Print Buyer: Karen Hunt
Permissions: Jeanne Bosschart
Interior Design: Detta Penna
Compositor: Omegatype Typography, Inc.

Credits: Illustrations from C. Starr and R. Taggart, *Biology,* Fourth and Fifth Editions, Wadsworth, 1987 and 1989: Root tip sketch by Marian Reeve (p. 226); buttercup photo by Chuck Brown (p. 226); cell diagrams after Weier et al., *Botany,* Sixth Edition, Wiley, 1982 (p. 226); brain and digestive system photos from C. Yokochi and J. Rohen, *Photographic Anatomy of the Human Body,* Second Edition, Igaku-Shoin Ltd., 1979 (p. 339 and p. 280). Other illustrations originally from C. Starr and R. Taggart, *Biology,* Fourth and Fifth Editions, Wadsworth, 1987 and 1989 (pp. 43, 68, 71, 82, 119, 192, 195, 202, 207, 224, 227, 243, 268, 271, 291, 297, 314, 323, 343, 353, 362, and 365).

Printed in the United States of America   49

4 5 6 7 8 9 10—94 93 92

ISBN 0-534-13369-X

CONTENTS

1 Introduction   1

Crossword Number One   10

2 Chemical Foundation for Cells   13

3 Cell Structure and Function   30

4 Ground Rules for Metabolism   46

5 Energy-Acquiring Pathways   55

6 Energy-Releasing Pathways   64

Crossword Number Two   75

7 Cell Division and Mitosis   79

8 Meiosis   87

9 Observable Patterns of Inheritance   94

10 Chromosome Variations and Human Genetics   104

11 DNA Structure and Function   115

12 From DNA to Proteins   123

13 Recombinant DNA and Genetic Engineering   135

Crossword Number Three   142

14 Microevolution   146

15 Macroevolution   155

16 Human Evolution: A Case Study   167

Crossword Number Four   173

17 Viruses, Monerans, and Protistans   176

18 Fungi and Plants   186

19 Animals   199

Crossword Number Five   216

20 Plant Tissues   219

21 Plant Nutrition and Transport   230

22 Plant Reproduction and Development   238

Crossword Number Six   252

23 Animal Tissues, Organ Systems, and Homeostasis   255

24 Protection, Support, and Movement   263

25 Digestion and Human Nutrition   274

26 Circulation   287

27 Immunity   301

28 Respiration   311

29 Solute-Water Balance   320

30 Neural Control and the Senses   329

31 Endocrine Control   347

32 Reproduction and Development   357

   Crossword Number Seven   372

33 Population Ecology   375

34 Community Interactions   384

35 Ecosystems   396

36 The Biosphere   406

37 Human Impact on the Biosphere   418

38 Animal Behavior   428

   Crossword Number Eight   442

Answers   445

PREFACE

This study guide is like a tutor; it increases the efficiency of your study periods; it condenses the major ideas of your text; it asks you to do a series of specific tasks to demonstrate your ability to recall key concepts and terms and relate them to life; it tests you on your understanding of the factual material and indicates what you might wish to reexamine or clarify; and it gives you a preliminary estimate of your next test score based on specific material. Most important, though, the study guide and text together help you make informed decisions about matters that affect your own well-being and the well-being of your environment. In the years to come our survival will increasingly depend on administrative and managerial decisions based on an informed biological background.

HOW TO USE THIS STUDY GUIDE

After this preface, you will find an outline that will show you how the study guide is organized and will help you use it efficiently. Each chapter begins with an outline of the topics discussed, to provide an overview of what follows. The content of each text chapter is then broken up into sections, which are labeled 1-I, 1-II, and so on. Each section has many parts. The *Summary* stresses important concepts and indicates the text pages covered; it is followed immediately by a list of *Key Terms*. To help you memorize the terms so that you can improve your grades, flash cards (with a term or name on one side and its definition on the reverse side) are extremely useful and well worth the time and effort. Some sections include a part called *Concept Aid* that provides conceptual assistance, memory aid, or learning shortcuts.

A series of learning *Objectives* follows each *Key Terms* section. These are tasks that you should be able to accomplish if you have understood the assigned reading in the text. There are generally three levels of difficulty: Some objectives require that you memorize the terms and concepts; some require that you understand the material; and others require that you apply your understanding of the terms and concepts to different situations.

So that you can immediately evaluate your mastery of the terms and concepts, each *Objectives* section is followed by *Self-Quiz Questions*, which are answered in the back of the study guide. Any wrong answers will show you which portions of the text you need to reexamine.

The last section of most of the chapters is followed by a *Chapter Test* and *Integrating and Applying Key Concepts*. The tests consist of multiple-choice (and some matching) questions; the answers to these are also in the back of the study guide. *Integrating and Applying Key Concepts* invites you to try your hand at applying the major concepts to situations in which there is not necessarily a single pat answer and so none is provided in the back of the study guide (except for problems in Chapters 9 and 10). Your text generally will provide enough clues to get you started on an answer, but these sections are intended to stimulate your thought and provoke group discussions.

Finally, *Crossword Puzzles* are included following Chapters 1, 6, 13, 16, 19, 22, 32, and 38, as another way to help you match terms with definitions. Solutions to the puzzles are in the *Answers* section in the back of the study guide.

We would like to thank the staff at Wadsworth, especially Jack Carey, Kathy Shea, Alan Noyes, Stephen Rapley, and Karen Garrison, for their help and support.

Structure of the Study Guide

The outline below indicates how each chapter in this study guide is structured.

1

Chapter Title ⟶ **ON THE UNITY AND DIVERSITY OF LIFE**

Chapter Outline ⟶ | **Chapter Outline** |

General Objectives ⟶ A List of Tasks to Be Accomplished

Section Title ⟶ **1-I ORIGINS AND ORGANIZATION**

These categories occur for each section within a chapter
- **Summary**
- **Key Terms**
- **Objectives**
- **Concept Aid (included in some chapters)**
- **Self-Quiz Questions**
  - **True-False**
  - **Matching**
  - **Fill-in-the-Blanks**

Chapter Tests ⟶ **UNDERSTANDING AND INTERPRETING KEY CONCEPTS**

INTEGRATING AND APPLYING KEY CONCEPTS

1
INTRODUCTION

SHARED CHARACTERISTICS OF LIFE
    DNA and Biological Organization
    Interdependency Among Organisms
    Metabolism
    Homeostasis
    Reproduction
    Mutation and Adapting to Change

LIFE'S DIVERSITY
    Five Kingdoms, Millions of Species
    An Evolutionary View of Diversity
THE NATURE OF BIOLOGICAL INQUIRY
    On Scientific Methods
    Commentary: Testing the Hypothesis
    Through Experiments
    Limitations on Science
SUMMARY

General Objectives

1. List features that distinguish living organisms from dead organisms. Then distinguish a dead organism from a rock.
2. Describe the general pattern of energy flow through Earth's life forms, and explain how Earth's resources are used again and again (cycled).
3. Explain what is meant by the term *diversity*, and speculate about what caused the great diversity of life forms on Earth.
4. Understand the theories that biologists have formulated to explain how life might have changed through time.
5. List the steps of the scientific approach to understanding and solving a problem.

1-I
(pp. 3–7)

SHARED CHARACTERISTICS OF LIFE
    DNA and Biological Organization
    Interdependency Among Organisms
    Metabolism
    Homeostasis
    Reproduction
    Mutation and Adapting to Change

Summary

At a very basic level, nonliving and living things are composed of the same fundamental subatomic particles that are organized into the energetic interactions of atoms and molecules. Organisms are constructed of one or more cells. Living beings tend to have specific patterns of chemical organization, and as part of the web of life, possess particular but interdependent means of obtaining and systematically using energy and materials. Almost all existing forms of life depend directly or indirectly on one another for materials and energy. There is a one-way flow of energy from the sun, through *producer* organisms, and on through *consumers* and *decomposers*. Organisms can generally adjust to many long-term and short-term environmental changes. Living forms can reproduce themselves and are committed to characteristic, inherited programs of growth, development, and reproduction; they have special molecules of inheritance called *deoxyribonucleic acid*, or DNA. DNA molecules are like blueprints for constructing new organisms from "lifeless" molecules. DNA molecules are capable of changes called *mutations*. Most genetic changes are not adaptive for a species; a few mutations provide benefits and so lend adaptive potential to organisms in their specific environments. Natural selection and other evolutionary forces have brought about a great diversity of organisms.

Key Terms

NOTE: The following underscored terms are especially important and are bold-faced in the text. Being able to define additional terms in the list greatly increases understanding of the written material.

life	multicelled organism	decomposers
adaptive	population	budget
energy	species	metabolism
subatomic particle	community	ATP
atom	ecosystem	homeostasis
molecule	biosphere	dynamic homeostasis
organelle	energy	reproduction
tissue	photosynthesis	inheritance
organ	aerobic respiration	DNA
organ system	producers	mutation
cell	consumers	trait

Objectives

After reading the section and thinking about the contents, you should be able to:

1. Compare the basic structures of (a) a frog and (b) a rock.
2. List some specific functions or examples of behavior carried out by a frog and a rock.
3. Define and contrast *energy* and *materials*. Explain how each is related to a living organism and to work.
4. Arrange in order, from smallest to largest, the levels of organization that occur in nature. Define each as you list it.
5. Explain how the basic processes of energy capture and energy release create interdependencies among organisms.
6. List and define basic characteristics of life ascribed to all organisms.

Self-Quiz Questions

True-False

If false, explain why.

____ (1) Cells and rocks are composed of the same fundamental chemical particles and types of molecules.

—— (2) Converting light into heat is an example of energy transformation.

—— (3) To stay alive, an organism must obtain energy from someplace else.

—— (4) Green plants absorb sugar, water, and minerals from the soil. The sugar absorbed in this manner is used to do cellular work.

—— (5) Trapping light energy in the chemical bonds of sugar is an example of an energy transformation.

—— (6) Photosynthesis and respiration involve energy transfers.

—— (7) Instructions that result in particular developmental and maintenance patterns being followed are encoded in the ATP molecule.

—— (8) Mutations are never advantageous to the organism in which they occur.

—— (9) Sweating is part of a homeostatic control system that keeps body temperature more or less constant in mammals.

—— (10) The instructions for production of each insect developmental stage exist before eggs are produced.

Matching

Choose the most appropriate answer to match with each of the following terms.

(11) __F__ atom

(12) __C__ cell

(13) __E__ community

(14) __G__ ecosystem

(15) __D__ molecule

(16) __B__ organelle

(17) __H__ population

(18) __A__ subatomic particle

(19) __I__ tissue

A. A proton, neutron, or electron
B. A well-defined structure within a cell, performing a particular function.
C. The smallest unit of life
D. Two or more atoms bonded together
E. All of the populations interacting in a given area
F. The smallest unit of a pure substance that has the properties of that substance
G. A community interacting with its nonliving environment
H. A group of individuals of the same species in a particular place at a particular time
I. A group of cells that work together to carry out a particular function

Fill-in-the-Blanks

(20) _____ is the chemical process whereby some organisms are able to convert sunlight energy to energy-rich molecules. Organisms release energy from energy-rich molecules for cellular work through a chemical process known as (21) _____ . (22) _____ is the capacity for acquiring and using energy for stockpiling, tearing down, building up, and eliminating materials in controlled ways. The four stages in the moth life cycle, in sequence, are (23) _____ , (24) _____ , (25) _____ , and (26) _____ . (27) _____ is the capacity to maintain internal conditions within some tolerable range, even when external conditions vary. (28) _____ are changes that occur in the kind, structure, or number of DNA's component parts.

LIFE'S DIVERSITY
    Five Kingdoms, Millions of Species
    An Evolutionary View of Diversity

Summary

The array of different organisms on Earth is the total of variations that have proved adaptive over time in obtaining available resources, in tolerating diseases and toxic substances, and in escaping from predators and parasites. Although Earth's organisms share the fundamental characteristics of life, different life forms present a bewildering array of variations. To organize this vast diversity, a classification system was devised that assigned a two-part (genus + species) name to each recognizably distinct kind of organism. This two-name system forms the basis for a concise hierarchical classification system, which allows placement of life forms in successively more inclusive groupings. The broadest groups are the five kingdoms of life.

Key Terms

unity	division	Animalia
diversity	phylum	evolution
species	kingdom	Darwin
genus	Monera	artificial selection
family	Protista	natural selection
order	Fungi	differential reproduction
class	Plantae	

Objectives

1. Explain how scientists can, at the same time, discuss the unity and diversity of life.
2. Explain the relationship of the *generic* name to the *specific* name.
3. Characterize the kingdoms in the five-kingdom system and distinguish each from the other four kingdoms. To do this, you may have to consult p. 7 and Figure 1.5.
4. Arrange, in order from greatest to fewest number of organisms included, the following categories of classification: class, family, genus, kingdom, order, phylum, and species.
5. Briefly explain how Darwin's explanation of the mechanisms of change helps account for the "diversity" of life.

Self-Quiz Questions

Fill-in-the-Blanks

The proposal that characteristics of organisms change through natural selection is primarily linked with the name of (1) _____ . Multicelled consumers are classified in the Kingdom (2) _____ . Single-celled producers or consumers of considerable internal complexity are placed in the Kingdom (3) _____ . Organisms that are, for the most part, multicelled decomposers that digest their food externally and then absorb it belong to the Kingdom (4) _____ . Kingdom (5) _____ holds relatively simple, single cells that can be producers or decomposers. Organisms that are multicelled producers are classified in the Kingdom (6) _____ .

Sequence

Arrange in correct hierarchical order with the largest, most inclusive category first and the smallest, most exclusive category last:

(7) _D_   A. Class
(8) _F_   B. Family
(9) _A._   C. Genus
(10) _E_   D. Kingdom
(11) _B_   E. Order
(12) _C_   F. Phylum
(13) _G_   G. Species

True-False

If false, explain why.

___ (14) The most inclusive (largest) taxonomic category is the phylum.

___ (15) There is a larger number of different species in a class than in an order.

___ (16) If some organisms in a population inherit traits that lend them a survival advantage, they would be less likely to produce offspring.

___ (17) Darwin described the selection occurring in pigeons as natural selection.

___ (18) All bacteria are single-celled and are therefore protistans.

1-III
(pp. 9–13)

THE NATURE OF BIOLOGICAL INQUIRY
   On Scientific Methods
   Commentary: **Testing the Hypothesis Through Experiments**
   Limitations on Science
SUMMARY

Summary

The *scientific method* of approaching questions is a commitment to systematic observation and testing. Various tools and experimental designs are used to record observations, test hypotheses, and draw conclusions. Most experiments are compared with a control group, and only one variable at a time is tested quantitatively. Discipline, objectivity, suspended judgment, testing, and repeatability of experimental results are the bases of experimental principles. Occasionally, accidents and intuition contribute to the success of the experimental process.

Key Terms

principle	experimental group	sampling error
hypothesis, -ses	control group	significant
induction	independent variable	theory
deduction	dependent variable	subjective
test, testable	controlled variable	supernatural
systematic observations	randomization	

Objectives

1. Define the term *principle* as applied by scientists.
2. Outline the principal steps generally used in the scientific method of investigating a problem.
3. Explain how observations differ from conclusions and how a *hypothesis* differs from a *theory*.
4. Explain why one or more control groups are used in an experiment.
5. Explain how the methods of science differ from answering questions by subjective thinking and systems of belief.

Self-Quiz Questions

True-False

If false, explain why.

___ (1) Making systematic observations means observing all information surrounding a problem, even if it has no bearing on the problem a scientist is involved with.

___ (2) Sampling errors are introduced into an experiment when a test group is not equivalent to a natural population.

___ (3) A control group helps to establish how far the value of a variable deviates from standard values.

___ (4) When any test group is not equivalent to a natural population, the test is said to be randomized.

Sequence

Arrange the following steps of the scientific method in correct chronological sequence from first to last:

(5) _b_   A. Carry out the tests. Repeat as often as necessary to find out whether or not results are consistent with predictions.

(6) _D_   B. Use trained judgment in selecting and summarizing the relevant preliminary observations from what could be nearly infinite observational trivia.

(7) _F_   C. Report objectively on the results of the tests and the conclusions drawn from them.

(8) _E_   D. Review all available preliminary observations. Be sure to note the range of conditions under which they have been made.

(9) _A_   E. Devise ways to test whether the explanation is valid. Think through how different but related conditions might affect the outcome. Be sure the test you devise addresses these conditions.

(10) _C_   F. Work out a hypothesis that seems in line with the observations.

Labeling

Assume that you have to determine what object is inside a sealed, opaque box. Your only tools to test the contents are a bar magnet and a triple-beam balance. Label each of the following with an O (for observation) or a C (for conclusion).

(11) _C_   The object has two flat surfaces.

(12) _O_   The object is composed of nonmagnetic metal.

(13) _C_   The object is not a quarter, a half-dollar, or a silver dollar.

(14) _O_   The object weighs x grams.

(15) _C_   The object is a penny.

CHAPTER TEST   **UNDERSTANDING AND INTERPRETING KEY CONCEPTS**

____ (1) About 12 to 24 hours after the last meal, a person's blood-sugar level normally varies from about 60 to 90 milligrams per 100 milliliters of blood, though it may attain 130 mg/100 ml after meals high in carbohydrates. That the blood sugar level is maintained within a fairly narrow range despite uneven intake of sugar is due to the body's ability to carry out _____ .

    (a) prediction
    (b) inheritance
    (c) metabolism
    (d) homeostasis

____ (2) As an eel migrates from seawater to freshwater, the salt concentration in its environment decreases from as much as 35 parts of salt per 1,000 parts of seawater to less than 1 part of salt per 1,000 parts of freshwater. The eel stays in the freshwater environment for many weeks because of its body's ability to carry out _____ .

    (a) adaptation
    (b) inheritance
    (c) puberty
    (d) homeostasis

____ (3) A boy is color-blind just as his grandfather was, even though his mother had normal vision. This situation is the result of _____ .

    (a) adaptation
    (b) inheritance
    (c) metabolism
    (d) homeostasis

____ (4) The digestion of food, the production of ATP by respiration, the construction of the body's proteins, cellular reproduction by cell division, and the contraction of a muscle are all part of _____ .

    (a) adaptation
    (b) inheritance
    (c) metabolism
    (d) homeostasis

____ (5) Which of the following does _not_ involve using energy to do work?

    (a) Atoms being bound together to form molecules
    (b) The division of one cell into two cells
    (c) The digestion of food
    (d) None of these

___ (6) To eliminate from consideration the influence of uncontrolled variables during experimentation, one should _____ .

   (a) increase the sampling error as much as possible and use suspended judgment
   (b) establish a control group identical to the experimental group except for the variable being tested
   (c) use inductive reasoning to construct a hypothesis
   (d) make sure the experiments are repeatable

___ (7) A hypothesis should *not* be accepted as valid if _____ .

   (a) the sample studied is determined to be representative of the entire group
   (b) a variety of different tools and experimental designs yield similar observations and results
   (c) other investigators can obtain similar results when they conduct the experiment under similar conditions
   (d) several different experiments, each without a control group, systematically eliminate each of the variables except one

___ (8) The principle point of evolution by natural selection is that _____ .

   (a) long-term heritable changes in organisms are caused by use and disuse
   (b) those mutations that adapt an organism to a given environment somehow always arise in the greatest frequency in the organisms that occupy that environment
   (c) mutations are caused by all sorts of environmental influences
   (d) survival of characteristics in a population depends on competition among organisms, especially among members of the same species

___ (9) A control group _____ .

   (a) is not subjected to experimental errors
   (b) is exposed to experimental treatments
   (c) is maintained under strict laboratory conditions
   (d) is treated exactly the same as the experimental group
   (e) varies from the experimental group only in the factor being tested by the experiment

___ (10) Statistical tests are designed to determine if differences between experimental and control groups are _____ .

   (a) valid
   (b) significant
   (c) qualitatively different
   (d) null
   (e) quantitatively different

___ (11) The least inclusive of the taxonomic categories listed is _____ .

   (a) family
   (b) phylum
   (c) class
   (d) order
   (e) genus

INTEGRATING AND APPLYING KEY CONCEPTS

(1) Humans have the ability to maintain body temperature very close to 37° C.
   (a) What conditions would tend to make the body temperature drop?
   (b) What measures do you think your body takes to raise body temperature when it drops?
   (c) What conditions would cause body temperature to rise?
   (d) What measures do you think your body takes to lower body temperature when it rises?

(2) Suppose you're an ecologist working in an African game preserve that includes elephants. You want to discover the precise migratory habits of the elephants so that they can encounter humans less frequently and thus have better chances to survive.
   (a) What sources would you go to in order to review all available observations?
   (b) Which hypotheses might you try to test?
   (c) What tools might you use in testing your hypotheses?
   (d) Name three variables that might affect the migratory movements of elephants.
   (e) How could you isolate each of these variables and study the effects of just one variable at a time so as to deduce its contribution to elephant migratory behavior?

(3) Do you think that all humans on Earth today should be grouped in the same species?

(4) What sorts of topics are usually regarded by scientists as untestable by the scientific method?

Crossword
Number One

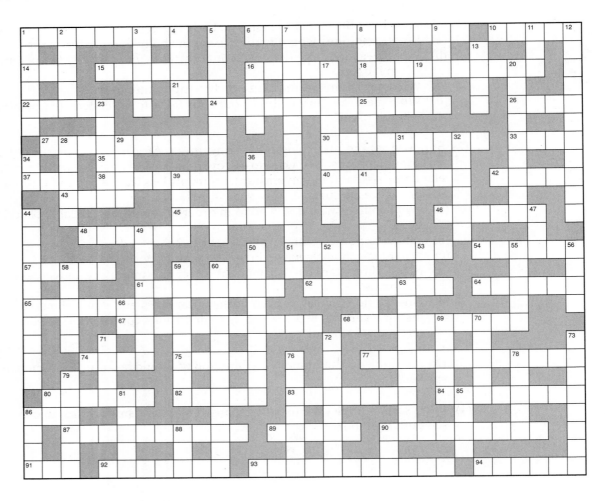

Across

1. a biological community interacting with its nonliving environment
6. unexplainable by ordinary laws
10. a group of closely related families
14. a unit of energy or work done by a force acting over a distance of one centimeter
15. a group of closely related orders
16. bunch or cluster
18. all of the members of a species living in a particular place defined in time
21. organ of light reception
22. a group of closely related species; the Latin word for "kind" or "sort"
24. creating a situation that has no specific pattern or objective
26. a desert in Asia
27. all of the physical and chemical processes involved in the maintenance of life
30. a group of different interacting populations defined in time and space
33. super_____ , exploding star
35. as
37. a dark, oily, thick mixture—mostly hydrocarbons—produced by distilling wood, coal, or peat

38. not requiring the help of others
40. the kingdom that includes single-celled organisms with membrane-bound organelles
42. Latin combining word that means "flat," as in _____ spiral
43. freedom from difficulty, hard work, or effort
45. kingdom of mostly green, multicelled photosynthesizers
46. a scheme that depicts expenditures and income of a resource, such as time, money, or energy
48. a group of similar cells working together to carry out a particular function
51. reasoning from a specific phenomenon to a more general set of circumstances
54. a coherent set of ideas that forms a general frame of reference for further studies in the field of inquiry
57. a coastal city in west-central Israel
61. subatomic _____ ; very small bit
62. that which is likely to change
64. proposed and described the theory of evolution by natural selection
65. used as a basis of comparison in experimental studies
67. important; not negligible
68. an organism that breaks down the remains of dead organisms
74. _____ Hsiao Ping, a Chinese leader of the 1980s
75. combining form that refers to the presence of iron; Latin
77. the process by which parental cells or organisms create offspring cells or organisms
80. a group of closely related classes of organisms
82. a protein-digesting enzyme released by the kidney to raise blood pressure
83. to sign the back of a check
84. a living individual
85. recline; also, to tell an untruth
87. orderly; proceeding according to a plan
89. the cardinal is a member of the _____ family, Fringillidae
90. the process of reasoning from the general to the specific
91. Greek goddess of the dawn
92. not artificial or man-made
93. the act of noting a phenomenon
94. the kingdom that includes bacteria

Down

1. the capacity to do work
2. a group of different tissues working together to carry out a specific function
3. characteristic
4. the lawgiver who led the Israelites out of Egypt
5. provisional, tentative, empirical; involving testing
7. the process that produces sugar (and other organic compounds) and oxygen from carbon dioxide, water, sunlight, and other helper molecules
8. the energy currency of living organisms
9. a tool that punches holes in leather and textiles
11. a color ranging from neutral brownish gray to dull grayish brown
12. the metabolic process by which an organism uses oxygen to break apart organic compounds, and yields carbon dioxide and other products of oxidation
13. the smallest amount of an element that still shows the physical and chemical properties of that element
16. departed; not here or there
17. a basic truth, law, or assumption

19. unknown spaceship
20. discrete structure that has a specific function in a cell
23. homeo-_____ ; the ability to return to a normal state
25. a mechanical device that carries out a variety of banking tasks; abbrev.
28. long-legged wading bird
29. part of a skeleton
31. state of being one
32. one sips a caffeinated beverage from this
34. in a place; located
36. experiment with; try
39. leader of the Roman Catholic Church
41. free from bias and prejudice
44. influenced by one's own cultural or emotional background
47. Lao-Tzu (6th century B.C. Chinese philosopher) founded a religion based on the _____ (= way)
49. selecting a portion regarded as representative of the whole
50. choice
52. the genetic molecule
53. wise bird
55. mistake
56. the male active cosmic principle
58. multicelled organisms that absorb food they have digested outside their own bodies
59. to take from one place (or thing) and give to another place (or thing)
60. not proceeding at the same rate
63. all of the regions on Earth that support self-sustaining and self-regulating ecological systems
66. bone or mouth in Latin
69. city in Utah
70. hurt by piercing by flying insect
71. the smallest unit of a living organism that is capable of independent function
72. organism that can synthesize its own food from appropriate inorganic substances
73. kingdom that includes multicelled organisms dependent on other organisms for their food
76. the category of classification of organisms that is least inclusive
78. a Latin American ballroom dance; means "I touch" in Latin
79. a dissertation that advances an original point of view
80. 3.14; π
81. the Latin word for "beyond"; surpassing a specified limit or range
85. a cereal grass, *Oryza sativa*, cultivated extensively in warm climates
86. not death
88. earth, _____ , fire, and water
90. point; a tiny, round spot

2
CHEMICAL FOUNDATIONS FOR CELLS

ORGANIZATION OF MATTER
  The Structure of Atoms
  Isotopes
  Commentary: Dating Fossils, Tracking
  Chemicals, and Saving Lives—Some Uses
  of Radioisotopes
BONDS BETWEEN ATOMS
  The Nature of Chemical Bonds
  Ionic Bonding
  Covalent Bonding
  Hydrogen Bonding
  Properties of Water
ACIDS, BASES, AND SALTS
  Acids and Bases

  The pH Scale
  Dissolved Salts
  Buffers
CARBON COMPOUNDS
  Families of Small Organic Compounds
  Functional Groups
  Condensation and Hydrolysis
  Carbohydrates
  Lipids
  Proteins
  Nucleotides and Nucleic Acids
SUMMARY

General Objectives

1. Understand how protons, electrons, and neutrons are arranged into atoms and ions.
2. Explain how the distribution of electrons in an atom or ion determines the number and kinds of chemical bonds that can be formed.
3. Know the various types of chemical bonds and the circumstances under which each forms.
4. Understand the essential chemistry of water and of some common substances dissolved in it.
5. Define *acids, bases, buffers,* and *salts;* relate these terms to the pH scale.
6. Understand why the chemistry of life is based on the element carbon.
7. Be familiar with the concept of *functional groups* and cite common examples listed in the text.
8. Understand how small organic molecules can be assembled into large biological molecules by condensation; understand how large biological molecules can be broken apart into their basic subunits by hydrolysis.
9. Know the general structure of a monosaccharide with six carbon atoms, glycerol, a fatty acid, an amino acid, and a nucleotide.
10. Know the main biological molecules into which the essential building blocks (cited in Objective 8) can be assembled by condensation.
11. Know where the main biological molecules tend to be located in cells or organelles and the activities in which they participate.

ORGANIZATION OF MATTER
The Structure of Atoms
Isotopes
Commentary: **Dating Fossils, Tracking Chemicals, and Saving Lives—Some Uses of Radioisotopes**

Summary

All forms of matter on Earth are composed of one or more *elements:* hydrogen, carbon, and silver are examples of elements. When two or more elements are combined in a fixed ratio, a compound is produced (one atom of sodium combined with one atom of chlorine produces the compound called table salt). There are about ninety-two naturally occurring elements on Earth. *Atoms* are the smallest units of elements that still retain the properties of elements. *Molecules* are groups of two or more atoms bonded together; the atoms may be of the same or different elements.

All events in the living world depend on the organization and behavior of atoms and molecules. All energy transfers and transformations within and between living things occur at the level of atoms and molecules. The shapes and behavior of cells are determined by molecules contained within them.

The structure of the atoms in a given element makes them different from atoms of other elements. The number and arrangement of their subatomic particles (protons, neutrons, and electrons) dictate the behavior of the atoms of a specific element. Uncharged neutrons in the nucleus, together with the protons (positive charges), account for the *mass number* of an element. The number of protons alone dictates the *atomic number* of the element and helps, indirectly, to establish how that atom reacts with other atoms. How the electrons (negative charges) are arranged outside the nucleus determines the ways in which the atom reacts with other atoms and the kinds of ions the atom forms. Electron activity between atoms is responsible for the flow of materials and energy in the world of life. The number of positively charged protons in an *atom* always equals the number of negatively charged electrons; thus, atoms have no net charge.

All atoms of an element have the same number of protons but can vary slightly in the number of neutrons they have. The variant forms are called *isotopes.* Carbon 12, carbon 13, and carbon 14 are examples. Unstable isotopes are radioactive and decay to a more stable form; energy given off in the form of particle emissions by *radioactive isotopes* can be detected by electronic devices. Thus they are useful as "tracers" of biochemical activity in organisms as well as in determining the age of fossil-bearing rocks.

Key Terms

matter	neutron	atomic number
elements	electron	mass number
compound	proton	isotopes
molecule	electric charge	radioactive isotopes
atom	nucleus	(tracers)

Objectives

1. Indicate the relationship between *atoms* and *elements* by comparing the definition of each term.
2. Compare the definitions of *atom* and *molecule.*
3. Define *electric charge* and state how (a) two identically charged particles react to each other and (b) two oppositely charged particles react to each other.
4. List and describe the three types of subatomic particles in terms of their location (inside or outside the nucleus) and charge.

5. Define and distinguish *atomic number* and *mass number*.
6. Show how an atom of carbon differs from atoms of oxygen, hydrogen, and nitrogen. (See Table 2.1.)
7. Explain what *isotopes* of an element are and how differences in the number of neutrons and in the mass numbers distinguish each isotope.
8. Name the four most abundant elements in the human body. (See Table 2.1 and Figure 2.2 of the main text.)

Self-Quiz Questions

True-False

If false, explain why.

___ (1) Conversion of light energy into the chemical energy of energy-rich sugars, as occurs in plants, is an energy transformation.

___ (2) The number of protons in the nucleus of an atom is equal to the number of electrons outside the nucleus.

___ (3) The number of protons in a nucleus helps, indirectly, to determine how that atom reacts with other atoms.

___ (4) If two particles bear the same electric charge, they are attracted to each other.

___ (5) Carbon 12 and carbon 14 are identical except for the number of electrons in each atom.

Matching

Choose the one most appropriate answer for each; not every letter may be matched with a number.

(6) ___ atom
(7) ___ atomic number
(8) ___ electric charge
(9) ___ electron
(10) ___ element
(11) ___ energy
(12) ___ isotope
(13) ___ matter
(14) ___ molecule
(15) ___ neutron
(16) ___ proton
(17) ___ radioactive

A. An uncharged subatomic particle
B. The number of protons in the nucleus of one atom of an element
C. The smallest neutral unit of an element that shows the chemical and physical properties of that element
D. That which occupies space and has mass
E. Two or more atoms linked together by one or more chemical bonds
F. Contributes the capacity to do work—that is, to move a certain amount of material across a specific distance
G. A positively charged subatomic particle
H. A term applied to unstable isotopes
I. The sum of protons + neutrons
J. A negatively charged subatomic particle
K. Ninety-two different types occur in nature
L. A form of an element, the atoms of which contain a different number of neutrons from other forms of the same element
M. Pushes away similar particles but attracts oppositely charged particles

BONDS BETWEEN ATOMS
   The Nature of Chemical Bonds
   Ionic Bonding
   Covalent Bonding
   Hydrogen Bonding
   Properties of Water

Summary

Atomic symbols are used to write formulas of chemical compounds. Chemical reactions among atoms and molecules are shown in balanced chemical equations using symbols and formulas. Chemical bonds occur when atoms give up, gain, or share electrons(s) with one or more other atoms. An *orbital* is a specific region of space outside the nucleus in a particular *energy level* (or shell), where an electron is most likely to be located. An orbital can contain no more than two electrons. Negatively charged *electrons,* arranged in orbitals outside the nucleus, tend to get as close as possible to the positively charged *protons* in the nucleus and stay as far away from each other as possible. As soon as the inner orbitals are filled with two electrons, orbitals of the next highest energy level are the next to fill. Atoms that have one or more orbitals containing only one electron tend to react chemically with other such atoms.

Electrons occupying the orbitals in the highest occupied (outermost) energy level of an atom are the ones that interact with other atoms to form either a molecule or some other group of associated atoms. Electrons in orbitals closest to the nucleus are at their lowest energy levels; electrons have higher energy levels if they are located in orbitals further from the nucleus. In a chemical bond, one or more orbitals of one atom become linked with one or more orbitals of another atom (or atoms), establishing an energy relationship between the participating atoms that holds the atoms at certain distances from each other and causes the group of atoms to assume a specific three-dimensional shape.

An electron can absorb certain amounts of incoming energy and as a result spends more of its time farther out from the pull of the nucleus. If the outside energy source is removed, the excited electron eventually gives off the extra energy it has absorbed and returns to whichever orbital closest to the nucleus can accommodate it. Such electron energy, for example, drives the chemistry of photosynthesis and visual perception.

In an *ionic bond*, a positive and a negative ion are linked by the mutual attraction of opposite charges. A *covalent bond* is formed when two different atoms share a pair of electrons. If the pair is shared equally, the covalent bond is *nonpolar;* if shared unequally, the bond is *polar.* Water is a molecule with polar covalent bonds. Covalent bonds may be double (two pairs of shared electrons) or triple (three pairs of shared electrons). A single covalent bond (a pair of shared electrons) is often represented by a single dash between the bonded atoms, for example, H-H or H_2. A double covalent bond (two pairs of shared electrons) is represented by two dashed lines between the bonded atoms, for example, $O = O$ or O_2.

In a *hydrogen bond*, an electronegative atom (such as oxygen) weakly attracts a hydrogen atom that is already participating in a polar covalent bond. Hydrogen bonds may form between two molecules or help stabilize the internal structure of many large biological molecules. Water owes some of its properties to hydrogen bonds formed by attractions of hydrogen of one water molecule to the oxygen of another. Most of the molecules involved in life's chemistry are held together by strong covalent bonds that involve a sharing of electron pairs and by weaker interactions (ionic bonds and hydrogen bonds) between atoms or ions having opposite charges.

Water molecules have negatively and positively charged ends. Certain substances dissolve in water because they can become dispersed among and form weak bonds with the polar water molecules. Such substances are *hydrophilic*. Substances that cannot form weak bonds with water molecules are *hydrophobic*. Hydrophobic molecules (such as oil) tend to cluster in a watery environment. Water dissolves ionic substances and most polar molecules. None of life's activities can proceed without water because: (1) water's cohesive nature resists penetration and allows whole columns of water to be moved about, (2) water stabilizes the cellular temperature so that chemical reactions can proceed in an orderly fashion; evaporation of water lowers the temperature of its surface; water resists freezing, and (3) solvent properties of water dissolve ionic substances and most polar molecules. The polar nature of water encourages *spheres of hydration* to be formed around ions, preventing ions from reacting with each other and forcing them to remain dispersed in the cytoplasm rather than concentrating in some part of the cell. Thus, water molecules help keep vital ions available for chemical reactions throughout the cell. Large molecules may have charged regions that tend to attract either the negative or the positive ends of water molecules (that is, to be hydrophilic) and other regions that bear no net charge (in other words, are hydrophobic).

The properties of water profoundly influence the organization and behavior of substances that make up the cellular environment; the chemistry of life proceeds in water.

Key Terms

chemical bond	electron excitation	hydrophobic
orbitals	ion	cohesion
energy levels	ionic bond	temperature stabilization
formula	covalent bond	evaporation
chemical equations	nonpolar covalent bond	solvent
reactant	polar covalent bond	solute
product	hydrogen bond	dissolved
law of conservation of mass	hydrophilic	spheres of hydration

Objectives

1. Understand what is meant by a *chemical bond*.
2. Explain what is meant by *atomic symbol* and *chemical formula* (see Figure 2.3).
3. Define *orbital* and indicate how electrons are related to orbitals.
4. Describe the relationships among energy levels, orbitals, and electrons. Explain what happens when an orbital is filled with two electrons.
5. Describe the conditions of atomic structure that enable atoms to become reactive.
6. Explain what happens to the behavior of one or more of an atom's electrons (a) when the atom absorbs energy and (b) when the atom releases energy.
7. Describe covalent, ionic, and hydrogen bonds fully, so as to distinguish each from the others.
8. Contrast *nonpolar* and *polar covalent* bonds.
9. State one way that living organisms use hydrogen bonding.
10. Distinguish between *hydrophilic* and *hydrophobic* substances.
11. Explain the role that hydrogen-bond formation plays in some of the physical and chemical properties of water.
12. Explain why an oxygen atom typically forms two bonds, why a nitrogen atom typically forms three (or five) bonds, and why a carbon atom typically forms four (occasionally two) bonds.
13. Give two specific examples of organisms that depend directly on the cohesive properties of water and describe the nature of that dependence. Speculate about what those organisms would do if water were not cohesive.
14. Explain what *to dissolve* means.
15. Explain why the fact that water is an outstanding solvent is of such importance to living organisms. State how you think life would be limited if water dissolved only a few different substances.

True-False

If false, explain why.

___ (1) In an atom, the volume of space that can accommodate (at most) two electrons is called an *energy level*.

___ (2) When an atom absorbs energy, electrons that are part of that atom move faster and tend to spend more of their time farther from the nucleus than they did before the energy was absorbed.

___ (3) For an atom to react with another atom, all orbitals of that atom must already contain two electrons.

___ (4) A chemical bond is formed when two atoms share a pair of electrons and is also formed by atoms losing or gaining one or more electrons.

___ (5) Covalent bonds typically occur within molecules; weaker bonds usually form between two or more different molecules but can also form within the same molecule.

___ (6) The atoms in table salt (sodium chloride) are linked together with covalent bonds.

___ (7) In order to form a single covalent bond, a pair of electrons is shared between the two nuclei.

___ (8) Hydrophilic substances generally dissolve in water.

___ (9) A polar covalent bond is formed when two atoms share a pair of electrons equally.

___ (10) Water molecules are polar, that is, the oxygen end becomes positive and the hydrogen end becomes negative.

___ (11) A solute is any fluid in which one or more substances can be dissolved.

___ (12) "Spheres of hydration" form around ions and prevent them from chemical interaction.

2-III
(pp. 25-27)

ACIDS, BASES, AND SALTS
    Acids and Bases
    The pH Scale
    Dissolved Salts
    Buffers

Summary

A substance whose molecules release hydrogen ions in solution is known as an *acid*. Any substance that combines with hydrogen ions in solution is a *base*. The greater the hydrogen ion concentration, the lower the *pH value* on the *pH scale*. pH values are a shorthand measure of the degree to which a solution is acidic, basic, or neutral. All biochemical reactions that usually take place in water are sensitive to changes in pH. It is fortunate that the pH of pure water is 7, or neutral, because that contributes to the stability of biochemical processes in cells. Each kind of cell is adapted to a particular pH range, usually not far from neutrality.

Inside cells, *buffer molecules* help maintain pH values within rather narrow tolerance ranges. Buffer molecules combine with or release hydrogen ions in response to changes in cellular pH. Cell functioning also depends on essential ions such as K^+ and Ca^{++}.

Salts are ionic compounds formed when acids react with bases. Salts can dissolve to form ions that are very important in many aspects of cellular chemistry.

Key Terms

hydrogen ion, H^+	base, basic	buffers
hydroxide ion, OH^-	neutral	dissolved salts
acid, acidic	pH scale	salt

Objectives

1. Explain what is meant by hydrogen ion and hydroxide ion.
2. Distinguish between an acid and a base; give three examples of acids and three examples of bases. (See Figure 2.12 in the main text.)
3. Describe the pH scale, specifying the acidic range, the basic range, and the point of neutrality.
4. Explain why a pH of 2 is 100 times more acidic than a pH of 4.
5. Explain why the pH of substances is important to living organisms.
6. Define *buffer* and give an example of how this type of substance functions in cells.
7. Define the term *salt* and state how a salt is related to ions; state how a salt may be related to acids and bases.
8. List two essential ions and give a cellular function of each.

Self-Quiz Questions

True-False

If false, explain why.

___ (1) Some substances dissolve in water and release hydrogen ions, OH^-.

___ (2) A pH scale is used to express the concentration of hydrogen ions in solutions.

___ (3) A pH of 4 is 10 times more acid than a pH of 3.

___ (4) Most living cells maintain a pH of about 4.

___ (5) Hydrochloric acid serves as a major buffer in cells.

Fill-in-the-Blanks

When acids dissolve in water, they release (6) _____ ions; when bases dissolve in water, they release (7) _____ ions. The (8) _____ _____ is used to express the (9) _____ _____ concentration of solutions. Most living cells maintain an H^+ concentration of pH (10) _____ . A pH of 3 has an H^+ concentration 10 times higher than pH (11) _____ . (12) _____ are molecules that combine with or release hydrogen ions to prevent rapid shifts in pH. A(n) (13) _____ is formed when an acid reacts with a base.

Matching

Choose the most appropriate answer for each.

(14) ___ acid

(15) ___ base

(16) ___ essential ions

(17) ___ hydrogen ion

(18) ___ hydroxide ion

(19) ___ neutral

(20) ___ salt

A. OH^-
B. Ca^{++}, K^+
C. H^+
D. NaCl
E. Combines with hydrogen ions in solution
F. Releases hydrogen ions in solution
G. 7

2-IV
(pp. 27–28)

CARBON COMPOUNDS
    Families of Small Organic Compounds
    Functional Groups
    Condensation and Hydrolysis

Summary

Oxygen, hydrogen, and carbon compose ninety-three percent of your body's weight; water accounts for much of the first two elements, but carbon is the most important structural element in the body because a carbon atom can form as many as four covalent bonds with carbon as well as other elements. Chains or rings of carbon atoms form the main structure of strandlike, globular, or sheet-like molecules in cells, some of which contain millions of atoms.

    Four families of small organic (= carbon-containing) compounds—simple sugars, fatty acids, amino acids, and nucleotides—serve as energy sources and as building blocks for the large molecules present in cells. *Functional groups* confer special properties to organic molecules. Large molecules such as polysaccharides, fats, proteins, and nucleic acids are assembled by *condensation*, which is the covalent linkage of small molecules. The condensation process is guided by specific *enzymes* and may also involve forming water as a product. Large molecules such as starch or proteins generally are split apart into two or more parts by reaction with water; this process of *hydrolysis*, too, is guided by specific enzymes.

Key Terms

carbon
organic compounds
inorganic compounds
hydrocarbons
functional groups
methyl group, $-CH_3$ $\left(-\overset{\displaystyle H}{\underset{\displaystyle H}{C}}-H \right)$
ethyl group, $-C_2CH_3$
aldehyde group, $-CHO$ $\left(-\overset{O}{\overset{\|}{C}}-H \right)$
ketone group $\left(-\overset{O}{\overset{\|}{C}}-\overset{}{C}-C \right)$

hydroxyl group, $-OH$ $(-OH)$
carboxyl group, $-COOH$ $\left(-\overset{O}{\overset{\|}{C}}-OH \right)$
amino group, $-N$; $-NH$; $-NH2$ $\left(-\overset{\displaystyle H}{\underset{}{N}}-H \right)$
phosphate group, $-PO_4(p_i)$ $\left(-\overset{O}{\overset{\|}{\underset{\displaystyle O^-}{P}}}-O^- \right)$
enzymes
condensation
hydrolysis

Objectives

1. Explain why carbon atoms are part of so many substances in living organisms.
2. Distinguish between the terms *organic* and *inorganic* compounds.
3. Illustrate, by drawing their carbon skeletons, a linear chain, a branched chain, and a ring structure.
4. List the four main families of organic compounds usually found dissolved in cellular fluid.
5. Define *functional group* and become familiar with some examples that identify the common large molecules involved in life's chemistry.
6. Understand how two subunits can be joined together through condensation by removing a water molecule. (See Figure 2.16 in the main text.)
7. State the role of enzymes in chemical reactions.
8. Explain why hydrolysis is said to be "condensation in reverse."

Self-Quiz Questions

True-False

If false, explain why.

___ (1) Carbon-based molecules are known as inorganic compounds.

___ (2) Units of smaller molecules are assembled by means of hydrolysis into larger molecules made of repeating subunits.

___ (3) Protein molecules that determine the structure and behavior of organic molecules are known as amino acids.

___ (4) All organic molecules contain carbon.

___ (5) Inorganic compounds have carbon chains or rings.

___ (6) In a branched chain, each carbon atom is never bonded to more than one other carbon atom. (Think!)

2-V
(pp. 28–29)

CARBON COMPOUNDS (cont.)
Carbohydrates

Summary

Carbohydrates form some of the main structural materials (including cellulose) and foods (such as sucrose, glucose, starch, and glycogen) that serve as reservoirs of energy. *Monosaccharides* are simple sugars that are the basic subunits of the larger carbohydrates, the polysaccharides. *Disaccharides* such as sucrose are composed of two covalently bonded simple sugars. *Polysaccharides* consist of more than two simple sugar units (the same or different) covalently bonded to form long, sometimes branching chains. Starch, for example, is a polysaccharide that may be digested by hydrolysis into simple sugars (such as glucose). Many simple sugars can be linked by condensation to form cellulose of cell walls, various plant starches, glycogen, or other polysaccharides.

Key Terms

carbohydrate	glucose	galactose
sugar	fructose	polysaccharide

monosaccharides	disaccharide	starch
simple sugar	sucrose	cellulose
ribose	lactose	glycogen
deoxyribose		

Objectives

1. Define *carbohydrate*. State what the basic molecular subunits of carbohydrates are.
2. Distinguish among *monosaccharides*, *disaccharides*, and *polysaccharides* and name as many of each as you can.
3. Illustrate how a carbohydrate is assembled from its basic subunits.
4. Indicate how monosaccharides, disaccharides, and polysaccharides are used in living organisms. Explain how the structure of each allows organisms to use that form of carbohydrate for a specific purpose.

Self-Quiz Questions

True-False

If false, explain why.

___ (1) Glucose and lactose are monosaccharides.

___ (2) Ribose and glucose are the building blocks of nucleic acids.

___ (3) Peptide bonds link monosaccharides into polysaccharides called *starches*.

___ (4) In living organisms, carbohydrates serve as structural supports and as food reserves.

___ (5) Polysaccharides can be broken down into simple sugars if the appropriate enzymes and water are present.

___ (6) Condensation is the process that assembles simple sugars into starches.

___ (7) Carbohydrates contain oxygen; hydrocarbons do not.

Matching

Match *all* applicable letters with the appropriate terms. A blank may contain more than one letter.

(8) ___ deoxyribose

(9) ___ cellulose

(10) ___ starch

(11) ___ fructose

(12) ___ glucose

(13) ___ glycogen

(14) ___ lactose

(15) ___ ribose

(16) ___ sucrose

A. Monosaccharide
B. Disaccharide
C. Polysaccharide
D. Used as a structural support
E. Used as a food reserve
F. Table sugar
G. Milk sugar
H. A building block of nucleic acids
I. A six-carbon sugar

CARBON COMPOUNDS (cont.)
  Lipids

Summary

Lipids are oily or waxy substances such as true fats, waxes, steroids (such as cholesterol and estrogen), and phospholipids. True fats are compact forms of stored energy. Waxes help to waterproof and protect surfaces from potentially damaging agents. Phospholipids and glycolipids are important features of cell membranes.

Lipids with fatty acid (a long, unbranched hydrocarbon with a —COOH group at the end) components include *glycerides* (for example, fat), *phospholipids*, and *waxes* (for example, beeswax and cutin, a plant waterproofing material). Lipids without fatty acid components include steroids (for example, testosterone, estrogen, other hormones, and cholesterol).

Key Terms

lipids	triglyceride	waxes
nonpolar solvents	fats	cutin
fatty acid	oils	steroids
glyceride	saturated	cholesterol
monoglyceride	unsaturated	hormones
diglyceride	phospholipid	

Objectives

1. List the major characteristics of lipids.
2. Understand what is meant by "lipids with fatty acids" and "lipids with no fatty acids."
3. Know the parts of a glyceride molecule.
4. Distinguish saturated fats from unsaturated fats; give examples of each.
5. State briefly how the following classes of lipids function in living organisms: glycerides; phospholipids; waxes; steroids.

Self-Quiz Questions

True-False

If false, explain why.

___ (1) Lipids contain oxygen, but carbohydrates lack oxygen.

___ (2) A triglyceride is broken down by hydrolysis into three fatty acid molecules and one molecule of glycerol.

___ (3) Saturated fats contain the maximum possible number of hydrogen atoms that can be covalently bonded to the carbon skeleton of their fatty acid tails.

___ (4) An unsaturated fat is usually an oil at room temperature and generally contains many double bonds in the fatty acid component.

___ (5) Waxes are also composed of long-chain fatty acids. They help organisms prevent water loss and predation.

___ (6) Some of the male and female sex hormones belong to the group of lipids known as *steroids*.

Matching

Choose *all* the appropriate answers for each.

(7) ___ cholesterol

(8) ___ cutin

(9) ___ phospholipid

(10) ___ saturated fat

(11) ___ unsaturated fat

A. Basic fabric for all membranes
B. Butter and bacon
C. Vegetable oil
D. Wax
E. Steroid

2-VII
(pp. 31–33)

CARBON COMPOUNDS (cont.)
 Proteins

Summary

Some proteins, such as enzymes, speed chemical reactions. Other proteins transport substances or cause cells (or parts of cells) to move. Yet other proteins constitute structural elements in bone and cartilage, transmit chemical information, or protect vertebrates from disease agents. Each protein may be broken down into its own particular assemblage of some twenty different *amino acids*. One amino acid is covalently bonded to the next in a chain by a *peptide bond* (formed by condensation). Three or more amino acids bonded in a chain by condensation form a *polypeptide chain*. The chemical nature, size, shape, and ordering of side groups projecting from amino acids dictate how a protein interacts chemically with other substances and, hence, what role the protein plays in a living system. The ultimate structure of a protein can be investigated on three (or four) levels: primary, secondary, tertiary, and quaternary. Disruption of the weak bonds on which the secondary, tertiary, and quaternary structures are based causes *denaturation*, a dramatic and irreversible loss of the all-important three-dimensional structure of the protein. Such a denatured protein becomes nonfunctional.

Key Terms

protein
amino acid
R group
peptide bond
polypeptide chain

primary structure
secondary structure
tertiary structure
quaternary structure

helical
hemoglobin
albumin
denaturation

Objectives

1. Define *protein* and identify the basic building blocks of proteins.
2. Describe the structure that all amino acids have in common and the structures that make each of the twenty amino acids different.
3. Explain how a peptide bond is formed.
4. Indicate what functions various proteins serve in a living organism.
5. Cite the names of two proteins mentioned in the text.
6. Illustrate how the primary, secondary, tertiary, and quaternary structures of a protein are formed; apply this to hemoglobin as an example (see p. 33 of the main text).
7. Define *denaturation* and explain how it occurs.
8. Explain what function enzymes serve in the chemical reactions occurring in cells.

True-False

If false, explain why.

___ (1) Amino acids are linked by hydrolysis, a process that splits molecules of water as the amino acid subunits are linked together.

___ (2) Bone and cartilage are constructed, in part, of specific proteins.

___ (3) R groups are identical on the different amino acids.

___ (4) The primary structure of a protein is formed principally by hydrogen bonds linking various amino acids.

___ (5) An amino group contains a nitrogen atom and two hydrogen atoms; a carboxyl group contains two oxygen atoms, a carbon atom, and a hydrogen atom.

___ (6) Enzymes are an important class of proteins whose subunits are simple sugars.

2-VIII
(pp. 33–35)

CARBON COMPOUNDS (cont.)
Nucleotides and Nucleic Acids
SUMMARY

Summary

Nucleotides are rather small organic molecules that serve chemical roles as nucleotide-based molecules and as the building blocks of much larger molecules, the *nucleic acids*. Each nucleotide contains at least a five-carbon sugar, a nitrogen-containing base, and a phosphate group. Nucleotide-based molecules serve as chemical messengers between cells (cAMP), as energy carriers (ATP), as transporters of hydrogen ions and electrons in metabolic reactions (NAD^+ and FAD). Covalently bonded chains of nucleotides form larger molecules that encode genetic instructions (DNA) or help translate these instructions (RNA) into the proteins of structure and function upon which all forms of life are based.

Key Terms

nucleotide	purine	NAD^+
five-carbon sugar	adenosine phosphates	FAD
ribose	cAMP	nucleic acids
deoxyribose	ATP	DNA
nitrogen base	nucleotide coenzymes	RNA
pyrimidine		

Objectives

1. Describe a *nucleotide* by discussing its components and the way the components are linked together.
2. List the three principal nucleotide-based molecules, give two examples of each, and tell how each functions in living organisms.
3. Define *nucleic acid* and describe its basic structure.
4. Explain the general role of DNA and RNA in cellular chemistry.

True-False

If false, explain why.

___ (1) ATP acts as an energy carrier.

___ (2) Nucleotide coenzymes transport the amino acids necessary in metabolism.

___ (3) Adenosine phosphates act as chemical messengers between cells and as energy carriers.

___ (4) Nucleic acids are long chains of covalently bonded nucleotides, with nitrogenous bases connecting the phosphates and with sugars sticking out to the side.

___ (5) DNA consists of two parallel nucleic acid strands twisted around each other and cross-linked between bases by hydrogen bonds.

___ (6) DNA contains the code for constructing one or more proteins.

___ (7) RNA molecules function in the processes in which DNA codes are used for constructing proteins.

Matching

Choose *all* the appropriate answers for each.

(8) ___ adenosine phosphates

(9) ___ nucleotide coenzymes

(10) ___ nucleic acid

A. Long; single-stranded or double-stranded
B. ATP
C. Transport hydrogen ions and their associated electrons
D. cAMP, a chemical messenger
E. NAD^+ and FAD
F. RNA and DNA

CHAPTER TEST

UNDERSTANDING AND INTERPRETING KEY CONCEPTS

___ (1) A molecule is _____ .

(a) a combination of two or more atoms
(b) less stable than its constituent atoms separated
(c) electrically charged
(d) a carrier of one or more extra neutrons

___ (2) A hydrogen bond is _____ .

(a) a sharing of a pair of electrons between a hydrogen and an oxygen nucleus
(b) a sharing of a pair of electrons between a hydrogen nucleus and either an oxygen or a nitrogen nucleus
(c) an attractive force that involves a hydrogen atom and an oxygen or a nitrogen atom that are either in two different molecules or within the same molecule
(d) none of the above

— (3) A mixture of sugar and water is an example of a(n) _____ .

    (a) compound
    (b) solution
    (c) suspension
    (d) colloid
    (e) ion

— (4) Radioactive isotopes have _____ .

    (a) excess electrons
    (b) excess protons
    (c) excess neutrons
    (d) insufficient neutrons
    (e) insufficient protons

— (5) The shapes of large molecules are controlled by _____ bonds.

    (a) hydrogen
    (b) ionic
    (c) covalent
    (d) inert
    (e) single

— (6) A pH solution of 10 is _____ times as basic as a pH of 7.

    (a) 2
    (b) 3
    (c) 10
    (d) 100
    (e) 1,000

— (7) Substances that are nonpolar and repelled by water are _____ .

    (a) hydrolyzed
    (b) nonpolar
    (c) hydrophilic
    (d) hydrophobic

— (8) Carbon is part of so many different substances because _____ .

    (a) carbon generally forms two bonds with a variety of other atoms
    (b) carbon generally forms four bonds with a variety of atoms
    (c) carbon ionizes easily
    (d) carbon is a polar compound

— (9) Proteins _____ .

    (a) include all hormones
    (b) are composed of nucleotide subunits
    (c) are not very diverse in structure and function
    (d) include all enzymes

— (10) Most of the chemical reactions in cells must have _____ present before they proceed.

    (a) RNA
    (b) salt
    (c) enzymes
    (d) fats

___ (11) Hydrolysis could be correctly described as the _____ .

    (a) heating of a compound in order to drive off its excess water and concentrate its volume

    (b) breaking of a long-chain compound into its subunits by adding water molecules (in the presence of the proper enzymes) to its structure between the subunits

    (c) linking of two or more molecules by the removal of one or more water molecules

    (d) constant removal of hydrogen atoms from the surface of a carbohydrate

___ (12) DNA _____ .

    (a) is one of the adenosine phosphates

    (b) is one of the nucleotide coenzymes

    (c) contains protein-building instructions

    (d) translates protein-building instructions into actual protein structures

___ (13) Amino acids are linked by _____ bonds to form the primary structure of a protein.

    (a) disulfide

    (b) hydrogen

    (c) ionic

    (d) peptide

___ (14) Lipids _____ .

    (a) serve as food reserves in many organisms

    (b) include cartilage and chitin

    (c) include fats that are broken down into one fatty acid molecule and three glycerol molecules

    (d) are composed of monosaccharides

INTEGRATING AND APPLYING KEY CONCEPTS

(1) Explain what would happen if water were a nonpolar molecule instead of a polar molecule. Would water be a good solvent for the same kinds of substances? Would the nonpolar molecule's specific heat likely be higher or lower than that of water? Surface tension? Cohesive nature? Ability to form hydrogen bonds? Is it likely that the nonpolar molecules could form unbroken columns of liquid? What implications would that hold for trees?

(2) Humans can obtain energy from many different food sources. Do you think this ability is an advantage or a disadvantage in terms of long-term survival? Why?

(3) If the ways that atoms bond influence molecular shapes, then do the ways that molecules behave toward one another influence the shapes of cellular organelles? Do the ways that organelles behave toward one another influence the structure and function of the cells?

(4) Proteins have the most diverse shapes and the most complex structure of all molecules. Why do you suppose that DNA, rather than proteins, is the code molecule used to construct new proteins?

3
CELL STRUCTURE AND FUNCTION

THE NATURE OF CELLS
    Basic Cell Features
    Cell Size and Microscopy
CELL MEMBRANES
    Membrane Structure and Function
    Diffusion
    Osmosis
    Available Routes Across Membranes
    Endocytosis and Exocytosis
PROKARYOTIC CELLS—THE BACTERIA

EUKARYOTIC CELLS
    Function of Organelles
    Typical Components of Eukaryotic Cells
    The Nucleus
    Cytomembrane System
    Mitochondria
    Specialized Plant Organelles
    The Cytoskeleton
    Cell Surface Specializations
SUMMARY

General Objectives

1. Appreciate the extremely small but highly ordered nature of cells and be able to discuss how such tiny objects are seen.
2. Understand the essential structure and functions of the cell membrane.
3. Know the forces that cause water and solutes to move across membranes passively (that is, without expending energy).
4. Understand which types of substances move by simple diffusion. Understand the importance of osmosis to all cells.
5. Know the mechanisms by which substances are moved across membranes against a concentration gradient.
6. Understand how material can be imported into or exported from a cell by being wrapped in membranes.
7. Contrast the general features of prokaryotic and eukaryotic cells.
8. Describe the organelles associated with the cytomembrane system, and tell the general function of each.
9. Describe the nucleus of eukaryotes with respect to structure and function.
10. Contrast the structure and function of mitochondria and chloroplasts.
11. Describe the cytoskeleton of eukaryotes and distinguish it from the cytomembrane system.
12. List several surface structures of cells and tell how they help cells survive.

3-I
(pp. 37–39)

THE NATURE OF CELLS
    Basic Cell Features
    Cell Size and Microscopy

Summary

The *cell* is the basic unit of life; most cells are so small they must be viewed with magnification. Such tiny cells are highly organized chemical systems we call "life." This chemical complexity must be maintained as the cell exchanges nutrients, water, and wastes with an often disorganized, harsh, external environment. The three basic cellular regions are the *plasma membrane*, the *nucleus* (lacking in bacteria), and the *cytoplasm*. Pathways of chemical communication link the plasma membrane, cytoplasm, and nuclear regions.

Scientists know that all organisms are composed of one or more cells, that the cell is the basic living unit of organization for all organisms, and that all cells arise from preexisting cells. Collectively, these three hypotheses are known as the *cell theory*. Crude systems of lenses developed early in the seventeenth century by Galileo, Hooke, and van Leeuwenhoek allowed observation of cells as small as bacteria and sperm. Since the 1820s, specific technological advances in lens design have enlarged our understanding of cellular structure and behavior. Various kinds of light and electron microscopes have their own uses. Minute cell structures are measured with very small metric units; a micrometer is $1/1,000,000$ (one-millionth) of one meter and a nanometer is $1/1,000,000,000$ (one-billionth) of one meter. Low-resolution (resolution is the ability of the microscope to discern between two closely adjacent objects) light microscopes are used to observe the cells of living organisms, whereas high-resolution electron microscopes form images of nonliving structures as small as 0.2 nanometers. Scanning electron microscopes have lower resolving power but remarkable depth of field; extremely minute surface features of cells and organisms can be examined.

Key Terms

Galileo	cytoplasm	compound light microscope
Hooke	nucleus	
van Leeuwenhoek	cell theory	resolution
Schwann	micrometers	transmission electron microscope
Schleiden	nanometers	
Virchow	micrograph	scanning electron microscope
plasma membrane		

Objectives

1. Understand that most cells are too small to be seen with the unaided eye.
2. Consider that cells are extremely complex and must maintain a highly ordered internal chemical environment separate from an often harsh external environment.
3. List the three basic cellular regions and describe their general functions.
4. Briefly describe the contributions made by each of the following scientists to the modern understanding of cell biology: Galileo, Hooke, and van Leeuwenhoek.
5. List the basic ideas of the cell theory stated by Schleiden, Schwann, and Virchow.
6. a. Describe how each of these microscopes works: conventional compound light microscope, transmission electron microscope, and scanning electron microscope.
   b. Tell whether living cells can be viewed by each type of microscope and state the limitations of each type.
7. Consult Appendix I (main text) and determine how you would explain the range of cell sizes to a person unfamiliar with the metric system.

Self-Quiz Questions

True-False

If false, explain why.

___ (1) The cell is the smallest independent living unit.

___ (2) The average cell is approximately as large as your thumbnail.

___ (3) A plasma membrane permits some substances to cross it and prevents other substances from crossing.

Matching

Choose the one best answer for each.

(4) ___ Galileo

(5) ___ Hooke

(6) ___ Schleiden and Schwann

(7) ___ van Leeuwenhoek

(8) ___ Virchow

A. Originated the term *cell*

B. Determined that all cells must come from preexisting cells

C. The first person to observe a great diversity of microscopic organisms

D. Said that all organisms are constructed of cells

E. The first person to record any biological observations under the microscope

Fill-in-the-Blanks

A (9) _____ is a photograph of an image formed with a microscope. If you wish to observe a living cell with a compound microscope, it must be small or thin enough for (10) _____ to pass through. (11) _____ is the property that dictates whether small objects close together can be seen as separate things. A(n) (12) _____ is one-billionth of a meter. With a(n) (13) _____ electron microscope, a beam of electrons is transmitted through a prepared, thinly sliced section of a cell or organism. With a(n) (14) _____ electron microscope, a narrow beam of electrons is played back and forth across a specimen's surface, which has been coated with a thin metal layer. The (15) _____ is a membrane-bound zone of hereditary control.

3-II
(pp. 40–42)

CELL MEMBRANES
    Membrane Structure and Function
    Diffusion
    Osmosis

Summary

The current model of cell membrane structure is the *fluid mosaic model*. Cell membranes are composed of lipids (predominantly phospholipids) and proteins. Membrane lipids have hydrophilic heads and hydrophobic tails; when surrounded by water they assemble spontaneously into a lipid bilayer. All heads are at the two outer faces of the lipid bilayer, and all tails are sandwiched between them. Lipid molecules move about within a bilayer, thereby contributing to membrane fluidity. This behavior maintains the integrity of the plasma membrane and cell organelles. The lipid bilayer provides the basic *structure* of all cell membranes and acts as a hydrophobic barrier between the fluid regions. Membrane functions are carried out largely by proteins associated with the bilayer. The combination of lipids and proteins in the membrane give it the "mosaic" quality. Some proteins act as open or gated channels and transport water-soluble substances across the membrane. Some pump substances across the bilayer while others transfer electrons. Some proteins are enzymes. Others are receptors for chemical signals from hormones or other substances that initiate changes in metabolism or cell behavior.

Maintaining internal concentrations of water and solutes depends on *passive transport mechanisms* such as diffusion and osmosis, which do not alter in any way the direction in which a substance is moving on its own. No direct energy outlay by the cell is required for these processes.

Diffusion is a random movement of like molecules down a concentration gradient from their region of greater concentration to a region of lesser concentration. It accounts for the greatest volume of substances that are moved into and out of cells, and it is also an important transport process within cells. *Osmosis* is the passive movement of *water* across a differentially permeable membrane in response to solute concentration gradients and/or pressure gradients. Osmotic movements across cell membranes are influenced by the concentrations of dissolved substances inside and outside of the membrane being considered. Red blood cells shrivel when placed in a *hypertonic* solution but swell and burst when immersed in a *hypotonic* solution. If solute concentrations are equal in the fluids inside and outside of the membrane, they are said to be *isotonic* and there is no water movement in either direction. *Turgor pressure* is the internal cell pressure on plant cell walls, which results from water moving into the cell by osmosis.

Key Terms

cell membrane	pump, or transport protein	differentially permeable
phospholipid	electron-transfer protein	osmosis
hydrophilic head	receptor proteins	solute
hydrophobic tail	concentration gradient	isotonic
lipid bilayer	simple diffusion	hypotonic
fluid mosaic model	pressure gradient	hypertonic
membrane proteins	temperature gradient	turgor pressure
channel protein	electric charge	plasmolysis
gate, gated channel		

Objectives

1. Describe the structure of a cell membrane as portrayed by the fluid mosaic model.
2. Explain how the formation of lipid bilayers occurs when phospholipid molecules are immersed in water.
3. List three functions carried out by cell membranes.
4. Explain what would occur if receptor proteins on membrane surfaces suddenly were to vanish from all of your cells.
5. Define *diffusion* and explain what causes diffusion of any sort.
6. Explain what is meant by *concentration gradient* and explain how such gradients are formed.
7. List three factors that influence the rate of diffusion and state what sorts of substances diffuse readily across plasma membranes.
8. Define *osmosis* and distinguish that process from *diffusion*.
9. Understand the effects of osmosis on living cells.

10. Distinguish between *permeable* and *differentially permeable.*
11. Describe the physical condition of a living cell in the following environments: *hypotonic solution, hypertonic solution,* and *isotonic solution.*
12. Define *turgor pressure* as the concept relates to plant cells.

Self-Quiz Questions

Fill-in-the-Blanks

The current model of membrane structure is the (1) _____ _____ model. A(n) (2) _____ _____ serves as a sort of fluid **sea** matrix in which diverse (3) _____ are suspended like icebergs. Together, the two components form the (4) "_____." (5) _____ have fatty acid tails (which repel water) and (6) _____ heads that are hydrophilic. (7) _____ _____ carry out most membrane functions. Some membrane proteins are open (8) _____ for passage of (9) _____ _____ substances. When activated, some membrane receptors bring about changes in cell (10) _____ or behavior. A(n) (11) _____ can exist between two regions that differ in (12) _____ , pressure, temperature, or net electric charge. Diffusion is driven by the (13) _____ inherent in all individual molecules as they move from a region of (14) _____ concentration to a region of (15) _____ concentration. A cell membrane is (16) _____ _____ ; some molecules travel rapidly across the membrane, others cross it more slowly, and some are kept from crossing it at all. Red blood cells shrivel and shrink when placed in a(n) (17) _____ [choose one] hypotonic ( ) isotonic ( ) hypertonic ( ) solution.

True-False

If false, explain why.

___ (18) In a plasma membrane, the hydrophilic tails point inward, tail to tail, and form a region that excludes water.

___ (19) The tails of phospholipids are hydrophobic.

___ (20) Diffusion accounts for the greatest volume of substances that are moved into and out of cells.

___ (21) Osmosis occurs in response to a concentration gradient that involves unequal concentrations of water molecules.

3-III
(pp. 42–44)

CELL MEMBRANES (cont.)
   Available Routes Across Membranes
   Endocytosis and Exocytosis

Summary

Simple diffusion moves small, electrically neutral molecules like H_2O, CO_2, and O_2 across the cell membrane. Large, electrically neutral molecules like glucose or small ions rarely cross the lipid bilayer. *Passive transport mechanisms* move solutes through proteins without expenditure of cell energy. In *facilitated diffusion*, proteins embedded in the plasma membrane assist passively in the passage of small molecules across the membrane in the direction simple diffusion would take them. Maintaining suitable internal concentrations of water and solutes depends on *active transport mechanisms* such as active transport, exocytosis, and endocytosis, all of which work to move a substance in a direction contrary to its spontaneous direction of movement. These mechanisms cannot operate without direct energy outlays by the cell. When ions and molecules required by a cell are scarce, the cell must expend energy in order to stockpile these nutrients by getting them to move against their concentration gradient. In *active transport,* proteins serve as carriers or as fixed channels for moving ions or molecules across the membrane. The sodium-potassium pump is an example of such an active transport system. Cells that must move relatively large amounts of solids or fluids across the plasma membrane enclose the substances to be transported in membrane-bound compartments known as *vesicles* and move them through the membrane to the opposite side. *Endocytosis* and *exocytosis* move substances in vesicles into and out of the cell, respectively.

Key Terms

membrane transport proteins	facilitated diffusion	vesicles
passive transport	active transport	exocytosis
	sodium-potassium pump	endocytosis

Objectives

1. Summarize the available routes across membranes.
2. Distinguish *active transport* from *passive transport;* cite an example of each.
3. Explain how a cell obtains the ions and molecules it requires if they are scarce.
4. Explain how a cell excretes ions and molecules it does not need when there are many more of them outside the cell than inside.
5. Describe the form of energy the cell expends in accomplishing active transport.
6. Distinguish *exocytosis* from *endocytosis;* try to cite an example of each.
7. State the purpose of an active transport system like the sodium-potassium pump.
8. Relate the role of *vesicles* in exocytosis and endocytosis.

Self-Quiz Questions

True-False

If false, explain why.

___ (1) Active transport depends on specialized proteins that transport substances across the membrane.

___ (2) The sodium-potassium pump is an example of a passive transport mechanism.

___ (3) Movement of cellular materials by exocytosis and endocytosis involves vesicles.

___ (4) The secretion of mucus is most likely accomplished by exocytosis.

___ (5) Facilitated diffusion is a type of diffusion that requires the cell to expend ATP molecules.

Fill-in-the-Blanks

(6) _____ , (7) _____ , (8) _____ _____ , a few other simple molecules, and some (9) _____ diffuse readily across plasma membranes.

(10) _____ _____ cannot operate without direct energy outlays by the cell.

3-IV
(pp. 45–46)

PROKARYOTIC CELLS—THE BACTERIA
EUKARYOTIC CELLS
   Function of Organelles
   Typical Components of Eukaryotic Cells

Summary

Bacteria are the smallest cells and structurally simple. Lacking a nucleus, they are said to be *prokaryotic* cells. Every prokaryotic cell has a plasma membrane, cytoplasm, an irregular area of cytoplasm containing DNA, and ribosomes; most also produce a cell wall and have a few internal membranes. Prokaryotic DNA does not coil into chromosomes.

   All cells except bacteria are *eukaryotic*. In eukaryotic cells, a conspicuous, true nucleus replaces the irregular area of prokaryotic cytoplasm containing DNA, and additional membrane-bounded organelles are distributed throughout the cytoplasm. Organelles separate often incompatible chemical reactions in space and time. Principal organelles include endoplasmic reticulum, Golgi bodies, lysosomes, transport vesicles, mitochondria, and ribosomes. Cells possess an internal network of protein filaments, called the *cytoskeleton*. Chloroplasts are found in photosynthetic cells. Plant and fungal cells often have central vacuoles. Unlike animals, many protistans, true fungi, and land plant cells have a cell wall surrounding the plasma membrane. Cell structure is extremely variable; there is no distinctly "typical" cell.

Key Terms

cell wall	Golgi bodies	cell wall
ribosomes	lysosomes	chloroplast
prokaryotic	transport vesicles	central vacuole
eukaryotic	mitochondria	protein filaments
organelles	cytoskeleton	plasma membrane
endoplasmic reticulum		

Objectives

1. Look at Figures 3.13 and 3.14 (main text) and describe how plant cells differ from animal cells. Then look at Figure 3.12 (main text) and describe how a generalized prokaryotic cell differs from plant and animal cells.
2. Explain how the term *prokaryote* is related to the Kingdom Monera and how the term *eukaryote* is related to the other four kingdoms.
3. Become familiar with the typical components of eukaryotic cells and their general functions.
4. List the types of organisms that produce cell walls and central vacuoles.

Matching

Select the single best answer. A letter can be used more than once.

(1) —— endoplasmic reticulum   A.   Photosynthesis occurs here

(2) —— Golgi complex

B.   Digestion and disposal

C.   Energy extraction

(3) —— lysosomes

D.   Material synthesis, modification, and distribution

(4) —— nucleus

E.   Hereditary instructions for synthesis and cell operation

(5) —— mitochondria

(6) —— chloroplast

F.   Links the nuclear envelope with the plasma membrane

Fill-in-the-Blanks

(7) _____ have a membrane-bound nucleus and other membrane-bound organelles; (8) _____ lack such structures. (9) _____ cells do not produce walls, although some secrete products to the surface layer of tissues in which they are formed. In most bacteria, a rather rigid (10) _____ _____ surrounds the cytoplasm.

3-V
(pp. 46–51)

PROKARYOTIC CELLS—THE BACTERIA (cont.)
EUKARYOTIC CELLS
    The Nucleus
    Cytomembrane System

Summary

All eukaryotic cells contain a membrane-bound compartment, the *nucleus*. DNA, containing the instructions for constructing proteins, is isolated within the nucleus. DNA and associated proteins actively regulate the metabolic behavior of the cell. Within cells, virtually all the chemical reactions involving carbohydrates, lipids, proteins, and nucleic acids depend on a special group of mediating proteins called *enzymes*. Thus, proteins are the basis for cell structure and function. A *nuclear envelope* encloses the *nucleoplasm* and is regularly traversed by pores; exchange of materials between nucleus and cytoplasm occurs across this boundary. Two or more *nucleoli*, sites of ribosome subunit construction, are found within the nucleoplasm. Eukaryotic DNA is long and thin with proteins (some are enzymes) attached along its length; this complex is a *chromosome*. A nucleus may contain a few to many different chromosomes, depending on the species of organism. Before a cell divides, the DNA is duplicated; during cell division, the copied and condensed forms of the chromosomes are efficiently distributed to daughter cells.

Prokaryotic DNA has few associated proteins and cannot condense into chromosomes.

Proteins manufactured in the cytoplasm are either stored or enter a processing and transport mechanism, the *cytomembrane system*. The cytomembrane system includes the *endoplasmic reticulum (ER)*, *Golgi bodies*, *lysosomes*, and various vesicles. These organelles in eukaryotic cells affect intracellular substances in different ways. *Ribosomes* are the cytoplasmic sites of protein synthesis. If ribosomes are on the *rough endoplasmic reticulum*, the proteins they synthesize will be exported as secretions from the cell or delivered to organelles. Carbohydrate groups are attached to protein chains in rough ER. The *smooth endoplasmic reticulum* lacks ribosomes but contains enzymes that help assemble lipids. The smooth endoplasmic reticulum also participates in isolating and transporting materials and in breaking down storage materials and potentially toxic materials such as drugs or harmful metabolic by-products. *Golgi bodies* make up a membrane system (generally located in the cytoplasm near the nucleus) that receives proteins and lipids from the ER, packages them, and transports them to specific destinations. *Exocytic vesicles* bud from the Golgi, move to the plasma membrane, and release their contents outside the cell. *Lysosomes*, budding from the Golgi, are vesicles of various digestive enzymes in animal cells; they sometimes merge with membrane-enclosed materials, bacteria, foreign particles, or worn-out cell parts and digest them. *Endocytic vesicles* bud from the ER and import extracellular substances into the cell; they later fuse with the Golgi. *Microbodies* are vesicles that bud from the ER; they degrade and convert substances such as hydrogen peroxide, fats, and oils.

Key Terms

nucleus	chromatin	lysosomes
nucleolus, -oli	cytomembrane system	exocytic vesicles
nuclear envelope	endoplasmic reticulum (ER)	endocytic vesicles
ribosomes	rough ER	transport vesicles
chromosomes	smooth ER	microbodies
nucleoplasm	Golgi bodies	

Objectives

1. Describe the basic organization of the nuclear envelope, the nucleolus, and the chromosomes.
2. State the processes carried out by each of the nuclear components mentioned in Objective 1.
3. Describe the different forms a chromosome can take during the life of a cell.
4. Distinguish between *rough endoplasmic reticulum* and *smooth endoplasmic reticulum* in terms of structure and particular functions.
5. Draw a diagram of a Golgi body and explain how Golgi bodies accomplish their functions.
6. Trace the pathway traveled by a protein that has just been exported from a cell. Start at the point where its constituent amino acids are in the cytoplasm (Figure 3.19, main text).
7. Explain how microbodies can be useful to the cells of your tissues.

Self-Quiz Questions

True-False

If false, explain why.

___ (1) The nuclear envelope consists of two lipid bilayers and is without pores.

___ (2) If the chromosome network did not condense during cell division, many more tangles and breaks would occur in the DNA material.

___ (3) The nucleolus produces the materials from which the Golgi complex is later constructed in the cytoplasm.

—— (4) Lysosomes contain digestive enzymes.

—— (5) The rough endoplasmic reticulum is a membrane system that has many ribosomes associated with it.

—— (6) A protein due to be exported from the cell is made by ribosomes located in the rough endoplasmic reticulum and exported in vesicles pinched off from the endoplasmic reticulum.

—— (7) Digestive enzymes that are usually stored in nucleoli are able to break down virtually every large molecule found in cells.

Fill-in-the-Blanks

The (8) _____ is the region where ribosomal subunits are synthesized.

Elongated, thin strands of DNA and its associated proteins extend throughout the nucleoplasm when it is not undergoing division; this material condenses during cell division and is called (9) _____ .

Matching

Choose the best answer for each.

(10) —— Golgi body

(11) —— lysosome

(12) —— microbody

(13) —— rough endoplasmic reticulum

(14) —— smooth endoplasmic reticulum

A. Synthesizes proteins to be exported from that cell

B. Formed as buds from the ER; contain enzymes that degrade substances

C. Helps assemble lipids

D. Assembles, packages, transports, exports substances

E. Small bags of digestive enzymes found in animal cells

3-VI
(pp. 52–55)

PROKARYOTIC CELLS—THE BACTERIA (cont.)
EUKARYOTIC CELLS
    Mitochondria
    Specialized Plant Organelles
    The Cytoskeleton
    Cell Surface Specializations

Summary

All eukaryotic cells contain *mitochondria*, which have membrane systems similar to those of chloroplasts except that the inner membrane folds into a form that resembles projecting shelves. Within two inner compartments, mitochondria use oxygen to convert energy stored in carbon compounds into forms that the cell can use—principally ATP.

In plants and in some protistans, the chemical reactions of photosynthesis occur in oval or disk-shaped organelles called *chloroplasts*. Each chloroplast has an outer membrane and a complex, much-folded inner membrane thrown into configurations that resemble stacks of coins (*grana*). Apparently the grana are the sites where some of the energy of sunlight is trapped by chlorophyll and passed on to be stored in the bonds of carbon compounds such as sugar molecules.

A fluid-filled *central vacuole* occupies a large portion of a living, mature plant cell and stores water. Vacuoles are storage organelles that concentrate materials such as amino acids, sugars, and wastes not intended for export in places that are out of the way of metabolic activity.

A *cytoskeleton* based on a three-dimensional cytoplasmic lattice composed of protein fibers and threads provides cell shape and internal organization. *Microtubules* (tubulin subunits), *microfilaments* (actin subunits), and *intermediate filaments* are the main components. Some cytoskeleton parts, such as the division spindle, are transient; other parts, such as muscle cell filaments, are permanent.

Cilia and *flagella* are microtubular structures that move cells through their environment. Flagella are generally longer and less numerous than cilia but they possess a similar internal structure, the "9 + 2 array" of microtubules. The cytoskeleton seems to be organized by small cytoplasmic masses of protein and other substances, the microtubule organizing centers (MTOCs). In animal cells, the MTOCs are the cylindrical, paired centrioles located near the nucleus; nine groups of three microtubules each circle the edge. Centrioles may control the plane of cell division, which may in turn influence embryo and adult form development. Centrioles help to fashion *basal bodies* found at the bases of cilia and flagella. Eukaryotic flagella and cilia are assembled from microtubules that find their source in basal bodies. Cilia and flagella also help attract food particles to nonmotile animals. Cilia move fluids and suspended particles along the surfaces of epithelium that line the digestive, respiratory, and reproductive tracts of many types of animals.

Cell walls occur among bacteria, protistans, fungi, and plants. Most cell walls have carbohydrate frameworks. These cellular surface products provide support, resist mechanical pressure, and confer tensile strength when cells absorb water and expand. Porous cell walls occur in all kingdoms except Animalia and are quite diverse structurally.

Animal cells and tissues are held together by a meshwork of collagen, fibrous proteins, glycoproteins, and specialized polysaccharides, the *extracellular matrix*. Substances move easily from cell to cell through the matrix.

The cells in multicelled organisms must interact with cellular neighbors and their environment. Interactions do occur because cells are tightly in contact with one another by various kinds of *cell junctions*.

Key Terms

mitochondrion, -dria	actin subunit	microtubule organizing
chloroplast	intermediate filaments	centers (MTOCs)
chlorophyll	permanent	centrioles
central vacuole	transient	basal bodies
cytoskeleton	flagella, flagellum (sing.)	cell walls
microtubules	cilia, cilium (sing.)	extracellular matrix
tubulin subunit	9 + 2 array	cell junctions
microfilament		

Objectives

1. Describe the basic structure of the mitochondrion. Indicate the processes that occur in mitochondria and the approximate size of mitochondria.
2. Describe the basic structure of the chloroplast, indicate how the grana differ from the stroma and what happens in each, and tell where the chlorophyll is located.
3. Explain how cells store materials.
4. Describe how plant cells increase their size and interior surface area.
5. Describe the cytoskeleton and explain why you think that it cannot be seen very well using an ordinary light microscope.
6. Distinguish *microfilaments* from *microtubules* in terms of structure and function. Mention actin and tubulins in your explanations.

7. Explain how microfilaments and microtubules influence cell shape, motion, and cell division.
8. Explain how the internal microtubule pattern differs when comparing cilia, flagella, and centrioles.
9. Describe the relationship between centrioles and basal bodies.
10. List the examples of organisms that have eukaryotic cilia and flagella and state some of the major functions provided by those organelles.
11. State the benefits and limitations imposed on a cell by the presence of cell coats, capsules, sheaths, and walls.
12. Describe the general structure of plant cell walls.
13. List the major groups of organisms that have walls surrounding their cells.
14. Describe the extracellular matrix, in what type of cells it is found, and its function.
15. Explain the major functions of cell junctions in multicelled organisms.

Self-Quiz Questions

True-False

If false, explain why.

___ (1) Muscle cells and other cells that demand high-energy output generally would have many more chloroplasts than less active cells have.

___ (2) Mitochondria and chloroplasts are organelles that have both inner and outer membranes.

___ (3) Enzymes and molecules involved in ATP formation are located on the outer membranes of mitochondria.

___ (4) Chloroplasts are found in animal cells.

___ (5) A plant's central vacuole can store microtubules and microfilaments.

Matching

Link each letter with its appropriate blank(s). Each blank should have only one letter. The same letter may be used in more than one blank.

(6) ___ vacuole

(7) ___ vesicle

(8) ___ mitochondrion

(9) ___ eukaryotic flagellum

(10) ___ basal body

(11) ___ plant cell wall

(12) ___ centriole

(13) ___ cilium

(14) ___ MTOC

(15) ___ microtubule

A. A tiny bag that transports substances through the cytoplasm
B. A large storage compartment that generally stores fluids
C. Site of photosynthesis
D. Tubulins
E. Provides the ATP necessary for rapid movements
F. Short, barrel-shaped organelle that organizes the interior of a cilium or flagellum
G. Cellulose
H. Assembled from microtubules in a 9 + 2 array within a sheath that is continuous with a plasma membrane
I. Exists in a pair that is pushed apart by spindle formation
J. Organizes the cytoskeleton

In photosynthetic prokaryotes such as bacteria and blue-green algae, light-trapping reactions and ATP formation occur on the (16) _____ _____ _____ . In photosynthetic eukaryotes these reactions occur in an organelle called the (17) _____ . An energy-rich molecule produced in the grana is (18) _____ . An energy-rich molecule produced by a mitochondrion is (19) _____ . The principal subcellular structures involved in establishing and maintaining the shapes of cells and extensions from cells are (20) _____ . (21) _____ are long, thin structural elements composed of contractile proteins. The (22) _____ is a three-dimensional network that pervades the cytoplasm. Microtubules are assembled from protein subunits called (23) _____ .

3-VII
(pp. 55–56)

SUMMARY

Summary

As time passed, cells became more complex. There were selective advantages for cells that increased their internal surface area with infoldings and outfoldings of membrane on which metabolic reactions could occur. Such developments have enabled cells to acquire energy and materials in highly controlled and specialized ways.

Objectives

1. List the basic minimal parts a cell must have in order to carry on metabolism and live.
2. Explain the advantages that accrue for cells that develop many internal compartments.
3. Consult Table 3.2 in your main text. Decide which kingdom has the least complex cells and which has the most complex cells. Indicate which other kingdom shares the most features with plants and which other kingdom most strongly resembles the animals.

Self-Quiz Questions

True-False

If false, explain why.

___ (1) Some cells can live if they contain only a plasma membrane, DNA molecules, and cytoplasm that lacks other compartmented cell structures.

___ (2) All living organisms are composed of one or more cells.

___ (3) On Earth today, new cells can arise only from cells that already exist.

___ (4) There are no essential differences in the ways that prokaryotes and eukaryotes acquire and process energy and materials.

Labeling   Identify each indicated part of the accompanying illustration.

(5) _____   (11) _____   (18) _____

(6) _____ _____   (12) _____ _____   (19) _____ _____

(7) _____ _____   (13) _____ _____   (20) _____

_____   (14) _____ _____   (21) _____

(8) _____   (15) _____   (22) _____

(9) _____   (16) _____   (23) _____

(10) _____   (17) _____ _____   (24) _____

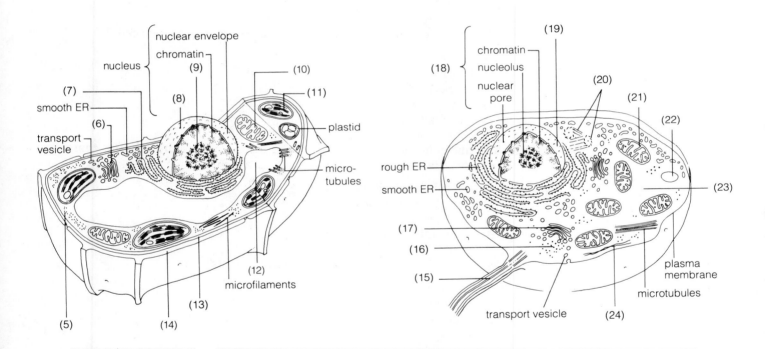

CHAPTER TEST   **UNDERSTANDING AND INTERPRETING KEY CONCEPTS**

___ (1) In a lipid bilayer, _____ tails point inward and form a region that excludes water.

(a) acidic
(b) basic
(c) hydrophilic
(d) hydrophobic

___ (2) A protistan adapted to life in a freshwater pond is collected in a bottle and transferred to a saltwater bay. Which of the following is likely to happen?

    (a) The cell bursts.
    (b) Salts flow out of the protistan cell.
    (c) The cell shrinks.
    (d) Enzymes flow out of the protistan cell.

___ (3) Which of the following is *not* a form of active transport?

    (a) Sodium-potassium pump
    (b) Endocytosis
    (c) Exocytosis
    (d) Facilitated diffusion

___ (4) Which of the following is *not* a form of passive transport?

    (a) Osmosis
    (b) Facilitated diffusion
    (c) Simple diffusion
    (d) Exocytosis

___ (5) Which of the following is *not* found as a part of prokaryotic cells?

    (a) Ribosomes
    (b) DNA
    (c) Nucleus
    (d) Cytoplasm
    (e) Cell wall

___ (6) The nucleolus is a dense region of _____ where the subunits that later will be constructed into _____ are made.

    (a) nucleoplasm, ribosomes
    (b) cytoplasm, vesicles
    (c) cytoplasm, chromosomes
    (d) nucleoplasm, chromatin

___ (7) Which of the following is *not* present in all cells?

    (a) Cell wall
    (b) Plasma membrane
    (c) Ribosomes
    (d) DNA molecules

___ (8) A nanometer is _____ of a meter.

    (a) one-ninth
    (b) one-tenth
    (c) one one-hundredth
    (d) one one-billionth

___ (9) Mitochondria convert energy stored in _____ to forms that the cell can use, principally ATP.

    (a) water
    (b) carbon compounds
    (c) O_2
    (d) carbon dioxide

___ (10) The _____ _____ is free of ribosomes, curves through the cytoplasm, and is the main site of lipid synthesis.

    (a) lysosome
    (b) Golgi body
    (c) smooth ER
    (d) rough ER

INTEGRATING AND APPLYING KEY CONCEPTS

(1) Name three aspects of the biological world that would be changed if water were not a polar molecule.

(2) If there were no such thing as active transport, how would the lives of organisms be affected?

(3) Which parts of a cell constitute the minimum necessary for keeping the simplest of living cells alive?

(4) How did the existence of a nucleus, compartments, and extensive internal membranes confer selective advantages on cells that developed these features?

4

GROUND RULES FOR METABOLISM

ENERGY AND LIFE
THE NATURE OF METABOLISM
    Energy Changes in Metabolic Reactions
    Metabolic Pathways
ENZYMES
    Enzyme Structure and Function
    Effects of Temperature and pH on Enzymes

Control of Enzyme Activity
COFACTORS
ATP: THE MAIN ENERGY CARRIER
    Structure and Function of ATP
    The ATP/ADP Cycle
ELECTRON TRANSPORT SYSTEMS
SUMMARY

General Objectives

1. Know two laws that govern the way energy is transferred from one substance to another.
2. Provide an example of a metabolic pathway and explain what kinds of substances regulate activity of the pathway.
3. Tell exactly what enzymes do and how they do it.
4. Know the full significance of the ATP/ADP cycle.
5. Explain how a molecule can "carry" energy.
6. Describe the role of electron transport systems in cellular chemistry.

4-I
(pp. 58–59)

ENERGY AND LIFE

Summary

All events in our universe are governed by two laws of energy. The first law of thermodynamics states that the total amount of energy in the universe never changes. Energy may be converted from one form into another, but it cannot be created or destroyed. Living organisms can channel the ways in which energy changes from one form into another, first hoarding it temporarily and then letting go of what is already there. The second law of thermodynamics says that, left to itself, any system and its surroundings spontaneously undergoes conversion to a less-organized form. Each time this happens, some energy is randomly dispersed in a form that is not readily available to do work: this is called *entropy*. Thus, although the total amount of energy in the universe stays the same, the amount available in *useful* forms is dwindling. The entropy of any local region can be lowered as long as that region is resupplied with usable energy that is being lost from some other place. Through energy transfusions from the sun, the universal trend toward increased entropy can be postponed here on Earth.

Key Terms

metabolism	low-quality energy	entropy
energy	second law of	system
first law of thermodynamics	thermodynamics	surroundings
high-quality energy	spontaneous	randomly dispersed

Objectives

1. Define *energy* and explain what is meant by the *quality of energy*.
2. State the first and second laws of thermodynamics.
3. Explain energy conversion by giving an example.
4. Explain what *spontaneous* means in relation to energy.
5. Trace the path of energy flow from the sun to your muscle movement.
6. Explain how, if the universe is becoming progressively disordered, a human embryo can grow into an infant and an infant can grow into an adult.

Self-Quiz Questions

True-False

If false, explain why.

___ (1) Energy is the capacity to accomplish work.

___ (2) The first law of thermodynamics states that entropy is constantly increasing in the universe.

___ (3) Your body steadily gives off heat equal to that from a 100-watt light bulb.

___ (4) When you eat a potato, some of the stored chemical energy of the food is converted into mechanical energy that moves your muscles.

___ (5) The amount of low-quality energy in the universe is decreasing.

4-II
(pp. 60–61)

THE NATURE OF METABOLISM
    Energy Changes in Metabolic Reactions
    Metabolic Pathways

Summary

A *metabolic reaction* is a form of internal energy change in the cell. How efficiently energy is used in any system depends on the precise route by which one energy form is transferred or converted to another form. If the route consists of a big explosive reaction that occurs in a single step, much of the potential energy becomes low-quality energy rather than energy that can be used to do work. Cells generally convert the potential chemical energy stored in nutrient molecules in a *series* of steps, each of which releases tiny amounts of kinetic energy. The energy from such *exergonic reactions* is used to drive energy-requiring (*endergonic*) reactions in the cell. By relying on such *metabolic pathways*, a cell temporarily conserves some energy that might otherwise be lost as entropy during its continual chemical conversions. Products from one or more reactions can serve as intermediates for subsequent reactions in the pathway.

    Chemical reactions are generally *reversible* under certain conditions. The greater the concentration of reactants, the faster the forward reaction. The greater the concentration of products, the faster the reverse reaction. An increase in concentration, temperature, or pressure may cause some *product* molecules to revert to reactant molecules. *Dynamic equilibrium* may be reached in which the reaction is proceeding as quickly in reverse as it is in forward, and there is no further change in the *net* concentration of reactants or products. At *equilibrium*, product concentrations may be greater or lower than reactant concentrations, depending on how much energy is fed into or released from the reaction. In cells the availability of raw materials (reactants) shifts, and requirements for different products may vary from minute to minute. In a changing environment, cellular homeostasis is maintained only through constant adjustments in the metabolic pathways that sustain life. *Degradative pathways* break apart larger molecules such as carbohydrates, lipids, and proteins; this releases smaller molecules, which may be stored or used in *biosynthetic pathways* to construct the large molecules of cell parts. There are limits to the number of compounds that can be formed during metabolic reactions. Compounds must be produced in concentrations high enough to let a reaction run to completion but low enough to prevent their use in unnecessary side reactions.

Key Terms

products	metabolic pathways	biosynthetic pathways
reactants	net change	reactants
exergonic reaction	rate of reaction	cofactors
endergonic reaction	chemical equilibrium	energy carriers
reversible reaction	degradative pathways	end products
dynamic equilibrium		

Objectives

1. Explain what might happen to cells if they tried to use the most direct route for releasing the potential chemical energy stored in their nutrient molecules.
2. In terms of reactant and product concentrations, explain what causes a chemical reaction to occur.
3. Distinguish between *exergonic* and *endergonic* reactions by defining and giving an example of each.
4. State what the requirements are for a system to achieve dynamic equilibrium. Then decide whether you think a living system can ever be in a true state of equilibrium.
5. Define *metabolic pathway* and list the various participants in such a system.
6. Contrast the functions of *degradative pathways* and *biosynthetic pathways*.
7. List some ways cells manage their metabolic pathways when the availability of raw materials shifts.

Self-Quiz Questions

Matching

Match the most appropriate letter to its number.

(1) ___ dynamic equilibrium

(2) ___ endergonic

(3) ___ biosynthetic pathways

(4) ___ energy carriers

(5) ___ exergonic

(6) ___ reactants

A. Mainly ATP
B. Reaction showing a net loss in energy
C. Small molecules are assembled into larger biological molecules
D. Rate of forward reaction = rate of reverse reaction
E. Substances able to enter into a reaction
F. Reactants show a net gain in energy as they are changed into products

ENZYMES
Enzyme Structure and Function
Effects of Temperature and pH on Enzymes
Control of Enzyme Activity

Summary

An *enzyme* is a protein catalyst that speeds the rate of a chemical reaction by lowering the *activation energy* to a level that must be reached before the reaction will occur. In living systems, metabolic reactions must occur within a certain range of low temperatures so that substances with shapes that depend on hydrogen bonding are not altered beyond redemption. An enzyme holds the reacting molecules in an orientation so favorable that collisions between the molecules are more likely to break certain bonds and cause atoms to become rearranged into the product molecules. In addition, an enzyme bonding to a reactant molecule can sometimes strain certain bonds so that they are forced to break apart.

Enzymes accelerate the rate at which a reaction approaches equilibrium, but they do not alter the proportions of reactants and products that will be present once equilibrium is reached. Enzymes can attach only to specific molecules (*substrates*); thereby they can cause high rates of required reactions even though overall concentrations of the reactants are low. The *induced-fit model* states that the *active site* of an enzyme and its *substrate* become fully complementary to each other.

Concept Aid

If we wish to obtain a chemical reaction between two or more different molecules in a test tube, we apply heat with a gas burner. The heat causes the molecules to vibrate and collide with one another at random; chemical reactions occur. How then do molecules in cells come together so they interact in the vital reactions of life's chemistry? In fact, the chemical reactions in human cells are so regularly controlled that a constant body temperature results. There are no "little flames" in cells that cause collisions between molecules! The answer is found in protein molecules called *enzymes* that mediate chemical reactions; reactant (substrate) molecules fit closely together on enzymes in order that they can interact with each other and chemical reactions can proceed. This is sometimes referred to as the "cold chemistry of life." Without enzymes to mediate chemical reactions, there would be no life. As the main text discusses various chemical reactions occurring in cells, enzymes are not always mentioned. This is because we understand that nearly all the chemical reactions in cells require enzymes.

Key Terms

enzymes	induced-fit model	heat-sensitive enzyme
catalyst, catalytic	transition state	inhibitors
substrates	activation energy	allosteric control
active site	energy hill diagram	feedback inhibition
enzyme-substrate complex		

Objectives

1. Define *enzyme* and *substrate*.
2. Explain how the *active site* of an enzyme molecule is related to specific chemical reactions.
3. Relate the *induced-fit model* to enzyme function.

4. Describe the precise role that enzymes play in speeding chemical reactions and explain why enzymes make particularly effective catalysts.
5. Understand the significance of the energy hill diagram in explaining the action of enzymes; relate this to *activation energy*.
6. Explain the effects of temperature and pH on enzymes.
7. Describe what happens when an enzyme is denatured.
8. Explain why parents might become very concerned when small children develop very high body temperatures as a result of severe infections.
9. Suppose that you switch to a diet that contains only fats, proteins, and nucleic acids and omits carbohydrates. Suppose also that your body can make the several different kinds of carbohydrates it needs from fats and proteins. Explain how *allosteric enzymes* might be of use to you in making the required carbohydrates.

Self-Quiz Questions

Fill-in-the-Blanks

(1) _____ are highly selective proteins that act as (2) _____ , which means they greatly enhance the rate at which specific reactions approach (3) _____ _____ . The specific substance upon which a particular enzyme acts is called its (4) _____ ; this substance fits into the enzyme's crevice, which is called its (5) _____ _____ . The (6) _____-_____ _____ describes how a substrate contacts the substrate without a perfect fit. Enzymes increase reaction rates by lowering (7) _____ _____ . (8) _____ and (9) _____ are two factors that influence the rates of enzyme activity. During severe viral infections, extremely high fevers can destroy the shape of (10) _____ , which leads to cell death. When denaturation occurs, (11) _____ _____ holding the enzyme in its three-dimensional shape break. Molecules that can bind with enzymes and interfere with their function as catalysts are called (12) _____ . When the end product binds to the first enzyme in a metabolic pathway and prevents product formation, this is known as (13) _____ _____ .

True-False

If false, explain why.

___ (14) Enzyme shape may change during the interaction between enzyme and substrate.

___ (15) The active site is a groove on the reactant molecule.

___ (16) For two reactant molecules to become product molecules, the reactant molecules must first collide with a certain minimum energy.

___ (17) Enzymes enhance reaction rates by increasing the activation energy required.

___ (18) High temperatures can denature enzymes and affect reaction rates; the pH level does not seem to affect enzymes and their action.

4-IV
(pp. 64–66)

COFACTORS
ATP: THE MAIN ENERGY CARRIER
  Structure and Function of ATP
  The ATP/ADP Cycle
ELECTRON TRANSPORT SYSTEMS
SUMMARY

Summary

Cofactors are nonprotein substances that either help enzymes to catalyze specific reactions or serve fleetingly as transfer agents of protons and electrons. Breaking of the high-energy phosphate bonds of ATP releases larger amounts of useful energy than does breaking of other kinds of covalent bonds. Reactions that do not proceed on their own can be driven indirectly by energy-releasing reactions such as the transfer of a phosphate group from ATP to another molecule; this is known as *phosphorylation*. ATP molecules are used over and over again in such reactions. ATP is sometimes called the *universal energy currency* of cells because it transfers usable energy to reactions concerned with energy metabolism, biosynthesis, active transport, and cellular movement.

Energy is embodied not only in phosphate bonds but also in some of the electrons associated with carrier molecules, such as NADH and NADPH, and the cytochromes. An *oxidation* reaction strips from an atom or molecule one or more electrons, which are simultaneously gained by other atoms or molecules in a *reduction* reaction. Sometimes hydrogen ions or atoms are transferred as well as the electron. Because the loss and gain happen simultaneously, these two events are regarded as being coupled in an oxidation-reduction reaction, which generates usable forms of energy. Free-moving electron carriers such as NAD^+ and $NADP^+$ accept electrons and hydrogen at one reaction site in the cell and transfer them to different reaction sites concerned with ATP production or biosynthesis. Sometimes electron carriers such as the cytochromes are arranged in a series known as an *electron transport system*, and *oxidation-reduction reactions* can occur one after another in organized sequence. These chains of membrane-bound electron carriers are the basis of transport systems that establish pH and electric gradients, which help drive ATP formation.

Concept Aid

The marvel of electricity is a familiar idea; electrons flow through wires. The flowing electrons represent energy; they encounter a resistance in the filament of a light bulb and it glows. This is a useful analogy when trying to understand the energy concepts of life's chemistry. Electrons and hydrogens are stripped off one compound (oxidation), which decreases the energy level of the compound. Another compound takes on the moving electrons and hydrogens (reduction) and its energy level increases. Thus, energy in the form of moving hydrogens and electrons is passed from compound to compound in biochemical pathways. Reactions involving such energy changes are catalyzed by proteins called *enzymes*.

Key Terms

cofactors	cytochromes	electron transport systems
coenzymes	ATP	oxidized
NAD$^+$	triphosphate	reduced
NADP$^+$	ATP/ADP cycle	oxidation-reduction reaction
metal ions	phosphorylation	bioluminescence

Objectives

1. Distinguish between *cofactor* and *coenzyme;* cite two examples of coenzymes and give their specific functions.
2. Identify the structure (name five parts) and function of ATP.
3. Explain why ATP is necessary to *all* metabolic pathways.
4. Explain how the hydrolysis of ATP coupled with a reaction in which two simple molecules are synthesized into one larger molecule resembles the connection of an electric motor to a sewing machine.
5. Describe the process of phosphorylation and explain how it is important to cells.
6. Name three energy-requiring activities of cells.
7. Distinguish an *oxidation* reaction from a *reduction* reaction; show how the two must be coupled into an oxidation-reduction reaction.
8. Explain how phosphorylation reactions and oxidation-reduction reactions are related to the transfer of energy.
9. Describe an electron transport system. Tell which types of molecules participate in such systems and describe how they participate.
10. State what electron transport systems accomplish for the cells that contain them.

Self-Quiz Questions

Fill-in-the-Blanks

Nonprotein substances that aid enzymes in their catalytic task are called (1) _____ . (2) _____ is an example of a coenzyme. Cells cannot extract energy (3) _____ for cellular activities from organic molecules. The energy must first be placed in the form of a molecule known as ATP, or (4) _____ _____ . ATP is constructed of (5) _____ , (6) _____ , and three (7) _____ groups. Through the linking of a phosphate group to ADP, an energy-rich molecule of (8) _____ is formed. Other molecules are primed for chemical reactions when they receive a phosphate group from (9) _____ through a process known as (10) _____ . ATP furnishes energy for almost all (11) _____ pathways. A(n) (12) _____ _____ system is a series of (13) _____ and (14) _____ bound in a cell membrane, that transfer electrons in an organized sequence. Electrons are transferred by (15) _____-_____ reactions. The first molecule in line accepts an excited (16) _____ from a donor molecule outside the chain. It gets transferred from the first molecule to a series of (17) _____ _____ molecules, with the potential of releasing usable (18) _____ at each step. At certain transfer points, the amount of energy given off is harnessed to do useful work, such as moving hydrogen (19) _____ across membranes to establish pH and electric gradients.

True-False

If false, explain why.

___ (20) A substance that has been oxidized has gained one or more electrons.

___ (21) Phosphorylation is a chemical reaction in which a phosphate group is attached to another molecule, thus increasing the potential chemical energy of that molecule.

___ (22) Reduction is a chemical reaction in which one or more electrons or hydrogen atoms is attached to a molecule, thus increasing its potential chemical energy.

CHAPTER TEST   **UNDERSTANDING AND INTERPRETING KEY CONCEPTS**

___ (1) The laws of thermodynamics state that _____ .

   (a) energy can be transformed into matter, and because of this, we *can* get something for nothing
   (b) energy can only be destroyed during nuclear reactions, such as those that occur inside the sun
   (c) if energy is gained by one region of the universe, another place in the universe also must gain energy in order to maintain the balance of nature
   (d) matter tends to become increasingly more disorganized

___ (2) Essentially, the first law of thermodynamics states that _____ .

   (a) one form of energy cannot be converted into another
   (b) entropy is increasing in the universe
   (c) energy cannot be created or destroyed
   (d) energy cannot be converted into matter or matter into energy

___ (3) Which is not true of enzyme behavior?

   (a) Enzyme shape may change while catalyzing a chemical reaction.
   (b) The active site of an enzyme orients its substrate molecules, thereby promoting interaction of their reactive parts.
   (c) All enzymes have an active site where substrates are temporarily bound.
   (d) An enzyme can catalyze a wide variety of different reactions.

___ (4) When NAD^+ combines with hydrogen, the NAD^+ is _____ .

    (a) reduced
    (b) oxidized
    (c) phosphorylated
    (d) denatured

___ (5) A substance that gains electrons is _____ .

    (a) oxidized
    (b) a catalyst
    (c) reduced
    (d) a substrate

___ (6) When a phosphate group is linked to ADP, the bond formed _____ .

    (a) releases a large amount of free energy
    (b) is transferred to other molecules
    (c) is usually found in each glucose molecule; that is why glucose is chosen as the starting point for glycolysis
    (d) can release a large amount of usable energy when the phosphate group is transferred to another molecule

___ (7) An allosteric enzyme _____ .

    (a) has an active site where substrate molecules bind and another site that binds with intermediate or end-product molecules
    (b) is an important energy-carrying nucleotide
    (c) carries out either oxidation reactions or reduction reactions but not both
    (d) raises the activation energy of the chemical reaction it catalyzes

---

INTEGRATING AND APPLYING KEY CONCEPTS

A piece of dry ice left sitting on a table at room temperature vaporizes. As the dry ice vaporizes into CO_2 gas, does its entropy increase or decrease? Tell why you answered as you did.

5
ENERGY-ACQUIRING PATHWAYS

FROM SUNLIGHT TO CELLULAR WORK:
PREVIEW OF THE MAIN PATHWAYS
PHOTOSYNTHESIS
    Simplified Picture of Photosynthesis
    Chloroplast Structure and Function
LIGHT-DEPENDENT REACTIONS
    Light Absorption

ATP and NADPH Formation
LIGHT-INDEPENDENT REACTIONS
    Calvin-Benson Cycle
    C4 Plants
SUMMARY

**General
Objectives**

1. Understand the main pathways by which energy from the sun or from specific chemical reactions enters organisms and passes from organism to organism and/or back into the environment.
2. Know the steps of the light-dependent and light-independent reactions. Know the raw materials needed to start each phase and know the products made by each phase.
3. Explain how C4 plants are able to continue carbohydrate construction even when the carbon dioxide/oxygen ratio is unfavorable.

5-I
(pp. 69–71)

**FROM SUNLIGHT TO CELLULAR WORK:
PREVIEW OF THE MAIN PATHWAYS
PHOTOSYNTHESIS**
    Simplified Picture of Photosynthesis
    Chloroplast Structure and Function

Summary

All organisms depend on energy contained in food molecules processed in the chemical pathways of cells. Energy from sunlight is captured by *photosynthetic autotrophs* and converted to the chemical energy of organic compounds. *Chemosynthetic autotrophs* obtain energy from electrons of inorganic compounds. Photosynthetic autotrophs are able to convert light energy to chemical energy. The three major pathways (photosynthesis, glycolysis, and cellular respiration) are linked by energy flowing through them. *Photosynthesis* consists of two sets of chemical reactions that occur in organelles called *chloroplasts;* in these reactions, energy-poor carbon dioxide and water from the environment are converted into energy-rich carbohydrates. The *light-dependent reactions* occur within chloroplasts on thylakoid membranes of grana disks. Here energy from sunlight is absorbed by pigment molecules such as chlorophyll and converted into potential chemical energy stored briefly in chemical bonds of ATP and NADPH. The *light-independent reactions* occur in the stroma of chloroplasts. It is in these reactions that ATP and NADPH are used to assemble glucose and other carbohydrates.

Key Terms

autotrophic organism
photosynthetic autotroph
chemosynthetic autotroph
heterotrophs
photosynthesis

glycolysis
aerobic respiration
light-dependent reactions
light-independent reactions
chloroplast

thylakoid membrane
thylakoid compartment
granum, grana
stroma

Objectives

1. Describe how autotrophs and heterotrophs differ in their means of obtaining energy.
2. Distinguish between photosynthetic autotrophs and chemosynthetic autotrophs.
3. Generally explain the relationship of photosynthesis to glycolysis and aerobic respiration.
4. Study the general equation for photosynthesis as shown on p. 70 of the main text until you can remember the reactants and products. Reproduce the equation from memory on another piece of paper.
5. Describe the structure of the chloroplast; identify thylakoid membrane, grana, and stroma.

Self-Quiz Questions

Fill-in-the-Blanks

(1) _____ _____ obtain energy from sunlight. (2) _____ _____ oxidize inorganic substances such as ammonium ions or iron or sulfur compounds to obtain energy. Photosynthetic autotrophs include all plants, some protistans, and some (3) _____ . Chemosynthetic autotrophs are limited to a few kinds of (4) _____ . (5) _____ organisms feed on autotrophs, each other, or organic wastes. Energy stored in organic compounds such as glucose may be released by the two interconnected pathways, (6) _____ and (7) _____ _____ . The two major sets of reactions of photosynthesis are the (8) _____ – _____ reactions, and the (9) _____ – _____ reactions. (10) _____ _____ and (11) _____ are the reactants of photosynthesis and the end product is usually given as (12) _____ . The internal membranes of the chloroplast are the (13) _____ , which appear as stacked disks, the (14) _____ . The area surrounding the stacks is known as (15) _____ .

LIGHT-DEPENDENT REACTIONS
Light Absorption

Summary

Photosynthesis occurs on thylakoid membranes, which are enclosed in the chloroplasts of plants. Embedded in the thylakoid membranes may be various *pigments* that absorb packets of light energy called *photons*. The variation in energy level of different photons corresponds to different wavelengths of light; we see most of these wavelengths as colors. The more abundant *chlorophylls* in leaves absorb blue and red wavelengths but reflect green—the reason leaves appear green. *Carotenoid pigments* absorb violet and blue wavelengths but reflect yellow, orange, and red. Taken together, the pigments of photosynthesis absorb nearly all the energy present in the visible light spectrum. A *photosystem* is a cluster of 200 to 300 pigment molecules on the thylakoid membranes, which ''harvest'' the energy of sunlight. The photosystem absorbs photon energy, which boosts an electron to a higher energy level. The electron instantly returns to a lower energy level while releasing energy that is trapped by special chlorophyll molecules. The trapped energy transfers in the form of an electron from a photosystem to an acceptor molecule embedded in the thylakoid membrane. This is the *first* event of photosynthesis and initiates the next series of photosynthetic reactions, the light-dependent reactions.

Key Terms

pigments	carotenoid	electromagnetic spectrum
light-trapping pigments	photosystems	nanometer
photon	absorb	T. Englemann
chlorophyll	reflect	

Objectives

1. Describe how the pigments found on thylakoid membranes are organized into photosystems and how they relate to photon light energy.
2. Describe the role chlorophylls and the other pigments found in chloroplasts play to initiate the light-dependent reactions. After consulting Figure 5.3 of the main text, state which colors of the visible spectrum are absorbed by (a) chlorophyll *a,* (b) chlorophyll *b,* and (c) carotenoids.
3. State what T. Englemann's 1882 experiment with *Cladophora* revealed.

Self-Quiz
Questions

Fill-in-the-Blanks

The light-capturing phase of photosynthesis takes place on (1) _____

_____ , which are arranged in stacks called (2) _____ . This system

forms a single compartment separate from the (3) _____ portion of the

chloroplast. A (4) _____ is a packet of light energy. Thylakoid

membranes contain (5) _____ , which absorb photons of light. The

principal pigments are the (6) _____ , which reflect green wavelengths

but absorb (7) _____ and (8) _____ wavelengths. (9) _____ are pigments that absorb violet and blue wavelengths but reflect yellow, orange, and red. A cluster of 200 to 300 of these pigment proteins is a (10) _____ . When pigments absorb (11) _____ energy, a(n) (12) _____ is transferred from a photosystem to a(n) (13) _____ molecule.

5-III
(pp. 73–76)

LIGHT-DEPENDENT REACTIONS (cont.)
    ATP and NADPH Formation

Summary

Most autotrophs carry out photosynthesis of one sort or another. Light absorbed by two types of photosystems activates the transfer of electrons from one or the other photosystem to a nearby acceptor molecule, which passes them on to the nearest electron transport system. As electrons are transferred down through the chain, some of the energy released is used to form ATP from ADP and inorganic phosphate. Eventually, the electrons lose their excess energy and return to a photosystem.

During *cyclic photophosphorylation*, excited electrons flow from *photosystem I* through an *electron transport system* and then reenter photosystem I. Every electron that enters the cyclic photophosphorylation pathway can cause one ATP molecule to be formed. The production of ATP is powered by the flow of hydrogen ions down concentration and electrical gradients established by hydrogen ions accumulating in the thylakoid disks during electron transfers through the transport system. Hydrogen ions moving through channel proteins into the stroma drives ATP formation. This process is known as *chemiosmosis*. The plant can then use this ATP to do various forms of biological work. Primitive organisms that use only the cyclic pathway *can* synthesize organic compounds, but their synthetic pathways are inefficient; hence, these organisms never evolved into large, complex, energy-demanding beings.

Most modern autotrophs carry out *noncyclic photophosphorylation*, which includes two photosystems (*photosystem II* and *photosystem I*) and two transport chains. When the two photosystems function together, electrons do not flow in a cycle. ADP is still phosphorylated to form ATP as electrons are transported from II to I. When the electrons are reexcited in photosystem I, they are sent to the second transport system, at the end of which they help to reduce NADP to NADPH. NADPH carries hydrogen ions and electrons (energy) to the light-independent reactions, where they help convert PGA to PGAL. Light absorption at the start of the pathway indirectly drives the splitting of water molecules; this is known as *photolysis*. Electrons released from water replace electrons being expelled from photosystem II. Every two new electrons from water molecules are used to make two ATP molecules via the first transport chain. Then they are reexcited in the second photosystem, and, as they pass down the second transport chain, the two electrons end up in a molecule of NADPH and are exported from the light reactions and used to drive one of the light-independent reactions. Because they are exported, the electrons do not follow a cyclic path. Some of the ATP is also used to drive one of the light-independent reactions. Oxygen atoms released from the photolysis of water are by-products that have accumulated in the atmosphere; oxygen provides the means for the chemistry of aerobic respiration to release energy from organic compounds.

Key Terms

electron transport system
photophosphorylation
cyclic pathway
cyclic
   photophosphorylation
P700
photosystem I

noncyclic pathway
noncyclic
   photophosphorylation
photosystem II
P680
photolysis

electric gradient
concentration gradient
channel proteins
chemiosmotic theory
ATP
NADPH

Objectives

1. Describe the general function of electron transport systems.
2. Contrast cyclic pathway and noncyclic pathways (photophosphorylations) in terms of the substances produced, the number of photosystems involved, and the number of transport chains used.
3. Explain what the water split during photolysis contributes to both cyclic and noncyclic pathways of the light-dependent reactions.
4. Name the two energy-carrier molecules produced during the noncyclic pathway and indicate how they will be used later.
5. Explain how the chemiosmotic theory is related to thylakoid compartments and the production of ATP.
6. Explain why atmospheric oxygen, a by-product of the noncyclic pathway, is essential for most life forms on Earth.

**Self-Quiz
Questions**

Fill-in-the-Blanks

Electrons ejected from chlorophyll molecules in photosystems on thylakoid membranes pass through one or two (1) _____ _____ systems. The formation of ATP molecules by electrons passing through electron transport systems is called (2) _____ . Pathways of this type are either (3) _____ or (4) _____ . The special chlorophyll found in photosystem I is referred to as (5) _____ . The (6) _____ pathway is the simplest one and operates with electrons traveling in a circle and producing only ATP. Today, land plants rely mostly on the (7) _____ pathway, which creates ATP and NADPH as energy carriers. This more complex pathway begins at photosystem II, which has a special chlorophyll molecule, (8) _____ . This molecule absorbs light energy and ejects an electron that passes over an electron transport system to be accepted by chlorophyll (9) _____ of photosystem I. (10) _____ absorbs light energy and boosts electrons to a higher energy level to a second (11) _____ _____ . Two electrons and one hydrogen (12) _____ are attached to (13) _____ , forming NADPH. (14) _____ splits water and releases oxygen, protons, and electrons. Electrons flow from split water to replace those given up by chlorophyll (15) _____ in photosystem II.

(16) _____ from split water accumulates in the atmosphere and makes aerobic respiration possible.

5-IV
(pp. 76–78)

LIGHT-INDEPENDENT REACTIONS
   Calvin-Benson Cycle
   C4 Plants
SUMMARY

Summary

During the *light-independent reactions,* a photosynthetic cell can use energy carriers such as ATP and NADPH to produce organic food molecules. In the first stage of these reactions, carbon dioxide fixation, carbon from atmospheric carbon dioxide is combined with (that is, fixed to) RuBP and incorporated into stable intermediate compounds (PGA). In the second stage, the Calvin-Benson cycle, ATP and NADPH supply chemical energy to convert PGA to PGAL. Some of this PGAL is then used to form glucose through sugar phosphate intermediates, but most is used to make new RuBP for fixation of more carbon. ADP, $NADP^+$, and leftover phosphates return to the chemistry of the light-dependent reactions where they again become NADPH and ATP. Some plants, particularly those of tropical origin, are known as C4 *plants;* they have an additional chemical means of fixing carbon that precedes the Calvin-Benson cycle. Photosynthetic bundle-sheath cells surround veins. Carbon dioxide fixation in these cells produces oxaloacetate, a four-carbon compound that is transferred to adjacent photosynthesizing leaf mesophyll cells; the carbon dioxide is released there and enters the Calvin-Benson cycle. On hot, dry days, the leaf stomata close and reduce water loss, but this also prevents carbon dioxide from entering the leaf. Photorespiration then occurs; oxygen rather than carbon dioxide attaches to RuBP used in the Calvin-Benson cycle. The result is less PGA formation, which results in lower food production. Growth of plants known as C3 (produce three-carbon PGA) plants suffer under such conditions. C4 plants such as crabgrass, sugarcane, and corn are adapted to high light levels, high temperatures, and limited water supply. C4 plants can continue to photosynthesize even on hot, dry days of intense sunlight when water must be conserved. C3 plants such as Kentucky bluegrass, wheat, and rice lack this additional means of CO_2 capture under stress. This explains why crabgrass thrives in the stress of late summer on what had earlier been a beautiful lawn of bluegrass.

Key Terms

Calvin-Benson cycle
carbon dioxide fixation
PGA
PGAL
RuBP

unstable six-carbon
   intermediate
sugar phosphate
photorespiration
C4 plants

stomates
mesophyll cells
bundle-sheath cells
C3 plants

Objectives

1. Explain why the light-independent reactions are called by that name.
2. Describe the process of carbon dioxide fixation by stating which reactants are necessary to get the process going and what stable products result from this process alone.

3. Describe the Calvin-Benson reaction series in terms of its reactants and products.
4. Explain what happens to each of the products of photosynthesis.
5. Describe the mechanism by which C4 plants thrive under hot, dry conditions; distinguish this CO_2 capturing mechanism from that of C3 plants.

Self-Quiz Questions

Fill-in-the-Blanks

The light-independent reactions can proceed without sunlight as long as (1) _____ and (2) _____ are available. The reactions begin when an enzyme links (3) _____ _____ to (4) _____ _____ , a five-carbon compound. The resulting six-carbon compound is highly unstable and breaks apart at once into two molecules of a three-carbon compound, (5) _____ . This entire reaction sequence is called carbon dioxide (6) _____ . ATP gives a phosphate group to each (7) _____ . This intermediate compound takes on H^+ and electrons from NADPH to form (8) _____ . It takes (9) _____ carbon dioxide molecules to produce twelve PGAL. Most of the PGAL becomes rearranged into new (10) _____ molecules—which can be used to fix more (11) _____ . Two (12) _____ are joined together to form a(an) (13) _____ _____ , primed for further reactions. The Calvin-Benson cycle yields enough RuBP to replace those used in carbon dioxide (14) _____ . ADP, NADP$^+$, and phosphate leftovers are sent back to the (15) _____ - _____ reaction sites where they are again converted to (16) _____ and (17) _____ . (18) _____ _____ formed in the cycle serves as a building block for the plant's main carbohydrates. When RuBP attaches to oxygen instead of carbon dioxide (19) _____ results and is typical of (20) _____ plants in hot, dry conditions. If less PGA is available, leaves produce a reduced amount of (21) _____ _____ . C4 plants can still construct carbohydrates when the ratio of carbon dioxide to (22) _____ is unfavorable because of the attachment of carbon dioxide to (23) _____ in certain leaf cells.

___ (1) The electrons that are passed to NADPH during noncyclic photophosphorylation were obtained from _____ .

    (a) water
    (b) CO_2
    (c) glucose
    (d) sunlight

___ (2) Cyclic photophosphorylation functions mainly to _____ .

    (a) fix CO_2
    (b) make ATP
    (c) produce PGAL
    (d) regenerate ribulose biphosphate

___ (3) Chemosynthesis involves the oxidation of such inorganic substances as _____ .

    (a) PGA
    (b) PGAL
    (c) sulfur
    (d) water

___ (4) The ultimate electron and hydrogen acceptor in noncyclic photophosphorylation is _____ .

    (a) $NADP^+$
    (b) ADP
    (c) O_2
    (d) H_2O

___ (5) C4 plants have an advantage in hot, dry conditions because _____ .

    (a) their leaves are covered with thicker wax layers than those of C3 plants
    (b) their stomates open wider than those of C3 plants, thus cooling their surfaces
    (c) special leaf cells possess a means of capturing CO_2 even in stress conditions
    (d) they also are capable of carrying on photorespiration

___ (6) Chlorophyll is _____ .

    (a) on the outer chloroplast membrane
    (b) inside the mitochondria
    (c) in the stroma
    (d) in the thylakoids

___ (7) Thylakoid disks are stacked in groups called _____ .

    (a) grana
    (b) stroma
    (c) lamellae
    (d) cristae

— (8) Plant cells produce O_2 during photosynthesis by _____ .

    (a) splitting CO_2
    (b) splitting water
    (c) degradation of the stroma
    (d) breaking up sugar molecules

— (9) Plants need _____ and _____ to carry on photosynthesis.

    (a) oxygen, water
    (b) oxygen, CO_2
    (c) CO_2, H_2O
    (d) sugar, water

INTEGRATING AND APPLYING KEY CONCEPTS

Suppose that humans acquired all the enzymes needed to carry out photosynthesis. Speculate about the attendant changes in human anatomy, physiology, and behavior that would be necessary for those enzymes actually to carry out photosynthetic reactions.

6
ENERGY-RELEASING PATHWAYS

ATP-PRODUCING PATHWAYS
AEROBIC RESPIRATION
    Overview of the Reactions
    Glycolysis
    Krebs Cycle
    Electron Transport Phosphorylation

ANAEROBIC ROUTES
    Lactate Fermentation
    Alcoholic Fermentation
ALTERNATIVE ENERGY SOURCES IN THE
HUMAN BODY
    Commentary: Perspective on Life
SUMMARY

General Objectives

1. Understand what kinds of molecules can serve as food molecules.
2. Know the relationship of food molecules to glucose and thus to glycolysis.
3. Understand the fundamental differences between glycolysis + fermentation and glycolysis + aerobic respiration. Know the factors that determine whether an organism will carry on fermentation or aerobic respiration.
4. Know the raw materials and products of each of these processes: glycolysis, fermentation, the Krebs cycle, and electron transport phosphorylation.

6-I
(p. 80)

ATP-PRODUCING PATHWAYS

Summary

Energy-rich carbohydrates and other food molecules are stockpiled in autotrophic organisms. Eventually, autotrophic cells and the cells of the organisms that dine on them will break down these molecules and use the stored energy to drive life processes. Although many different types of molecules are stored and eventually used, glucose breakdown and the associated energy transfers leading to *ATP* formation are central to the functioning of almost all prokaryotic and eukaryotic cells. The main degradative pathways that release energy from carbohydrates, lipids, or proteins for ATP formation are *aerobic respiration* (requires free oxygen) and *fermentation* (does not require free oxygen).

Key Terms

ATP
phosphate group

aerobic respiration

fermentation

Objectives

1. Explain why much of the world of life (including humans) depends largely on plants and other autotrophic organisms for energy sources.

2. Give the reasons for this statement: "At the biochemical level, there is an undeniable unity among organisms."
3. Cite the importance of adenosine triphosphate, or ATP.
4. Explain how ATP plays a role in priming other molecules for entering chemical reactions.
5. List the main pathways by which plants, as well as all other organisms, produce ATP.

Self-Quiz Questions

Fill-in-the-Blanks

Virtually all forms of life depend on a molecule known as (1) _____ as their primary energy carrier. This molecule donates a(n) (2) _____ _____ to other potential reactant molecules as a primer for entering chemical reactions. Plants produce adenosine triphosphate during (3) _____ , but plants and all other organisms produce ATP through chemical pathways that degrade food molecules. One degradative pathway requires free oxygen and is called (4) _____ _____ ; another pathway, (5) _____ , does not require free oxygen.

6-II
(pp. 80–81)

AEROBIC RESPIRATION
Overview of the Reactions

Summary

Aerobic respiration produces the highest yield of ATP of all possible degradative chemical pathways in organisms. Anaerobic respiration has a net yield of two ATP molecules while the aerobic route yields thirty-six (or sometimes more) ATPs. On this basis it is easier to understand why simpler organisms like bacteria rely on anaerobic ATP production and larger, more complex organisms such as humans utilize aerobic means to supply greater ATP requirements. Degrading chemical pathways that produce ATP prefer carbohydrate molecules if they are available; glucose serves as the starting molecule in most instances. The aerobic pathway of ATP production proceeds in three stages: (1) all respiratory activity begins with *glycolysis,* in which glucose is partly broken down anaerobically to two pyruvate molecules with a small yield of ATP; (2) the *Krebs cycle,* where the pyruvate is broken down to produce carbon dioxide while electrons and protons stripped from the molecule are delivered by carriers to a transport system—a small ATP yield occurs; and (3) carriers deliver the electrons and protons to *electron transport phosphorylation,* which produces a high yield of ATP as oxygen accepts electrons and protons to form water.

Key Terms

glucose	glycolysis	electron transport
degradative pathway	pyruvate	phosphorylation
ATP yield	Krebs cycle	

Objectives

1. Be familiar with which degradative chemical pathways produce low and high yields of ATP.
2. Explain why larger, more complex organisms utilize aerobic respiration for ATP production.
3. Memorize the equation for aerobic respiration as illustrated with the art on page 81 of the main text.
4. Summarize the reactions of glycolysis, the Krebs cycle, and electron transport phosphorylation; state which have relatively low and high ATP yields.
5. Explain, in general terms, the role of oxygen in aerobic respiration.

Self-Quiz Questions

Fill-in-the-Blanks

(1) _____ _____ is a degradative pathway requiring free oxygen in which the breakdown of one glucose molecule produces (2) _____ - _____ molecules of ATP. (3) _____ is an anaerobic degradative pathway in which the partial degrading of one glucose molecule yields (4) _____ ATP molecules. The summary equation for aerobic respiration states that one molecule of glucose plus six molecules of (5) _____ yields six molecules of carbon dioxide and six molecules of (6) _____ . In the reactions of respiration, (7) _____ precedes the Krebs cycle, which then precedes (8) _____ _____ phosphorylation. Electrons from electron transport phosphorylation are accepted by (9) _____ .

6-III
(pp. 81–86)

AEROBIC RESPIRATION (cont.)
    Glycolysis
    Krebs Cycle
    Electron Transport Phosphorylation

Summary

During *glycolysis,* which occurs in the cytoplasm, one molecule of glucose is broken down into two molecules of *pyruvate.* In the process, four ATP molecules are produced by *substrate level phosphorylation* for every two ATPs used, and two NADH molecules bear away from glycolysis the hydrogen ions and electrons recently acquired from an intermediate of glucose breakdown, *PGAL.* Glycolysis would quickly halt if the process ran out of NAD^+, which serves as the hydrogen and electron acceptor. When sufficient oxygen is present in eukaryotic cells, pyruvate from glycolysis is completely oxidized to become carbon dioxide during two phases called *acetyl-coenzyme A formation* and the *Krebs cycle* (Figure 6.5). Enzymes catalyze each step, and the intermediate produced serves as a substrate for the next enzyme in the series. In these reactions, which occur in the matrix within the inner membrane of the mitochondrion, NAD^+ and FAD serve as temporary acceptors for hydrogen ions and electrons released during the oxidations and are reduced to NADH and FADH$_2$. One ATP molecule is produced by substrate-level phosphorylation with each turn of the Krebs cycle (two ATPs for each pyruvate molecule). NADH and FADH$_2$, in turn, feed their hydrogen ions and electrons into the process known as *electron transport phosphorylation*, which occurs on the surface of the inner mitochondrial membrane, where NADH and FADH$_2$ are stripped of hydrogen ions and the high-energy electrons associated with those ions. The hydrogen ions are pumped into a reservoir, the outer mitochondrial compartment; thus, a hydrogen ion gradient is established, and the energy from the flow of these ions through the membrane into the inner mitochondrial compartment drives the phosphorylation of ADP, producing ATP. This idea is known as *chemiosmotic theory*. Most NADH molecules that enter the oxidative-phosphorylation sequence yield three ATP molecules; every FADH$_2$ molecule yields two. In heart and liver cells, each incoming NADH generates three ATP molecules. In other eukaryotic cells, NADH molecules made during glycolysis must be transported into the mitochondrion in order to enter the electron transport chain at the same point as FADH$_2$, and so each NADH generates only two ATP molecules (like FADH$_2$).

Key Terms

glycolysis

pyruvate

coenzymes

energy-requiring

PGAL

energy-releasing

NAD$^+$/NADH

electron transport
  phosphorylation

substrate-level
  phosphorylation

ADP/ATP

acetyl-coenzyme A

Krebs cycle

oxaloacetate

FAD/FADH$_2$

mitochondrial matrix

inner mitochondrial
  membrane

inner compartment

outer compartment

outer mitochondrial
  membrane

chemiosmotic theory

net yield

Objectives

1. Explain the purpose served by molecules of ATP reacting first with glucose and then with fructose-6-phosphate in the early part of glycolysis (see Figure 6.3).

2. Explain why your text says that four ATP molecules are made for every two used during glycolysis. Consult Figure 6.3 in your main text.

3. Consult Figure 6.5 in your main text. State the events that happen during acetyl-coenzyme A formation and explain how the process of acetyl-CoA formation relates glycolysis to the Krebs cycle.

4. State the factors that cause pyruvic acid to enter the acetyl-CoA formation pathway.

5. State what happens to the CO$_2$ produced during acetyl-CoA formation and the Krebs cycle.

6. Consult Figure 6.5 in the main text and predict what will happen to the NADH produced during acetyl-CoA formation and the Krebs cycle.

7. Calculate the number of ATP molecules produced during the Krebs cycle for each glucose molecule that enters glycolysis.

8. Summarize the basic ideas of the chemiosmotic theory and tell what the theory helps to explain.

9. Briefly describe the process of electron transport phosphorylation by stating what reactants are needed and what the products are. State how many ATP molecules are produced through operation of the transport system.

10. Account for the total *net yield* of thirty-six ATP molecules produced through aerobic respiration; that is, state how many ATPs are produced in glycolysis, the Krebs cycle, and the electron transport system.

Self-Quiz Questions

Fill-in-the-Blanks

If sufficient oxygen is present, the end product of glycolysis enters
(1) _____ pathways (acetyl-CoA formation, the (2) _____ cycle + (3)
_____ _____ _____), during which processes (4) _____
additional (5) _____ molecules are generated. In (6) _____
_____ , the food molecule fragments are further broken down into
(7) _____ _____ . During the reactions, hydrogen atoms (with their
(8) _____) are stripped from the fragments and are transferred to the
energy carriers (9) _____ and (10) _____ . The electrons are then sent
down a(n) (11) _____ system; hydrogen ions are pumped into the outer
mitochondrial compartment. The hydrogen ions accumulate and then follow
a gradient to flow into the inner compartment. The energy of the hydrogen
ion flow across the membrane is used in forming (12) _____ . Electrons
leaving the electron transport system combine with hydrogen ions and
(13) _____ to form water. These reactions occur only in (14) _____ .

Labeling

In problems 15 to 20, identify the structure or location; in problems 21 to 24,
identify the chemical substance involved.

(15) _____ _____ of mitochondrion   (20) _____

(16) _____ _____ of mitochondrion   (21) _____

(17) _____ _____ of mitochondrion   (22) _____

(18) _____ _____ of mitochondrion   (23) _____

(19) _____   (24) _____ _____ _____

ANAEROBIC ROUTES
  Lactate Fermentation
  Alcoholic Fermentation

Summary

In *anaerobic pathways*, something besides oxygen is the final electron acceptor. Glucose is partially dismantled by the glycolytic pathway; during this process, some of its stored energy ends up in two ATP molecules. At the end of glycolysis, glucose has been converted to two molecules of pyruvate; also, substrates give up electrons and hydrogen ions to two NAD^+ coenzymes, which yields two molecules of the electron carrier NADH. If oxygen—O_2—is not present in sufficient amounts, pyruvate enters *fermentation pathways,* in which pyruvate is converted into fermentation products such as lactate or ethanol and carbon dioxide. Fermentation pathways ensure that the essential carrier molecule NAD^+ is regenerated and ATP production by glycolysis can continue. In *alcoholic fermentation,* pyruvate from glycolysis is broken down to acetaldehyde. Acetaldehyde accepts electrons from NADH and becomes ethanol. Carbon dioxide is the gaseous product of alcoholic fermentation. Yeasts carry on alcoholic fermentation; they are valuable in alcohol and bread production. In *lactate fermentation,* pyruvate from glycolysis is converted to a three-carbon compound, lactate. Some bacteria and animal muscle cells carry on lactate fermentation.

Key Terms

anaerobic pathway   alcoholic fermentation   lactate fermentation
fermentation pathways

Objectives

1. Contrast the anaerobic and aerobic pathways of glucose metabolism in terms of (a) the organisms that use the pathways, (b) the number of ATP molecules generated by each, (c) the names of the major components of each type of pathway, and (d) the products of each type of pathway.
2. Explain why glucose breakdown and ATP formation are so important to the functioning of almost all prokaryotic and eukaryotic cells.
3. List some places where there is very little oxygen present and where anaerobic organisms might be found.
4. Describe what happens to pyruvate in anaerobic organisms. Then explain the necessity for pyruvate to be changed to a fermentative product.
5. State which factors determine whether the pyruvate (pyruvic acid) produced at the end of glycolysis will enter into the alcoholic fermentation pathway, the lactate fermentation pathway, or the acetyl-coenzyme A formation pathway.

**Self-Quiz
Questions**

Fill-in-the-Blanks

(1) _____ organisms can synthesize and stockpile energy-rich carbohydrates and other food molecules from inorganic raw materials.

(2) _____ is partially dismantled by the glycolytic pathway, during which process some of its stored energy ends up in two (3) _____ molecules.

Some of the energy of glucose is released during the breakdown reactions and is used in forming the energy carriers (4) _____ and (5) _____ . These reactions take place in the cytoplasm. If (6) _____ is not present in sufficient amounts, the end product of glycolysis enters (7) _____ pathways; in some bacteria and muscle cells it is converted into such products as (8) _____ or in yeast cells, it is converted into (9) _____ and (10) _____ _____ .

ALTERNATIVE ENERGY SOURCES IN THE HUMAN BODY

Summary

Many energy sources other than glucose can be fed into the glycolysis, acetyl-CoA formation, or Krebs cycle. Controls over enzymes that catalyze key steps in glycolysis and the Krebs cycle govern whether molecules are degraded as an ATP-generating energy source or converted to intermediate forms for use in biosynthesis. Carbohydrates can be converted into glucose or some other compound and thus enter glycolysis. Fats are first digested into glycerol and fatty acids, which enter glycolysis and acetyl-CoA formation. Proteins are digested into a variety of amino acids; some enter glycolysis, but many enter the Krebs cycle at various points by being converted into Krebs cycle intermediates. Metabolizing a gram of fat generally yields about twice as much ATP as metabolizing either a gram of carbohydrate or a gram of protein, so fats are efficient long-term energy-storage molecules. Cells generally use simple sugars, amino acids, nucleotides, fatty acids, and glycerol to synthesize as many of the more complex molecules as they need them; the remainder of the nutrients taken in are sent down respiratory pathways to make ATP and waste products or are converted into lipids or starches for longer-term energy storage.

Objectives

1. List some sources of energy (other than glucose) that can be fed into the respiratory pathways.
2. Explain what cells do with simple sugars, amino acids, fatty acids, and glycerol that exceed what the cells need for synthesizing their own assortments of more complex molecules.
3. Predict what your body would do to synthesize its needed carbohydrates and fats if you switched to a diet of 100 percent protein.

Self-Quiz Questions

True-False

If false, explain why.

___ (1) Glucose is the only carbon-containing molecule that can be fed into the glycolytic pathway.

___ (2) Fats are efficient long-term energy-storage molecules.

_____ (3) Simple sugars, fatty acids, and glycerol that remain after a cell's biosynthetic needs have been met are generally sent to the cell's respiratory pathways for energy extraction.

_____ (4) Carbon dioxide and water, the products of aerobic respiration, generally get into the blood and are carried to gills or lungs, kidneys, and skin, where they are expelled from the animal's body.

Labeling

Identify the process or substance indicated in the accompanying illustration.

(5) _____ _____

(6) _____

(7) _____

(8) _____ _____

(9) _____ _____

(10) _____ - _____

(11) _____

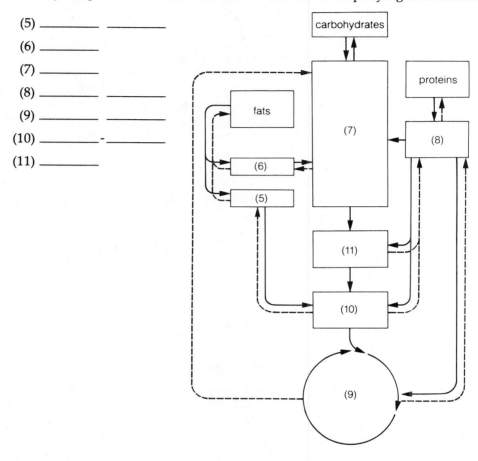

6-VI
(pp. 90–91)

ALTERNATIVE ENERGY SOURCES IN THE HUMAN BODY (cont.)
  Commentary: **Perspective on Life**
SUMMARY

Summary

Life apparently originated when the environment was rich in molecules containing carbon, hydrogen, oxygen, and nitrogen. Presumably, the first forms of life were heterotrophic prokaryotes that utilized glycolysis to extract energy from the early carbon compounds. As increasing demands were made on these carbon resources, alternative forms of harnessing energy, such as photosynthesis and chemosynthesis, evolved among the prokaryotes. Glycolysis followed by fermentation is an inefficient method of energy extraction that produces lactic acid or alcohol as accumulating by-products. Photosynthesis produces oxygen as a waste product.

Advances in the complexity of cells caused cells to become compartmentalized, and a variety of complex processes could occur in these well-organized regions. Cellular respiration was one such complex process that used the waste products of glycolysis and photosynthesis to extract much more energy and produce carbon dioxide and water, the raw materials of photosynthesis. Thus, the cycling of carbon, hydrogen, and oxygen came full circle, and life became more self-sustaining and balanced with the *carbon cycle.* Similar cycles have come to exist for other essential elements such as nitrogen and phosphorus.

Once molecules are assembled into the cells of organisms, their organization must be sustained by outside energy derived from food, water, and air. All living forms are part of an interconnected web of energy use and materials cycling that permeates all levels of biological organization. Should energy fail to reach any of these levels, life will become increasingly disordered. Energy flows only from forms rich in potential energy to forms having fewer and fewer usable stores of it. Life is no more and no less than a wonderfully complex system of prolonging order. It can do so because hereditary instructions are passed from generation to generation; thus, even though individual organisms die, life and order continue.

Objectives

1. Outline the supposed evolutionary sequence of energy-extraction processes.
2. Study the carbon cycle shown in the *Commentary* carefully; reproduce this from memory.

Self-Quiz Questions

True-False

If false, explain why.

___ (1) Energy is recycled along with materials.

___ (2) The first forms of life on Earth were most probably photosynthetic eukaryotes.

___ (3) Photosynthesis produces molecular oxygen as a by-product.

___ (4) Energy flows only from forms rich in potential energy to forms with fewer usable stores of energy.

CHAPTER TEST

UNDERSTANDING AND INTERPRETING KEY CONCEPTS

___ (1) Glycolysis would quickly halt if the process ran out of _____ , which serves as the hydrogen and electron acceptor.

(a) $NADP^+$
(b) ADP
(c) NAD^+
(d) H_2O

___ (2) The ultimate electron acceptor in aerobic respiration is _____ .

(a) NADH
(b) carbon dioxide (CO_2)
(c) oxygen ($1/2\ O_2$)
(d) ATP

___ (3) When glucose is used as an energy source, the largest amount of ATP is generated by the _____ portion of the entire respiratory process.

(a) glycolytic
(b) acetyl-CoA formation
(c) Krebs cycle
(d) electron transport phosphorylation

___ (4) The process by which about ten percent of the energy stored in a sugar molecule is released as it is converted into two small organic-acid molecules is _____ .

(a) photolysis
(b) glycolysis
(c) fermentation
(d) the dark reactions

___ (5) During which of the following phases of aerobic respiration is ATP produced directly by substrate-level phosphorylation?

(a) Glucose formation
(b) Ethyl-alcohol production
(c) Acetyl-CoA formation
(d) The Krebs cycle

___ (6) What is the name of the process by which reduced NADH transfers electrons along a chain of acceptors to oxygen so as to form water and in which the energy released along the way is used to generate ATP?

(a) Glycolysis
(b) Acetyl-CoA formation
(c) The Krebs cycle
(d) Electron transport phosphorylation

___ (7) Pyruvic acid can be regarded as the end product of _____ .

(a) glycolysis
(b) acetyl-CoA formation
(c) fermentation
(d) the Krebs cycle

___ (8) Which of the following is *not* ordinarily capable of being reduced at any time?

(a) NAD
(b) FAD
(c) Oxygen, O_2
(d) Water

INTEGRATING AND APPLYING KEY CONCEPTS

How is the "oxygen debt" experienced by runners and sprinters related to aerobic and anaerobic respiration in humans?

Crossword
Number Two

Across

1. organelle of photosynthesis
5. pigment that absorbs light energy
9. cellular material between the nucleus and cell membrane
13. negatively charged subatomic particle
15. starches and sugars are in this group of macromolecules
17. a functional group (of atoms) associated with energy transfer from molecule to molecule
19. pimple (slang)
20. acetyl _____
21. matter with very low density and viscosity; this type of matter fills its container
22. nutrient $(C_6H_{10}O_5)_n$ found in the storage areas of plants

24. incline, slope
25. unit of illumination (= 1 lumen per square meter)
26. the most abundant reduced coenzyme in living organisms
28. center from which microtubules are organized
33. not bold
34. cellular organelle of aerobic respiration
36. domesticated canine
37. the _____ cycle, a metabolic pathway that converts pyruvate to CO_2 and also transfers energy (as electrons and hydrogen atoms) to NAD^+ and FAD
38. a painter or sculptor
40. the major energy carrier common to all living organisms
41. the end-product of glycolysis: CH_3COCOO^-
43. a substance composed of atoms that have an identical number of protons
47. usually derived from living organisms; carbon-containing
48. a chemical reaction in which water or another simple substance is released by the combination of two or more molecules
50. next-to-last Greek letter, Ψ
55. adenine + ribose; a structural component of some nucleic acids
56. to escape as a gas
58. health food substitute for chocolate
59. nothing; naught
61. CFCs in the upper atmosphere are depleting the _____ layer, which blocks harmful ultraviolet radiation
62. one billionth of a meter unit
64. a process that uses CO_2 as the carbon source to synthesize food nutrients using the energy released from chemical reactions
66. having a tendency to combine with water
67. a process used by C_4 plants that fixes O_2 to RuBP rather than CO_2 to RuBP
68. molecules that have localized regions of positive and negative charge are said to be _____
70. building blocks of proteins; _____ acids
72. $C_3H_5O_3^-$; an end product of a particular fermentative pathway found in milk-souring organisms
75. a bluish white, lustrous metal; _____ oxide is used as a sunblock
80. a process that attaches a phosphate group to another molecule
81. same concentration of dissolved salts on both sides of a membrane
84. seed borne in a fruit that has a hard shell and tends to be one-seeded
86. a substance that releases a hydrogen ion in solution
88. if a cell is placed in a(n) _____ solution, water will flow out of the cell
89. any of several forces by which atoms or ions are bound in a molecule or crystal
90. a positively charged subatomic particle
92. the loss, by a molecule or atom, of one or more electrons
93. a metabolic pathway in which smaller molecules are joined together creating larger molecules
96. everything outside the grana of a chloroplast
97. two people; a pair
98. the emission of visible light by living organisms such as the firefly, various fish, and bacteria
99. one of the two products of ATP breakdown
101. the normal fullness or tension produced by the fluid content of plant or animal cells
102. concentration and electric gradients across a membrane drive ATP formation; _____ theory

103. one of several yellow or orange-colored pigments that absorb light energy with wavelengths shorter than 500 nm
104. an organelle in the nucleus where ribosomal subunits are produced

Down

1. large organic molecules that serve as cofactors; FAD and NAD$^+$ are examples
2. canal system within a eukaryotic cell
3. organism that lacks a membrane-bound nuclear region
4. at the crest of an energy hill, the reactants are in an activated, intermediate condition called the _____ state.
6. the acceptance, by a molecule or atom, of one or more electrons
7. PGA after it has been phosphorylated and reduced
8. glycerides, phospholipids, waxes, and steroids are all examples of the class of organic compounds called _____
9. the most common steroid in animal tissues and in cell membranes
10. the liquid portion of blood
11. _____ CoA is the molecule that enters the Krebs cycle from the glycolytic reactions
12. a surrounding substance within which something originates, develops, or is contained
14. a measure of the randomness, disorder, or chaos in a system
16. a _____ body gives rise to the microtubular core of cilia and flagella
18. the first stable intermediate of the light-independent reactions of photosynthesis; the product of carbon dioxide fixation
23. relating to the smallest portion of an element that still retains the properties of the element
26. a coenzyme that participates in glycolysis, fermentation, and aerobic respiration
27. pathways that break down large molecules into smaller ones
29. supplying the necessary energy to reach the transition state of a chemical reaction
30. sticking together
31. ability to change direction 180°
32. one of a pair of short, barrel-shaped structures that gives rise to the microtubule system of a cilium or flagellum; in animal cells
35. information organized for analysis
38. organism capable of synthesizing its own food from inorganic compounds
39. a _____ disk is part of a granum in a chloroplast
42. a vase used especially as a receptacle for cremation ashes
44. the crime of maliciously burning the property of another
45. place where something is located
46. a metal ion or coenzyme that helps make substrates bind to the active site of an enzyme or that makes them more reactive
49. organ of hearing
51. an organism that must consume other organisms to survive
52. a degradative process that continues from glycolysis and ends with the "spent" electrons being transferred to lactic acid or ethanol
53. a class of organic compounds that are assembled by linking together amino acids
54. preventing a process from happening
57. a colorless, volatile, flammable liquid often obtained by fermentation of sugars and starches; one kind is a widely used intoxicant

60. affirm; to state positively and definitely
63. pleasant and kind
65. the physics of the relationship between heat and other forms of energy
68. phosphoenol pyruvate, or_____ , for short
69. a combination of glycerol and fatty acids
71. of or relating to charged atoms
73. waxlike, water-repellant material that covers the epidermis of many plants
74. inflammation of the oil glands characterized by pimples
76. iron-containing protein molecules that occur in electron transport systems used in photosynthesis and aerobic respiration
77. an acid + a base $\longrightarrow$ a _____ + water
78. enzymes with both an active site and one or more regulatory sites
79. a DNA molecule and the proteins intimately associated with it
82. the molecule upon which an enzyme acts and to which an enzyme binds
83. a sleeveless garment fastened at the throat and worn hanging over the shoulders
85. a protein capable of speeding up a chemical reaction by lowering its activation energy
86. requiring the presence of O_2
87. one of two or more atoms, the nuclei of which have the same number of protons but different numbers of neutrons
91. a coenzyme reduced in the Krebs cycle and oxidized in the electron transport system that follows
94. the last letter of the Greek alphabet, Ω
95. an atom with unequal numbers of protons and electrons
100. a Middle Eastern organization devoted to reestablishing the land between the Mediterranean Sea and the Jordan River as Arab territory

7

CELL DIVISION AND MITOSIS

DIVIDING CELLS: THE BRIDGE BETWEEN GENERATIONS
    Overview of Division Mechanisms
    Some Key Points About Chromosome Structure
    Mitosis, Meiosis, and the Chromosome Number
MITOSIS AND THE CELL CYCLE

STAGES OF MITOSIS
    Prophase
    Metaphase
    Anaphase
    Telophase
CYTOKINESIS
SUMMARY

General Objectives

1. Understand the significance of cell division to reproduction.
2. Describe the general functions of both mitosis and meiosis in eukaryotic life cycles.
3. Define *chromosome* in terms of chemical construction and the different physical forms during cell divisions.
4. Describe, in terms of chromosome number, how mitosis and meiosis maintain a constant chromosome number in the life cycle of a particular species.
5. Understand what is meant by *cell cycle* and be able to visualize where mitosis fits into the cell cycle.
6. Describe each phase of mitosis.
7. Explain how the apportioning of cytoplasm to the daughter cells follows mitosis, which is a nuclear event.

7-I
(pp. 94–96)

DIVIDING CELLS: THE BRIDGE BETWEEN GENERATIONS
    Overview of Division Mechanisms
    Some Key Points About Chromosome Structure
    Mitosis, Meiosis, and the Chromosome Number

Summary

The point on which all life cycles pivot is *reproduction*, wherein organisms produce copies of themselves. Instructions for producing each new cell reside in the DNA, in which the sequence of nucleotide bases represents instructions for assembling proteins. Enzymes are proteins that give the cell a means to control the assembly of organic molecules from simple building blocks present in its surroundings. Thus, directly or indirectly, DNA governs which organic materials will occur in a cell and, in many cases, how those materials will become arranged as the cell grows and develops.

The continuity of life depends on the replication of a cell's set of DNA molecules before that cell reproduces. When a cell divides, copies of parental DNA are distributed to each daughter cell. In eukaryotes, hereditary instructions are usually divided among a set of two or more different DNA molecules, which vary in length and shape. There can be one, two, or more of these sets in a cell. With *mitosis* in somatic (body) cells, the number of DNA sets (chromosomes) in the nucleus does not change. Mitosis functions in the growth, repair, and asexual reproduction of some organisms. With *meiosis* in germ cells, the number of DNA sets in the daughter nucleus of each gamete or spore is exactly half of what it was in the parent nucleus. When two gametes (as with sperm and egg) fuse in fertilization to form a *zygote*, the number of DNA sets is restored to the number found in the body cells of a particular species. Through *cytokinesis*, each daughter cell receives a portion of the parental cell cytoplasm that contains necessary organelles and "start-up" enzymes involved in reading DNA and RNA instructions. Each unduplicated chromosome contains a DNA molecule and different proteins grouped together as a single, intact fiber. A constricted region, the *centromere*, occurs somewhere along the length of the chromosome and serves as a point of microtubule attachment for chromosome movement during cell division. During cell construction and maintenance activities, the chromosome is unfolded and uncoiled (long and thin) so that replication of DNA and synthesis of proteins can occur in accordance with DNA's instructions. Following DNA replication, the chromosome takes a different form (the duplicated chromosome) in which two threads of DNA and proteins remain attached as sister *chromatids*. During nuclear division, the DNA and proteins of the chromosome are folded and coiled into a condensed form.

The number of chromosomes differs from one species to the next, but within one species of multicellular organism, each cell (except the sex cells) contains the same number of chromosomes as the other. This constancy of chromosome number in a species is possible because in eukaryotic, sexually reproducing organisms chromosomes occur in *homologous pairs*. The homologues in a pair are physically similar in size and shape (except the pair of sex chromosomes) and contain DNA base sequences that code for the same characteristics of an organism; the expression of those characteristics may vary as in blue or brown eyes in humans. When both members of every pair of homologous chromosomes is present in a cell, that cell is said to be *diploid*. When only one member of each pair of homologous chromosomes is present in a cell, that cell is said to be *haploid*. Mitosis is a nuclear division that maintains the same number of chromosomes (haploid or diploid) in the daughter cells as in the parent cell. Meiosis is a nuclear division of diploid cells that separates or reduces the members of homologous pairs of chromosomes, sending one member of each pair of homologous chromosomes to one haploid sex cell and the other to another haploid sex cell.

Key Terms

reproduction	cytokinesis	centromere
daughter cell	germ cells	somatic cells
parental DNA	gametes	homologous chromosomes
mitosis	zygote	haploid
meiosis	chromosome	fertilization
nuclear division	sister chromatids	diploid

Objectives

1. Define the term *reproduction* and use it in an example.
2. Explain how DNA and its copying process is related to cellular reproduction and protein synthesis.

3. Describe how cells will use the instructions in DNA to synthesize molecules such as carbohydrates and lipids.
4. Explain why DNA must copy itself before a cell reproduces and relate this to the term *duplicated chromosome*.
5. Explain why it is important that daughter cells receive a significant portion of the parent cell cytoplasm through cytokinesis.
6. Define *mitosis* and *meiosis* and relate these processes to somatic cells, germ cells, gametes, and zygote.
7. Describe a chromosome and the two forms it may take during nuclear division. Briefly tell how the nuclear divisions, mitosis and meiosis, differ in their mechanisms of distributing chromosomes to somatic cells and gametes.
8. Relate the concept of homologous chromosomes to the haploid and diploid chromosome numbers; explain why gametes must be haploid and a zygote is diploid.

Self-Quiz Questions

Fill-in-the-Blanks

Making a copy of oneself is called (1) _____ , and the process begins at the molecular level. Instructions for producing each new cell and its important chemicals reside in (2) _____ . (3) _____ and (4) _____ are nuclear division mechanisms. (5) _____ is the process of dividing parental cell cytoplasm between two daughter cells. In animal cells, mitosis occurs in (6) _____ cells and meiosis occurs in (7) _____ cells and leads to the production of (8) _____ . Sperm nucleus and egg nucleus fuse in fertilization to form a(n) (9) _____ . The combination of DNA and proteins form an elongated structure, the (10) _____ . The two attached threads of a duplicated chromosome are known as (11) _____ _____ . The microtubule attachment site on a chromosome is the (12) _____ . A pair of chromosomes that are physically similar and whose genetic information deals with the same traits are said to be (13) _____ . Chromosome numbers of (14) _____ cells do not change during the process of (15) _____ . Chromosome numbers of (16) _____ cells are reduced by one-half during the process of (17) _____ . A cell possessing one-half the chromosomes of a diploid cell of its species is said to be (18) _____ .

True-False

If false, explain why.

___ (19) Mitosis and meiosis are cytoplasmic division mechanisms.

___ (20) In animals, mitosis occurs in germ cells.

___ (21) In animals, meiosis results in the formation of haploid gametes.

___ (22) A zygote is the first cell of a diploid individual.

___ (23) The two elongated portions of a duplicated chromosome are also known as chromosomes.

___ (24) Homologous chromosomes look alike but contain genetic information that controls different traits.

7-II
(p. 97)

MITOSIS AND THE CELL CYCLE

Summary

Mitosis, the division of the nucleus, is only a small slice of a larger series of events known as the eukaryotic *cell cycle.* Interphase, about ninety percent of a cell's life, consists of two phases of protein synthesis (G_1 and G_2) separated by one phase of DNA replication (the S phase).

Key Terms

cell cycle   gap phases, G_1 and G_2 S phase
interphase

Objectives

1. Scrutinize Figure 7.3 in the main text; then reproduce the entire figure of the cell cycle from memory.
2. List the events of interphase and briefly state the importance of each.

Self-Quiz Questions

Labeling

Identify the stage in the cell cycle indicated by each number.

(1) _____
(2) _____
(3) _____
(4) _____
(5) _____
(6) _____
(7) _____
(8) _____
(9) _____

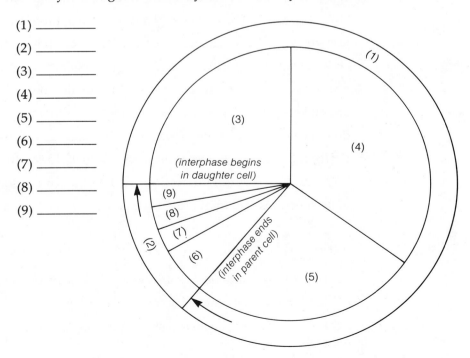

Link each time span identified below with the most appropriate number in the preceding labeling section.

___ (A) Period after replication of DNA during which the cell prepares for division by further growth and protein synthesis

___ (B) Period of asexual cell division

___ (C) DNA replication occurs now

___ (D) Period of cell growth before DNA replication

___ (E) Period when chromosomes are not visible

7-III
(pp. 97–102)

STAGES OF MITOSIS
   Prophase
   Metaphase
   Anaphase
   Telophase
CYTOKINESIS
SUMMARY

Summary

Parceling out DNA without errors to a new generation is a remarkable feat. Unwinding, unfolding, untwisting, assembling a duplicate strand, rewinding, refolding, and separating must occur in long, easily tangled DNA molecules. The physical isolation of DNA in the nucleus, chromosome packaging, and a microtubular system for moving many chromosomes simultaneously form the basis of reproduction in single-celled eukaryotes, as well as of physical growth in multicelled eukaryotes.

DNA replication precedes mitosis and is followed at some point by the sister chromatids coiling and condensing into chromosomes. The stages of mitosis (prophase, metaphase, anaphase, and telophase) do not have distinct boundaries; nevertheless, the chromosomes ultimately are guided into separate daughter cells in two equivalent parcels. During *prophase* a spindle is constructed of microtubules and the nuclear membrane disappears. The already duplicated chromosomes (each chromosome composed of two chromatids) condense, attach to spindle microtubules, and move to the spindle equator in early *metaphase*. The microtubules forming the *spindle apparatus* are responsible for separating *sister chromatids* (now independent chromosomes) from each other during *anaphase*. Once the two parcels of separated chromosomes arrive at opposite poles, newly forming nuclear membranes surround each parcel, and two nuclei eventually form during *telophase*. The chromosome content of each daughter nucleus is identical with that of the original nucleus. As soon as nuclear division (mitosis) is complete, *cytokinesis* (apportioning of cytoplasm to each of the daughter nuclei) generally occurs. Cytokinesis is the basis for reproduction in single-celled eukaryotes and is the basis for physical growth and repair in multicellular eukaryotes. Cytokinesis usually occurs during the latter part of mitosis. In animal cells, contractile microfilaments form a progressively smaller *cleavage furrow*, which eventually separates the cytoplasm into two daughter masses. In plant cells, a *cell plate* is formed from vesicles of materials and microtubule remnants. Growing outward from the center of the parent cell, the cell plate eventually grows to become a wall of cellulose that separates the cytoplasm into the two daughter masses.

Key Terms

mitosis	microtubules	spindle equator
prophase	anaphase	microfilaments
centriole	telophase	cell plate formation
metaphase	cytokinesis	vesicles
spindle apparatus	cleavage furrow	

Objectives

1. Describe the process of mitosis and what it accomplishes.
2. Describe the appearance of hereditary material (a) immediately before and during nuclear division and (b) when the cell is carrying out construction and maintenance activities.
3. List the events of each phase of mitosis.
4. Describe microtubule behavior during metaphase and anaphase as chromatids are lined up and then pulled apart.
5. Contrast the way cytokinesis occurs in animals with the way it occurs in plants.

Self-Quiz Questions

True-False

If false, explain why.

___ (1) In plant cells, a cleavage furrow forms around each cell's midsection.

___ (2) Cell plate formation occurs as part of mitosis in most land plants.

___ (3) Chromosomes move to the poles under their own power during anaphase.

___ (4) The stages of mitosis, in order, are prophase, anaphase, metaphase, and telophase.

___ (5) During anaphase, the microtubules that connect the centromere to its pole shorten.

Matching

Choose the one most appropriate answer for each.

___ (6) anaphase
___ (7) centromere
___ (8) chromatid
___ (9) cytokinesis
___ (10) G₁ phase
___ (11) G₂ phase
___ (12) metaphase
___ (13) prophase
___ (14) S phase
___ (15) telophase

A. First phase of protein synthesis
B. Cytoplasm allotted to and divided between two nuclei
C. Chromosomes condense; mitotic spindle starts to form
D. DNA replication occurs
E. Final phase of mitosis
F. Connects two sister chromatids
G. Second phase of protein building
H. Elongate half of a chromosome
I. Sister chromatids separate and move to opposite spindle poles
J. Chromosomes line up along equator of cell

UNDERSTANDING AND INTERPRETING KEY CONCEPTS

—— (1) The replication of DNA occurs _____ .

   (a) between the growth phases of interphase
   (b) immediately before prophase of mitosis
   (c) during prophase of mitosis
   (d) during prophase of meiosis

—— (2) In the cell life cycle of a particular cell, _____ .

   (a) mitosis directly precedes S
   (b) mitosis directly precedes G_1
   (c) G_2 directly precedes S
   (d) G_1 directly precedes S and G_2
   (e) mitosis and S directly precede G_1

—— (3) In eukaryotic cells, which of the following can occur during mitosis?

   (a) Two mitotic divisions to maintain the parental chromosome number
   (b) The replication of DNA
   (c) A long growth period
   (d) The nuclear envelope and nucleolus disappear

—— (4) *Haploid* refers to _____ and *diploid* refers to _____ .

   (a) half the parental chromosome number; having two chromosomes of each type in somatic cells
   (b) having two chromosomes of each type in somatic cells; twice the parental chromosome number
   (c) having two chromosomes of each type in somatic cells; half the parental chromosome number
   (d) twice the parental chromosome number; having two chromosomes of each type in somatic cells

—— (5) If a parent cell has sixteen chromosomes and undergoes meiosis, the resulting cells will have _____ chromosomes.

   (a) sixty-four
   (b) thirty-two
   (c) sixteen
   (d) eight
   (e) four

—— (6) If a parent cell has sixteen chromosomes and undergoes mitosis, the resulting cells will have _____ chromosomes.

   (a) sixty-four
   (b) thirty-two
   (c) sixteen
   (d) eight
   (e) four

___ (7) Chromosomes are duplicated during the _____ phase.

   (a) M
   (b) D
   (c) G_1
   (d) G_2
   (e) S

___ (8) The correct order of the stages of mitosis is _____ .

   (a) prophase, metaphase, telophase, anaphase
   (b) telophase, anaphase, metaphase, prophase
   (c) telophase, prophase, metaphase, anaphase
   (d) anaphase, prophase, telophase, metaphase
   (e) prophase, metaphase, anaphase, telophase

INTEGRATING AND APPLYING KEY CONCEPTS

Runaway cell division is characteristic of cancer. Imagine the various points of the mitotic process that might be sabotaged in cancerous cells in order to halt their multiplication. Then try to imagine how one might discriminate between cancerous and normal cells in order to guide the methods of sabotage most effective in combating cancer.

8
MEIOSIS

ON ASEXUAL AND SEXUAL REPRODUCTION
OVERVIEW OF MEIOSIS
    Think "Homologues"
    Overview of the Two Divisions
STAGES OF MEIOSIS
    Prophase I Activities
    Separating the Homologues

Separating the Sister Chromatids
MEIOSIS AND THE LIFE CYCLES
    Gamete Formation
    More Gene Shufflings at Fertilization
MEIOSIS COMPARED WITH MITOSIS
SUMMARY

General Objectives

1. Contrast asexual and sexual types of reproduction that occur on the cellular and multicellular organism levels.
2. Understand the effect that meiosis has on chromosome number.
3. Describe the events that occur in each meiotic phase.
4. Compare mitosis and meiosis; cite similarities and differences.
5. Show where meiosis generally occurs in plant life cycles and contrast this with where it generally occurs in animal life cycles.

8-I
(pp. 104–110)

ON ASEXUAL AND SEXUAL REPRODUCTION
OVERVIEW OF MEIOSIS
    Think "Homologues"
    Overview of the Two Divisions
STAGES OF MEIOSIS
    Prophase I Activities
    Separating the Homologues
    Separating the Sister Chromatids

Summary

Prokaryotic fission and eukaryotic mitosis are examples of *asexual reproduction* in which offspring cells are virtually identical to each other and to the parent cell; there is little variation in the *genes* (hereditary instructions). Most forms of *sexual reproduction* distribute hereditary instructions in the form of *alleles* (alternative forms of genes) into offspring by means of meiosis, a nuclear division occurring in diploid *germ cells* located in reproductive organs. The variation produced by the regrouping of genes that is associated with meiosis and gamete formation provides a testing ground for agents to select which groups work best in a particular environment; this serves as a basis for evolutionary change. In organisms that reproduce sexually, the diploid parent organisms must, at some point, produce haploid reproductive cells (gametes), which find each other and fuse (an event known as *fertilization*), thus forming a new diploid cell, the *zygote*.

Mitosis occurs in diploid as well as haploid organisms and maintains whatever number of chromosomes the parent cell had (either 2N or 1N), but *meiosis* reduces the parental diploid number (2N) to the haploid state (1N). That is, the diploid chromosome number is reduced by one-half as each member of each pair of *homologous chromosomes* is distributed to a different gamete. If there were no such reduction, fertilization would produce a zygote with twice the amount of hereditary information required to build a new individual.

Complete meiosis is composed of two sequential events (meiosis I and meiosis II) that follow an interphase of a diploid germ cell in which DNA replication and chromosome duplication occur. The duplicated chromosomes remain attached to each other at their centromere as *sister chromatids*. In *prophase I* of the first sequence, each pair of homologous chromosomes (two duplicated chromosomes or four chromatids) align closely, gene for gene. At this time, changes in the way genes are arranged on chromosomes occur through *crossing over* and *genetic recombination*. Now many of the duplicated chromosomes are composed of a mixture of paternal and maternal genes. Thus the chromosomes come to the spindle equator of *metaphase I* in the duplicated and recombined form. During *anaphase I* each homologue (composed of two attached sister chromatids) is moved to opposite ends of the spindle. The second division sequence, occurring in each of two cells, then separates the chromatids by means of a division process that resembles mitosis mechanically—although four haploid daughter cells form, rather than the two diploid cells that result from mitosis.

Eukaryotes rely on both sexual and asexual reproductive processes at various stages of the life cycle. The sexual process requires both gamete formation and fertilization, but the timing of these separate events varies considerably during the life cycles of different species. Most animals form haploid gametes; some time thereafter, one male and one female gamete generally fuse during fertilization.

Concept Aid

Confusion often exists as to what constitutes a "chromosome" during meiosis. The eukaryotic chromosome in the unduplicated state is composed of one DNA molecule coursing through it from end-to-end; this is associated with proteins of various types as a wrapping. Following the interphase of a germ cell, the "chromosome" is then composed of two attached "sister chromatids." Note that this duplicated structure is still referred to as one chromosome with a changed form. And so the term "chromosome" is correctly applied to the DNA thread and its protein wrapping in both the unduplicated and the duplicated state.

Key Terms

asexual reproduction	anther	genetic recombination
genes	ovary, ovaries	metaphase I
sexual reproduction	testis, testes	anaphase I
alleles	n and 2n	telophase I
meiosis	centromere	zygote
germ cells	sister chromatids	meiosis II
genetic variation	meiosis I	prophase II
gamete	prophase I	metaphase II
chromosomes	crossing over	anaphase II
homologous chromosomes	nonsister chromatids	telophase II

Objectives

1. Compare the characteristics of *asexual reproduction* with those of *sexual reproduction*.
2. Define *allele* correctly.
3. Explain the relationships among the following terms: *germ cells; somatic cells; reproductive organs*.
4. Explain why the phrase, *pair of homologous chromosomes,* is necessary.
5. List the differences between prophase I of meiosis and prophase of mitosis. State how the events of meiotic prophase I increase diversity within a species.
6. List the events that occur in the eight stages (meiosis I and meiosis II) of meiosis.
7. Scrutinize Figure 8.3 (main text) and note the stage in which crossing over and genetic recombination occur.
8. Study Figure 8.5 (main text) and be able to state why a chance, random alignment of homologous chromosomes at metaphase I adds to genetic variation in gametes.

Self-Quiz Questions

Matching

Assume for the following linkages that the cell starts with two pairs of already-duplicated chromosomes and that each chromosome consists of two chromatids. Choose the one most appropriate answer for each term.

(1) ___ anaphase I

(2) ___ anaphase II

(3) ___ metaphase I

(4) ___ metaphase II

(5) ___ prophase I

(6) ___ prophase II

(7) ___ telophase I

(8) ___ telophase II

A. Meiosis I is completed by this stage.
B. One member of each of the homologous pairs is separated from its mate; both are guided to opposite poles.
C. Two pairs of chromosomes are lined up at the spindle apparatus equator.
D. Chromosomes become clearly visible; nuclear region has two pairs of already-duplicated chromosomes.
E. Each chromosome splits; sister chromatids are separated and moved to opposite poles.
F. Each daughter cell ends up with one chromosome of each type and may function as a gamete.
G. Spindle apparatus reforms; each cell has two chromosomes, each of which consists of two chromatids.
H. Two chromosomes are lined up at spindle apparatus equator.

True-False

If false, explain why

___ (9) The first diploid cell formed after fertilization is called a *gamete*.

___ (10) Most eukaryotic cells are diploid.

___ (11) New plants sprouting from strawberry plant runners is an example of sexual reproduction.

___ (12) Homologous chromosomes have the same length, the same shape, but different genes.

8-II
(pp. 110–115)

MEIOSIS AND THE LIFE CYCLES
    Gamete Formation
    More Gene Shufflings at Fertilization
MEIOSIS COMPARED WITH MITOSIS
SUMMARY

Summary

Life cycles of multicellular animals advance from meiosis, gamete formation, fertilization, and then growth of the zygote through mitosis to form a mature animal. Plants differ from animals in certain features of their life cycles. The body of the more highly evolved plants is diploid, with multicellular organs in which meiosis occurs. But in plants, meiosis does not result directly in the formation of gametes. Instead, haploid spores are produced. Gamete formation in animals occurs only in special sex cells in male and female reproductive organs. In male animals, *spermatogenesis* (meiosis and gamete formation) produces from each spermatogonium four tiny, equal-sized haploid spermatids, which are transformed to mature *sperm*. In female animals, *oogenesis* (meiosis and gamete formation) produces one larger functional egg (*ovum*) and three tiny nonfunctional polar bodies from each oogonium.

Following the recombination occurring in prophase I and random chromosome alignments at metaphase I, the occurrence of fertilization adds to the "shuffling" of genes to create additional new genetic combinations. *Fertilization is a random event!* Which sperm bearing recombined genes will meet which egg-bearing recombined genes—by chance?

The two eukaryotic nuclear divisions we have studied are mitosis and meiosis. The functions of mitosis are asexual reproduction of single-celled organisms to produce genetic clones as well as growth and repair of multicelled forms. Meiosis is the basis of sexual reproduction in that this process produces new allele combinations and reduces the diploid chromosome number to haploid in the production of gametes. Fertilization of gametes from two parents restores the diploid number. The genetic variation that is achieved in offspring arising from sexual reproduction is due to: the genetic recombination occurring in prophase I; random spindle equator alignments of homologous chromosomes in metaphase I; and random fertilization.

Key Terms

animal life cycle	secondary spermatocyte	secondary oocyte
plant life cycle	spermatids	polar bodies
gamete formation	sperm	egg, ovum
spermatogenesis	oogenesis	fertilization
spermatogonium	oogonium	zygote
primary spermatocyte	primary oocyte	

Objectives

1. Describe and distinguish between the two forms of gamete formation. Explain why daughter cells of unequal size (one large egg and three polar bodies) are produced during oogenesis.

2. Contrast the generalized life cycle for plants with that for animals; describe how the products of meiosis differ in these two types of life cycles.
3. List all the events in sexual life cycles that lead to increased variation of gene combinations arising in diploid offspring.
4. State the functions of mitosis as compared with the functions of meiosis.

Self-Quiz Questions

Sequence

Arrange the following entities in correct order of development, putting a 1 by the stage that appears first and a 5 by the stage that finishes the process of spermatogenesis. Refer to Figure 8.8 in the main text.

(1) ___ primary spermatocyte

(2) ___ sperm

(3) ___ spermatid

(4) ___ spermatogonium

(5) ___ secondary spermatocyte

True-False

If false, explain why.

___ (6) The stage that ends meiosis II in male animals is the spermatid stage.

___ (7) The spore in a plant life cycle is the haploid product of meiosis.

___ (8) Oogenesis in female, multicellular animals results in four equal-sized eggs.

___ (9) All your cells are diploid except for any gametes you may produce.

___ (10) The exact gene combinations you now possess could have arisen from the meeting of any sperm from your father with any egg from your mother.

CHAPTER TEST

UNDERSTANDING AND INTERPRETING KEY CONCEPTS

___ (1) Which of the following does *not* occur in prophase I of meiosis?

(a) A cytoplasmic division
(b) A cluster of four chromatids
(c) Homologues pairing tightly
(d) Crossing over

___ (2) Crossing over is one of the most important events in meiosis because
_____ .

    (a) it produces new combinations of alleles on chromosomes
    (b) homologous chromosomes must be separated into different daughter cells
    (c) the number of chromosomes allotted to each daughter cell must be halved
    (d) homologous chromatids must be separated into different daughter cells

___ (3) Crossing over _____ .

    (a) generally results in pairing of homologues and binary fission
    (b) is accompanied by gene copying events
    (c) involves breakages and exchanges being made between sister chromatids
    (d) alters the composition of chromosomes and results in new combinations of alleles being channeled into the daughter cells

___ (4) The appearance of chromosome ends lapped over each other in meiotic prophase I provides evidence of _____ .

    (a) meiosis
    (b) crossing over
    (c) chromosomal aberration
    (d) fertilization
    (e) spindle fiber formation

___ (5) Which of the following does not increase genetic variation?

    (a) Crossing over
    (b) Random fertilization
    (c) Prophase of mitosis
    (d) Random homologue alignments at metaphase I

___ (6) Which of the following is the most correct sequence of events in animal life cycles?

    (a) Meiosis, fertilization, gametes, diploid organism
    (b) Diploid organism, meiosis, gametes, fertilization
    (c) Fertilization, gametes, diploid organism, meiosis
    (d) Diploid organism, fertilization, meiosis, gametes

___ (7) In sexually reproducing organisms, the zygote is _____ .

    (a) an exact genetic copy of the female parent
    (b) an exact genetic copy of the male parent
    (c) unlike either parent genetically
    (d) a genetic mixture of male parent and female parent

INTEGRATING AND APPLYING KEY CONCEPTS

A few years ago, the actual cloning of a human being was claimed to have been accomplished. Later, this claim was admitted to be fraudulent. Supposing that sometime in the future the cloning of humans becomes possible, speculate about the effects of reproduction without sex on human populations.

9

OBSERVABLE PATTERNS
OF INHERITANCE

MENDEL'S INSIGHTS INTO THE PATTERNS
OF INHERITANCE
    Mendel's Experimental Approach
    Some Terms Used in Genetics
    The Concept of Segregation
    Testcrosses
    The Concept of Independent Assortment

VARIATIONS ON MENDEL'S THEMES
    Dominance Relations
    Interactions Between Different Gene Pairs
    Multiple Effects of Single Genes
    Environmental Effects on Phenotype
    Continuous Variation in Traits
SUMMARY

**General
Objectives**

1. Know Mendel's principles of dominance, segregation, and independent assortment.
2. Know (very well) the terms used in inheritance studies (p. 118, main text) and concepts illustrated in Figure 9.3.
3. Understand how to solve genetics problems that involve monohybrid and dihybrid crosses.
4. Understand the variations that can occur in observable patterns of inheritance such as incomplete dominance, codominance, epistasis, pleiotropy, and continuous variation; be able to solve problems dealing with these hereditary variations.
5. Describe how the environment can affect the phenotype.

**9-I
(pp. 117–121)**

MENDEL'S INSIGHTS INTO THE PATTERNS OF INHERITANCE
    Mendel's Experimental Approach
    Some Terms Used in Genetics
    The Concept of Segregation

Summary

In 1865 Gregor Mendel presented the results of his now-famous experiments establishing the physical and mathematical rules that govern inheritance. Through his combined talents in plant breeding and mathematics, he perceived patterns in the expression of traits from one generation to the next. Mendel examined seven different traits (flower color, for instance), each of which had two different strains (white-flowered and purple-flowered), in sexually reproducing garden pea plants. Not only can garden pea flowers fertilize themselves but also the gametes from one individual can fertilize the gametes of another individual (in a process known as *cross-fertilization*). Mendel obtained seeds from strains in which all self-fertilized offspring displayed the same form of a trait, generation after generation; these are known as *true-breeding* strains. When two true-breeding organisms displaying different forms of a trait are crossed, the offspring are called *hybrids* because the two genetic units inherited, one from each parent, are unlike each other. Because Mendel first concentrated on studying a *single* trait, his initial experiments were called *monohybrid* crosses. The results of monohybrid cross experiments demonstrated that, in sexually reproducing organisms, one member from each pair is separated (segregated) from the other, and each ends up in a different gamete. This statement is the Mendelian *principle of segregation*.

Biochemical experiments and improvements in microscopy have advanced our understanding of the segregation process since Mendel's work in the mid-1800s, and the terms used to describe genetic events have changed accordingly. *Genes* (really linear segments of DNA coding) are distinct units of heredity that dictate the traits of a new individual. A particular gene has a location on a chromosome, its *locus*. If a trait is controlled by two genes, there will be one gene for that trait in a gamete (which is haploid). Following the fusion of the gametes (egg and sperm), the new individual has two genes for each trait; that is, the individual is diploid.

Alternative forms of genes for a given trait are known as *alleles*. In a diploid individual, two alleles for a given trait may be identical (in which case the individual is said to be *homozygous* for that trait) or two alleles may be different (in which case the individual is *heterozygous*). In heterozygous individuals, expression of one allele may mask expression of the other. The masking allele is *dominant* and the masked allele is *recessive*. Even so, both alleles retain their physical identity, and all the pairs of alleles constitute the individual's *genotype* (genetic makeup) for a particular assemblage of traits. The term *phenotype* (appearance makeup) refers to the collection of structural and behavioral features that are expressed, not masked.

The outcomes of genetics crosses are understood in terms of *probability* (chance events). The *Punnett square* is a device drawn on paper to simulate genetics crosses clearly and display the products of meiosis and fertilization on paper.

Concept Aid

Although the Punnett-square (checkerboard) method is a favored method for solving genetics problems, there is a quicker way. Six different outcomes are possible from monohybrid crosses. Studying the following *relationships* allows the result of any monohybrid cross to be obtained by *inspection*.

1. AA x AA = all AA
   (Each of the four blocks of the Punnett square would be AA.)
2. aa x aa = all aa
3. AA x aa = all Aa
4. AA x Aa
         or = 1/2 AA; 1/2 Aa
   Aa x AA
   (Two blocks of the Punnett square are AA, two blocks are Aa.)
5. aa x Aa
         or = 1/2 aa; 1/2 Aa
   Aa x aa
6. Aa x Aa = 1/4 AA; 1/2 Aa; 1/4 aa
   (One block in the Punnett square is AA, two blocks are Aa, and one block is aa.)

Key Terms

trait
the "blending hypothesis"
Mendel
true-breeding organism
cross-fertilization
genes
locus
gene pair

alleles
homozygous, homozygote
heterozygous, heterozygote
homozygous recessive
homozygous dominant
dominant
recessive
genotype

phenotype
homologous chromosomes
hybrids
monohybrid crosses
P, F$_1$, F$_2$
probability
Punnett-square method
segregation

Objectives

1. Explain why the blending hypothesis is not regarded as an acceptable explanation of genetic behavior today.
2. Distinguish self-fertilizing organisms from cross-fertilizing organisms. Explain the relation of true-breeding strains to self-fertilization.
3. Describe Mendel's monohybrid cross experiments with different true-breeding strains of garden peas (Figure 9.5 in main text). Give the numbers and ratios he obtained in his F$_1$ and F$_2$ generations when he crossed a round-seeded parent with a wrinkled-seeded parent. Recall that Mendel began his experiments with true-breeding plants.
4. Distinguish between the terms *dominant* and *recessive* and state what happens to the recessive gene in a hybrid.
5. Contrast the number of chromosomes in a gamete with the number of genes in a zygote.
6. Define and distinguish between *homozygous* and *heterozygous*.
7. State the Mendelian principle of segregation and explain what would happen if segregation did not occur.
8. Symbols are used as shorthand to represent genes in diagramming genetic crosses. The alphabet letter chosen for the trait is the first letter of the phenotype that is less common or is not considered normal. *That* letter in lowercase is used for the recessive allele and the capital represents the dominant allele. Assume the genotypes are homozygous and write the genotype (the pair of letter symbols) for each of the following traits:
   a. normal (dominant) vs. albinism (recessive) in humans
   b. freckled (dominant) vs. unfreckled (recessive) in humans
   c. normal (dominant) vs. chondrodystrophic dwarfism (recessive) in humans
9. Define *probability* and explain what it has to do with genetics.
10. Explain why, as a result of monohybrid crosses, Mendel always obtained an approximate 3:1 phenotypic ratio of dominant to recessive traits in the F$_2$ generation.
11. In one of Mendel's studies, the gene for green pod color was found to be dominant over its allele for yellow pods. Cross a homozygous green-podded plant with a yellow-podded plant. Show by a diagram (with the letter symbols) both the genotype and the phenotype for the P, F$_1$, and F$_2$ generations. Use the Punnett-square method or solve by inspection.

Self-Quiz Questions

Fill-in-the-Blanks

Mendel perceived patterns in the expression of (1) _____ from one generation to the next. He examined seven different traits, each with two different (2) _____ in sexually reproducing garden (3) _____ plants, which can carry out both self-fertilization and (4) _____-_____ . Self-fertilized plants that all display the same form of a trait generation after generation are (5) _____-_____ strains. When two such strains that display different forms of a trait are crossed, the offspring are called (6) _____ because the two genetic units inherited, one from each parent, are unlike each other. If two of the hybrid offspring are mated, one-fourth of their offspring are (7) _____ _____ ; that is, they again show the trait that seemed to have disappeared in their parents' generation. Through such studies, Mendel demonstrated that the widely believed (8) _____ theory was untrue. Mendel's initial experiments were called (9) _____ crosses because he concentrated his examinations on a single trait. With such experiments, he was able to demonstrate that, in sexually reproducing organisms, two units of heredity control each trait. During gamete formation, the two units of each pair are separated from each other and end up in different gametes; the statement of this fact is the Mendelian principle of (10) _____ .

9-II
(pp. 121–122)

MENDEL'S INSIGHTS INTO THE PATTERNS OF INHERITANCE (cont.)
  Testcrosses
  The Concept of Independent Assortment

Summary

To determine whether an individual that shows the dominant form of a trait is heterozygous or homozygous, a *testcross* is performed: the individual is mated with an individual that possesses the homozygous *recessive* genotype. If the unknown individual is homozygous, all offspring will show the dominant trait; if it is heterozygous, half the offspring will show the recessive trait. The *Punnett-square method* is perhaps the favored way to predict the probable ratio of traits in offspring of monohybrid and dihybrid crosses. *Dihybrid cross* experiments deal simultaneously with contrasting forms of two traits. Through experiments with dihybrid crosses, Mendel began to perceive how sexual reproduction might foster genetic diversity. The Mendelian principle of *independent assortment* states that, when gametes are formed, distribution of alleles for a given trait that have been segregated into one gamete or another is independent of, and does not interfere with, distribution of alleles for other traits that have been segregated.

Concept Aid

The inspection method for solving monohybrid crosses illustrated in 9-I (p. 95) may also be applied to an algebraic method for solving dihybrid crosses, although some writing is required. One must keep in mind the six possible outcomes from monohybrid crosses. The ratios are combined, as in the following example, when all the factors of one ratio are multiplied by all the factors of the other.

The dihybrid cross, Aabb x AaBb, may be solved as follows:

$$
1AA \begin{cases} 1bb = 1AAbb \\ \\ 1Bb = 1AABb \end{cases}
$$

$$
2Aa \begin{cases} 1bb = 2Aabb \\ \\ 1Bb = 2AaBb \end{cases}
$$

$$
1aa \begin{cases} 1bb = 1aabb \\ \\ 1Bb = 1aaBb \end{cases}
$$

The answer (right vertical column) may be read as: 1/8 AAbb (or 2/16 blocks in a 16-block Punnett square); 1/8 AABb; 2/8 Aabb; 2/8 AaBb; 1/8 aabb; 1/8 aaBb.

Key Terms	testcross	dihybrid crosses	independent assortment

Objectives

1. Define *testcross* and give an example.
2. Define *dihybrid cross* and distinguish it from *monohybrid cross*.
3. Explain why Mendel obtained a 9:3:3:1 phenotypic ratio of characteristics in the F_2 generation as a result of a dihybrid cross.
4. State the principle of independent assortment as formulated by Mendel.
5. Solve the following problems—both of which deal with two different traits on different chromosomes—by using the Punnett-square method or by multiplying the separate probabilities.
   a. The abilities to bark and to erect their ears are dominant traits in some dogs; keeping silent while trailing prey and droopy ears are recessive traits. If a barking dog with erect ears (heterozygous for both traits) is mated with a droopy-eared dog that keeps silent while trailing, what kinds of puppies can result?
   b. A spinach plant with straight leaves and white flowers was crossed with another spinach plant with curly leaves and yellow flowers. All of the F_1 generation have curly leaves and white flowers. If any two individuals of the F_1 are crossed, what phenotypic ratio will result? Show all genotypes.

Self-Quiz Questions

True-False

If false, explain why.

___ (1) The results of Mendels testcross supported his principle of independent assortment.

___ (2) The phenotypic ratio in the F_1 generation of a testcross that involves monohybrids is 1:1 if the dominant-appearing parent is heterozygous.

___ (3) The phenotypic ratio of 9:3:3:1 in the F_2 generation obtained in a dihybrid cross supports Mendels principle of independent assortment.

(4) Albinos cannot form the pigments that normally produce skin, hair, and eye color, so albinos have white hair and pink eyes and skin (because the blood shows through). To be an albino, one must be homozygous recessive for the pair of genes that codes for the key enzyme in pigment production. Suppose that a woman of normal pigmentation with an albino mother marries an albino man. State the possible kinds of pigmentation possible for this couple's children and specify the ratio of each kind of children the couple is likely to have. Show the genotype(s) and state the phenotype(s).

Multiple Choice

___ (5) In heterozygous individuals, _____ .

(a) the expression of one allele may mask the expression of the other for any given trait
(b) both alleles for a given trait retain their identity throughout the individual's life cycle
(c) if one allele masks the expression of the other, the former allele is dominant and the latter is recessive
(d) all of the above

___ (6) When independent assortment of alleles does occur, _____ .

(a) the alleles that code for the differing traits of an individual are parceled out independently of one another into separate gametes
(b) the process of meiosis retains the diploid number of chromosomes
(c) the alleles that code for a single trait are parceled out into the same gamete
(d) both (a) and (c)

___ (7) A true-breeding red-flowered, tall plant is crossed with a true-breeding dwarfed, white-flowered plant. When two of the resulting F_1 plants are crossed, the phenotypic ratio in the F_2 offspring is approximately a _____ ratio. Red flowers and tall plants are dominant.

(a) 1:1:1:1
(b) 1:2:1
(c) 1:3:1
(d) 9:3:3:1

___ (8) If AaBb is mated to AaBb, the probability is _____ that their offspring will express both dominant A and dominant B.

(a) 3/16
(b) 1/16
(c) 9/16
(d) 5/16

9-III
(pp. 122–124)

VARIATIONS ON MENDEL'S THEMES
    Dominance Relations
    Interactions Between Different Gene Pairs
    Multiple Effects of Single Genes

Summary

Since Mendel's time, studies have shown that interactions between genes can lead to many phenotypic possibilities in addition to the clear-cut fully dominant or fully recessive ("all-or-nothing") examples investigated by Mendel. Cases of *incomplete dominance* and *codominance* tell us that there can be a range of dominance when only two alleles are under consideration. Moreover, for one trait, dominance can also be expressed in a variety of forms if the trait is governed by a system involving more than two alleles. Blood types that result from the ABO blood group are an example of such a codominant system. *Epistasis* is a situation in which one gene pair masks the effect of one or more other pairs; in calico cats the spotting allele masks either black or yellow, producing white patches in whatever region of the cat's coat the spotting allele is behaving epistatically to the alleles for coat color.

Single genes also may have major or minor influences on the expression of more than one trait—a situation known as *pleiotropism*. Sickle-cell anemia and mutations in genes that code for proteins necessary in a variety of structures and/or processes are examples of pleiotropy.

Key Terms

incomplete dominance	ABO blood typing	albino
codominance	epistasis	pleiotropy
multiple allele system	melanin	sickle-cell anemia

Objectives

1. Define *dominance* and *codominance;* cite examples of each.
2. Explain why inheritance of many traits is no longer viewed as being under the control of purely dominant or purely recessive agents.
3. List and explain several examples of inheritance in which expression of a given trait occurs as a spectrum of intermediate aspects of the trait, rather than just as either of the two extremes (dominant or recessive).
4. State whether each of the examples you gave in Objective 3 is controlled by one pair of genes or by more than one pair.
5. Define *epistasis* and cite an example.
6. Explain the meaning of *pleiotropy* and discuss how it is related to sickle-cell anemia. Name some of the phenotypic characteristics that are part of the sickle-cell syndrome.

Self-Quiz Questions

True-False

If false, explain why.

___ (1) If a true-breeding, red-flowered snapdragon is crossed with a white-flowered snapdragon, all of the F_1 generation will be red-flowered.

___ (2) If two of the F_1 red-flowered snapdragons are crossed, three-fourths of their offspring (the F_2) will be red-flowered, and one-fourth will be white-flowered.

___ (3) Hemophilia is an example of a disorder caused by incomplete dominance.

___ (4) The expression of flower color in snapdragons is an example of codominance.

___ (5) A child of blood type AB has a mother with type A blood. The child's father could not have type O.

___ (6) Human skin color is an example of incomplete dominance that demonstrates the validity of the "blending" hypothesis.

9-IV
(pp. 124–126)

VARIATIONS ON MENDEL'S THEMES (cont.)
   Environmental Effects on Phenotype
   Continuous Variation in Traits
SUMMARY

Summary

Throughout any individual's life cycle, genes must interact with the environment in the expression of phenotype. Genes provide the chemical messages for growth and development, but neither growth nor development can proceed without environmental contributions to the living form. The environment profoundly influences the expression of all genes and their contributions to phenotype.

For the sake of simplicity, most examples of heritable traits mentioned in texts are controlled by a single gene locus, but there are many traits (human skin color, for instance) that are governed by several active alleles at different loci; such traits show continuous variation and are explained by several genes that each has an "additive" effect on the phenotype.

Key Terms

environmental effect   continuous variation

Objectives

1. Cite several examples of environmental alteration of an organism's basic phenotype.
2. Describe the difficulty of evaluating the influence of heredity and environment on a particular phenotype.
3. Explain continuous variation; list some examples of traits that continuously vary and give the cause.
4. Explain why frequency distribution diagrams (see Figure 9.12 in the main text) are used to portray continuous variation in populations.

Self-Quiz Questions

True-False

If false, explain why.

___ (1) A person's becoming thirty pounds overweight is one way that the external environment can influence the phenotypic expression of human genotype.

___ (2) The fact that it is possible to change the extent to which genotypic potential is expressed in the phenotype demonstrates that acquired phenotypic characteristics can be inherited.

___ (1) Mendel's principle of independent assortment states that _____ .

    (a) one allele is always dominant to another
    (b) hereditary units from the male and female parents are blended in the offspring
    (c) the two hereditary units that influence a certain trait separate during gamete formation
    (d) each hereditary unit is inherited separately from other hereditary units

___ (2) One of two or more alternative forms of a gene for a single trait is a(n)_____.

    (a) chiasma
    (b) allele
    (c) autosome
    (d) locus

___ (3) In the F_2 generation of a monohybrid cross involving complete dominance, the expected *phenotypic* ratio is _____ .

    (a) 3:1
    (b) 1:1:1:1
    (c) 1:2:1
    (d) 1:1

___ (4) In the F_2 generation of a cross between a red snapdragon (homozygous) and a white snapdragon, the expected phenotypic ratio of the offspring is _____ .

    (a) three-fourths red, one-fourth white
    (b) 100 percent red
    (c) one-fourth red, one-half pink, one-fourth white
    (d) 100 percent pink

___ (5) The results of a testcross reveal that all of the offspring resemble the parent being tested. That parent necessarily is _____ .

    (a) heterozygous
    (b) polygenic
    (c) homozygous
    (d) recessive

___ (6) When gametes from one individual undergo sexual fusion with gametes from another, this is called _____ .

    (a) blending
    (b) cross-fertilization
    (c) true-breeding
    (d) independent assortment

—— (7) A single gene that affects several seemingly unrelated aspects of an individual's phenotype is said to be _____ .

    (a) pleiotropic
    (b) epistatic
    (c) mosaic
    (d) continuous

—— (8) Suppose two individuals, each heterozygous for the same characteristic, are crossed. The characteristic involves complete dominance. The expected genotypic ratio of their progeny is _____ .

    (a) 1:2:1
    (b) 1:1
    (c) 100 percent of one genotype
    (d) 3:1

—— (9) If the two homozygous classes in the F_1 generation of the preceding cross are allowed to mate, the observed genotypic ratio of the offspring will be _____ .

    (a) 1:1
    (b) 1:2:1
    (c) 100 percent of one genotype
    (d) 3:1

—— (10) Assume that genes A, B, and C are located on different chromosomes. How many different types of gametes could an individual of genotype AaBbCc produce during meiosis if crossing over did not occur?

    (a) Two
    (b) Eight
    (c) Six
    (d) Twelve

CHAPTER TEST   **INTEGRATING AND APPLYING KEY CONCEPTS**

Solve the following genetics problems. Show all setups, genotypes, and phenotypes.

(1) Holstein cattle have spotted coats due to a recessive gene; a solid-colored coat is caused by a dominant gene. If two spotted Holsteins are mated, what types of offspring would be likely? Explain your statement with a diagram.

(2) Give the genotypes of the parents in this problem, labeling appropriately. Cocker spaniels are black due to a dominant gene and red due to its recessive allele. Solid color depends on a dominant gene, and the development of white spots depends on its recessive allele. A solid red female was mated with a black-and-white male, and five puppies resulted: one black, one red, two black-and-white, and one red-and-white.

10

CHROMOSOME VARIATIONS AND HUMAN GENETICS

RETURN OF THE PEA PLANT
    Autosomes and Sex Chromosomes
    Linkage and Crossing Over
CHROMOSOME VARIATIONS AND HUMAN
GENETICS
    Autosomal Recessive Inheritance
    Autosomal Dominant Inheritance

X-Linked Recessive Inheritance
Changes in Chromosome Structure
Changes in Chromosome Number
Commentary: Prospects and Problems in
Human Genetics
SUMMARY

General Objectives	1. Know that each karyotype is composed of autosomes and sex chromosomes and that genes are carried on all chromosomes.
	2. Relate the significance of linkage and crossing over to inheritance.
	3. Distinguish *autosomal recessive inheritance* from *sex-linked recessive inheritance;* define *autosomal dominant inheritance.*
	4. Understand how changes in chromosome structure and number can affect the outward appearance of organisms.
	5. Explain how knowing about modern methods of genetic screening and genetic counseling can minimize potentially tragic events.

10-I
 (pp. 128–131)
	RETURN OF THE PEA PLANT
	     Autosomes and Sex Chromosomes
	     Linkage and Crossing Over

Summary

During the late 1800s, several cytologists described the events that compose mitosis (Flemming) and meiosis (Weismann, Flemming). Mendel's paper on pea plant hybridization was published in 1866 but remained unnoticed until 1900. In that year, researchers rediscovered Mendel's paper, which stated that diploid cells have two units of instruction (genes) for each trait; the two units segregate when sex cells are formed. Mitotic and meiotic phenomena were then linked with Mendel's concepts of segregation and independent assortment, and it was suggested that Flemming's "threads" were likely the carriers of Mendel's "units of instruction."

Just after 1900, microscopists began to examine chromosomes very closely; a difference was noted when comparing the chromosome pairs of the two sexes. Most of the chromosomes could be matched for size, other physical characteristics, and number. These chromosomes are known as *autosomes*. However, one pair was found to differ; that is, they are not the same in males and females; these two chromosomes are called *sex chromosomes*. In many species, females have two identical sex chromosomes; they are designated "X." In males, the two sex chromosomes differ physically; one matches the female X chromosome and is designated "X" while the other is known as "Y."

Chromosomes at metaphase are studied by staining them to produce visual bands where DNA is most concentrated. In this way, the chromosomes can be matched by size, centromere location, and band thickness; they are arranged and numbered from largest to smallest. The appearance of the chromosomes as they appear at metaphase is known as the *karyotype*. They are often photographed in this organized form.

When segregated in meiosis, XX females produce only X-bearing eggs whereas males produce X- and Y-bearing sperms. If an X sperm fertilizes an X egg, a female is produced; if a Y sperm fertilizes an X egg, a male is produced. The Y chromosome seems to carry far fewer genes than the X chromosome. One recently discovered gene on the Y is the *sex-determining gene*, which produces a substance that is crucial in determining whether the new organism will develop testes or ovaries. The X chromosome may carry as many as 200 genes; most affect nonsexual characteristics. Genes carried on the X or Y chromosome may be said to be sex-linked. Now that scientists are able to identify more genes on chromosomes, it is preferred to be more precise and refer to genes either as "X-linked" or "Y-linked." X-linked genes were discovered in the early 1900s in the laboratory of Thomas Hunt Morgan. The discovery was made during breeding experiments with fruit flies. Their work supported what was then a new concept: genes are carried on chromosomes (study Figure 10.3 and the accompanying legend in the main text).

Early researchers also discovered that some groups of genes do not assort independently—as Mendel had said—because they are physically linked as part of the same chromosome. These and later studies with other organisms have made it clear that the number of *linkage groups* (genes on the same chromosome form a linkage group) corresponds to the number of chromosomes characteristic of the species, which further confirms the hypothesis that chromosomes are the vehicles that transport the genes through generations. Different genes physically located on the same chromosome tend to end up together in the *same* gamete; they do *not* assort independently. Different genes located on *nonhomologous* chromosomes usually assort independently of each other into gametes.

Crossing over happens during the first division of meiosis, in which two nonsister chromatids of homologous pairs of chromosomes exchange corresponding segments of breakage points. Crossing over results in genetic recombination—the formation of new combinations of alleles in a chromosome, hence in gametes, and finally in zygotes. The description of the formation of crossovers gave biologists the idea that locations of genes on their chromosomes could be mapped because the farther apart on the chromosome two linked genes may be, the more likely it is that crossing over will disrupt the original combination of linked alleles. The probability that crossing over and recombination will occur at a point somewhere between two genes located on the same chromatid is directly proportional to the distance that separates them. Genes are carried in linear array on chromosomes.

Key Terms

Flemming	sex chromosomes	testis determining factor
Weismann	X chromosome	sex determination
chromosomal theory	Y chromosome	sex-linked
of inheritance	karyotype	X-linked gene
autosome	sex-determining gene	Y-linked

Thomas Hunt Morgan   linkage, linked   reciprocal cross
fruit fly, Drosophila   crossing over

Objectives

1. Identify the researcher who first described threadlike bodies in the nuclei of dividing cells; state what was so useful about this contribution.
2. State the significant proposal made by August Weismann in 1887.
3. Relate the significance of the year 1900 to genetics knowledge.
4. Define *sex chromosomes* and *autosomes*; then distinguish the types of alleles found on each, if possible.
5. Define *karyotype*; state why it is useful; describe how karyotyping is done.
6. Distinguish between *sex-linked genes* and *sex-determining genes*.
7. Describe two significant contributions the Morgan research team made to our early understanding of genetics.
8. Explain how the meiotic segregation of sex chromosomes to gametes and subsequent random fertilization determines sex in many organisms.
9. Define gene *linkage* and state the relationship between the probability of crossing over (and subsequent recombination) and the distances between two genes located on the same chromosome.

Self-Quiz Questions

Fill-in-the-Blanks

(1) _____ first described the behavior of threadlike bodies (chromosomes) during mitosis. (2) _____ first proposed that chromosomal number must be reduced by half during gamete formation; the process became known as (3) _____ . (4) _____ chromosomes resemble each other in length, shape, and gene sequence.

Matching

Choose the single most appropriate letter.

(5) ___ crossing over

(6) ___ linkage group

(7) ___ karyotype

(8) ___ independent assortment

(9) ___ autosomes

(10) ___ segregation

(11) ___ Y-linked

A. All the chromosomes in a cell except the sex chromosomes
B. Homologous chromosomes are channeled into different gametes
C. Chromosomes of a cell arranged in order from largest to smallest
D. Refers to all the genes found on a chromosome
E. Exchange of parts of chromatids between nonsister chromatids during prophase I of meiosis
F. Genes carried on the smaller sex chromosome
G. How one pair of chromosomes segregates into different gametes does not influence how a different pair of chromosomes segregates into the same gamete

Multiple Choice

___ (12) Most animal and some plant species have sex chromosomes that _____ .

    (a) are called autosomes
    (b) always consist of one X and one Y
    (c) differ in number or in kind between males and females
    (d) all of the above

___ (13) Normally, all eggs from a human female contain _____ .

    (a) one X chromosome
    (b) one Y chromosome
    (c) one X chromosome and one Y chromosome
    (d) only autosomes

___ (14) _____ is a consequence of crossing over.

    (a) A dihybrid cross
    (b) Linkage
    (c) Dependent assortment
    (d) Genetic recombination

___ (15) The closer together two genes are on one chromosome, _____ .

    (a) the fewer times there will be crossing over and genetic recombination occurring between them
    (b) the more times there will be crossing over and genetic recombination occurring between them
    (c) the farther apart they will appear in a linkage group
    (d) the greater the chance will be that point mutations will happen to either of them

___ (16) Genetic recombination occurs as a result of crossing over between nonsister chromatids of homologous chromosomes during _____ .

    (a) metaphase of mitosis
    (b) prophase I of meiosis
    (c) syngamy (fusion of gametes)
    (d) prophase II of meiosis
    (e) a one percent frequency of recombination between the genes involved

___ (17) Which of the following did Morgan and his research group not do?

    (a) They isolated and kept under culture fruit flies with the sex-linked recessive white-eyed trait.
    (b) They developed the technique of amniocentesis.
    (c) They discovered X-linked genes.
    (d) Their work reinforced the concept that each gene is located on a specific chromosome.

___ (18) In most species, males transmit the Y chromosome to _____ .

    (a) their daughters but not to their sons
    (b) both sons and daughters
    (c) their sons but not to their daughters
    (d) neither sons nor daughters

CHROMOSOME VARIATIONS AND HUMAN GENETICS
    Autosomal Recessive Inheritance
    Autosomal Dominant Inheritance
    X-Linked Recessive Inheritance

Summary

The study of human genetics has proven difficult because humans live in a variety of environments, have erratic reproductive patterns, rear typically small families, and live long. *Pedigree charts* are constructed to study genetic relationships. An abnormal *trait* is one that deviates from normal; an abnormal trait may or may not be a medical problem. Abnormal traits considered to be medical problems are termed *disorders*. Genetic disorders arising from *mutations* are of considerable importance to families and society.

Autosomal recessive inheritance is characterized by normal but heterozygous parents; offspring have a twenty-five percent chance of being homozygous recessive by receiving an abnormal, autosomal recessive allele from each parent. *Galactosemia* is a disorder caused by a recessive allele carried on an autosome. The disorder appears to be transmitted according to the rules of simple Mendelian inheritance, because it is expressed in homozygotes of both sexes. Symptoms of galactosemia arise from the inability to produce an enzyme responsible for metabolizing the sugar galactose, a breakdown product of lactose (milk sugar). Accumulated high levels of galactose damage the brain, liver, and eyes.

Autosomal dominant inheritance is characterized by an abnormal trait being expressed in both heterozygotes and homozygotes; an affected individual will nearly always have an affected parent (with exception of a new mutation). Expectation would be that one-half the sons and one-half the daughters of an affected parent would display the abnormality. *Achondroplasia* is caused by a dominant allele carried on an autosome. The disorder is characterized by a failure of the long bones of the arms and legs to develop normally, producing a form of dwarf that is fertile and otherwise normal. *Huntington's disorder* is also a dominant genetic disorder that causes a progressive deterioration of the nervous system; symptoms first appear about age forty or after with eventual death.

X-linked recessive inheritance is characterized by the mutated gene occurring only on the X chromosome, not the Y. Heterozygous females may carry the trait but not express it; males, having only one X chromosome, are afflicted with the disorder, as are homozygous recessive females, and express the trait. Hemophilia A, an X-linked trait in which a blood-clotting mechanism fails, is an example.

Key Terms

family pedigrees
polydactyly
abnormal
disorder
autosomal recessive
  inheritance

galactosemia
autosomal dominant
  inheritance
achondroplasia

Huntington's disorder
X-linked recessive
  inheritance
hemophilia A

Objectives

1. List reasons it is difficult to study inheritance patterns in humans; describe the purpose of a *pedigree chart*.
2. Distinguish *abnormality* from *disorder*.
3. Describe the characteristics of *autosomal recessive inheritance*; summarize the characteristics of *galactosemia* as an example.

4. Describe the characteristics of *autosomal dominant inheritance;* summarize the characteristics of *achondroplasia* and *Huntington's disorder* as examples.
5. Describe the characteristics of *X-linked inheritance;* summarize the characteristics of *hemophilia A* as an example.

Self-Quiz Questions

True-False

If false, explain why.

___ (1) Inheritance patterns may be tracked by pedigree charts.

___ (2) Polydactyly is caused by an autosomal recessive gene.

___ (3) An abnormal condition always requires medical attention.

___ (4) Galactosemia is caused by a buildup of lactose in the blood.

___ (5) Huntington's disorder is caused by an autosomal recessive gene.

Fill-in-the-Blanks

(6) _____ are very difficult to study as a genetics subject. The human trait exhibiting onset after age forty is (7) _____ _____ and is caused by an autosomal (8) _____ gene. The type of inheritance having symptom-free heterozygotes but affected homozygotes is (9) _____ _____ . The type of inheritance where every affected individual usually has at least one affected parent is (10) _____ _____ .

10-III
(pp. 135–141)

CHROMOSOME VARIATIONS AND HUMAN GENETICS (cont.)
    Changes in Chromosome Structure
    Changes in Chromosome Number
    Commentary: **Prospects and Problems in Human Genetics**
SUMMARY

Summary

Rapid advances in microscopy enabled cytogeneticists to visualize the structure of chromosomes. Most often, crossing over proceeds normally during gamete formation. On very rare occasions, rearrangements in chromosome structure occur during crossing over—*deletions, duplications, translocations,* and *inversions.* These mistakes invariably affect fairly large chromosomal segments of the total chromosome number characteristic of the species. The *cri-du-chat* (cat-cry) disorder results from a deletion of human chromosome 5. Duplications of genes may have been important in evolution. Inversions reverse the position and sequence of genes in a chromosome. Some types of human cancer have been traced to a translocation.

Besides these rare internal arrangements, from time to time whole chromosomes or chromosome sets go astray. Changes in chromosome number sometimes occur in which a particular type of chromosome is either absent entirely (*monosomy*) or present three (*trisomy*) or more times in the diploid chromosome set. Such abnormal chromosome numbers arise when *nondisjunction* occurs, the failure of homologous chromosomes to segregate properly during anaphase I or anaphase II of meiosis. In humans, nondisjunction during gamete formation is the cause of *Down syndrome* (trisomy 21), *Turner syndrome* (XO), *Klinefelter syndrome* (XXY), and the *XYY condition*.

Because genetic disorders have no permanent cures, a variety of *phenotypic treatments* include diet modification, environmental adjustment, surgical correction, and chemical modifications of gene products. Early detection is important to establish effective treatment. As examples, diet control is used to treat galactosemia and phenylketonuria; albinos can avoid sunlight; and those with sickle-cell anemia avoid activity under oxygen stress. Surgery is used to correct the appearance and function of cleft lip. Wilson's disorder is treated with a drug to avoid excess copper. Preventive measures include prenatal diagnosis, genetic screening, genetic counseling, and induced abortion. Prenatal diagnosis through *amniocentesis* (the sampling of the fluid in the mother's uterus) makes it possible to determine whether the developing fetus has a severe genetic disorder. If a severe disorder is discovered, the parents may elect to induce abortion of the fetus, if specific requirements are met. *Chorionic villi analysis* uses chorion cells for genetic analysis. *Genetic screening* involves large-scale detection of persons who are affected or carriers of genetic disease. *Genetic counseling* deals with family pedigree analysis and risk prediction. Parents of embryos diagnosed with severe disorders can be given information as a basis to consider induced abortion; ethics play a part in such a decision.

Key Terms

deletion	Turner syndrome	chorionic villi analysis
duplication	sex chromosome	chorion
cri-du-chat disorder	abnormality	amnion
inversion	Klinefelter syndrome	cleft lip
translocation	XYY condition	Wilson's disorder
nondisjunction	phenotypic treatments	genetic screening
trisomy	phenylketonuria, PKU	genetic counseling
monosomy	prenatal diagnosis	induced abortion
Down syndrome	amniocentesis	

Objectives

1. Describe the following types of chromosome structural changes: *deletion, duplication, inversion,* and *translocation.*
2. Define *nondisjunction;* relate this to monosomy and trisomy.
3. Practice sketching a diagram of nondisjunction similar to Figure 10.13 in the main text.
4. State the abnormal chromosome constitutions and describe the outstanding phenotypic characteristics of Down syndrome, Turner syndrome, Klinefelter syndrome, and the XYY condition.
5. Distinguish between *autosomal nondisjunction* and *sex chromosome nondisjunction.*
6. Define *phenotypic treatment* and describe one example.
7. Explain the procedures used in two types of *prenatal diagnosis, amniocentesis,* and *chorionic villi analysis;* compare the risks.
8. Explain the benefits of *genetic screening* and *genetic counseling* to society.
9. Discuss some of the ethical considerations that might be associated with a decision of *induced abortion.*

Matching

Choose the single most appropriate letter. You can use the same letter more than once.

(1) ___ Down syndrome

(2) ___ galactosemia

(3) ___ hemophilia A

(4) ___ Huntington's disorder

(5) ___ XYY condition

(6) ___ Turner syndrome

(7) ___ *cri-du-chat* disorder

(8) ___ achondroplasia

(9) ___ XXY syndrome

A. Autosomal recessive inheritance
B. X-linked recessive inheritance
C. Nondisjunctive inheritance involving autosomes
D. Autosomal dominant inheritance
E. Nondisjunctive inheritance involving sex chromosomes
F. An abnormal phenotype due to chromosome structure

Matching

Choose the one most appropriate answer for each. A letter may be used more than once or not at all.

(10) ___ albinism

(11) ___ amniocentesis

(12) ___ chorionic villi analysis

(13) ___ cleft lip

(14) ___ galactosemia

(15) ___ phenylketonuria

(16) ___ karyotype analysis

(17) ___ sickle-cell anemia

(18) ___ Wilson's disorder

(19) ___ pedigree chart

A. A phenotypic defect that can be helped by diet modification
B. A phenotypic defect that can be helped by environmental adjustments
C. A phenotypic defect that can be helped by surgical correction
D. A phenotypic defect that can be helped by chemotherapy
E. Used to arrive at a diagnosis
F. Used to determine the location of a specific gene on a specific chromosome
G. Used to depict genetic relationships among the members of families

Fill-in-the-Blanks

A (20) _____ is the loss of a piece of a chromosome during gamete formation. Segments of chromosomes occasionally break off and move to a nonhomologous chromosome; such an event is known as a(n) (21) _____ . A(n) (22) _____ occurs when a deleted piece of chromosome gets turned around and rejoins the chromosome at the same place, but backward. A(n) (23) _____ occurs when a segment of genes is repeated and present in more than one copy in the chromosomes. (24) _____ causes (25) _____ , in which three chromosomes of the same kind are present in the chromosome set. In (26) _____ , an individual is $2n - 1$. Homologous

chromosomes can exchange parts as a result of (27) _____-_____ .
The XO, XXY, and XYY chromosome conditions are classified as types of
(28) _____ _____ _____ . (29) _____ is a genetic condition
that was once mistakenly thought to be genetically predisposed to criminal
behavior.

CHAPTER TEST

UNDERSTANDING AND INTERPRETING KEY CONCEPTS

___ (1) All the genes located on a given chromosome compose a _____ .

   (a) karyotype
   (b) bridging cross
   (c) wild-type allele
   (d) linkage group

___ (2) Chromosomes other than those involved in sex determination are
       known as _____ .

   (a) nucleosomes
   (b) heterosomes
   (c) alleles
   (d) autosomes

___ (3) If two genes are very close to each other on the same chromatid,
       _____ .

   (a) crossing over and recombination occurs between them quite often
   (b) they act as if they are assorted independently into gametes
   (c) they will most probably be in the same linkage group
   (d) they will be segregated into different gametes when meiosis occurs

___ (4) Karyotype analysis is _____ .

   (a) a means of detecting and reducing mutagenic agents
   (b) a surgical technique that separates chromosomes that have failed
       to segregate properly during meiosis II
   (c) used in prenatal diagnosis to detect chromosomal mutations and
       metabolic disorders in embryos
   (d) a process that substitutes defective alleles with normal ones

___ (5) A color-blind man and a woman with normal vision whose father was
       color-blind have a son. Color blindness, in this case, is caused by a
       sex-linked recessive gene. If only the male offspring are considered, the
       probability that their son is color-blind is _____ .

   (a) .25 (or 25 percent)
   (b) .50 (or 50 percent)
   (c) .75 (or 75 percent)
   (d) 1.00 (or 100 percent)
   (e) none of the above

___ (6) A woman heterozygous for color blindness (a sex-linked recessive allele) marries a man with normal color vision. What is the probability that their first child will be color-blind?

(a) 25 percent
(b) 50 percent
(c) 75 percent
(d) 100 percent

___ (7) Red-green color blindness is a sex-linked recessive trait in humans. A color-blind woman and a man with normal vision have a son. What are the chances that the son is color-blind? If the parents ever have a daughter, what is the chance for each birth that the daughter will be color blind? (Consider only the female offspring.)

(a) 100 percent, 0 percent
(b) 50 percent, 0 percent
(c) 100 percent, 100 percent
(d) 50 percent, 100 percent
(e) none of the above

___ (8) Suppose that a hemophilic male (sex-linked recessive allele) and a female carrier for the hemophilic trait have a nonhemophilic daughter with Turner syndrome. Nondisjunction could have occurred in _____ .

(a) both parents
(b) neither parent
(c) the father only
(d) the mother only

___ (9) Nondisjunction involving the X chromosome occurs during oogenesis and produces two kinds of eggs, XX and O (no X chromosome). If normal Y sperm fertilize two types, which genotypes are possible?

(a) XX and XY
(b) XXY and YO
(c) XYY and XO
(d) XYY and YO

___ (10) Of all phenotypically normal males in prisons, the type once thought to be genetically predisposed to becoming criminals was the _____ .

(a) XXY disorder
(b) XYY disorder
(c) Turner syndrome
(d) Down syndrome

___ (11) Amniocentesis is _____ .

(a) a surgical means of repairing deformities
(b) a form of chemotherapy that modifies or inhibits gene expression or the function of gene products
(c) used in prenatal diagnosis to detect chromosomal mutations and metabolic disorders in embryos
(d) a form of gene-replacement therapy

(1) The parents of a young boy bring him to their doctor. They explain that the boy does not seem to be going through the same vocal developmental stages as his older brother. The doctor orders a common cytogenetics test to be done, and it reveals that the young boy's cells contain two X chromosomes and one Y chromosome. Describe the test that the doctor ordered and explain how and when such a genetic result—XXY—most logically occurred.

(2) Solve the following genetics problem. Show all setups, genotypes, and phenotypes.

A husband sues his wife for divorce, arguing that she has been unfaithful. His wife gave birth to a girl with a fissure in the iris of her eye, a sex-linked recessive trait. Both parents have normal eye structure. Can the genetic facts be used to argue for the husband's suit? Explain your answer.

11

DNA STRUCTURE AND FUNCTION

DISCOVERY OF DNA FUNCTION
DNA STRUCTURE
    Components of DNA
    Patterns of Base Pairing
DNA REPLICATION
    Assembly of Nucleotide Strands

    Replication Enzymes
ORGANIZATION OF DNA IN CHROMOSOMES
SUMMARY

General Objectives

1. Understand how experiments using bacteria and viruses demonstrated that instructions for producing heritable traits are encoded in DNA.
2. Know the parts of a nucleotide and know how they are linked together to make DNA.
3. Understand how DNA is replicated and what materials are needed for replication.
4. Describe how DNA is organized into chromosomes.

11-I
(pp. 143–147)

DISCOVERY OF DNA FUNCTION
DNA STRUCTURE
    Components of DNA
    Patterns of Base Pairing

Summary

During its life span, each cell deploys thousands of different enzymes in translating its hereditary instructions (its genotype) into specific physical traits (its phenotype). Each cell comes into the world with a limited set of specific instructions that promote survival in a particular kind of habitat. Hereditary instructions must be of a sufficiently stable chemical nature to ensure constancy in structure and function and yet must allow room for subtle change. In the search for the hereditary molecule, two principal candidates were considered as likely: *proteins*, because they are diverse and complex, and *nucleic acids*, because—although they have only a simple repeating-unit structure—they are abundant in the nuclear regions of cells where the hereditary instructions were thought most likely to reside. After a variety of experiments, in almost all organisms, DNA (deoxyribonucleic acid) was demonstrated to be the molecule that contains the genetic code.

Only four different kinds of nucleotides are found in a DNA molecule, and each has the same phosphate group and the same five-carbon sugar (deoxyribose). Each nucleotide has one of four different nitrogen-containing bases.

Cytosine (C) and thymine (T) are single-ring pyrimidines; adenine (A) and guanine (G) are double-ring purines. In a DNA molecule, nucleotides are strung together into a long chain, with each phosphate group of one nucleotide linking with the sugar of the next nucleotide; when nucleotides are thus assembled, the purines and pyrimidines project to one side. Franklin and others, by using a variety of physical and chemical experimental techniques, provided Watson and Crick with enough clues for them to build a scale model, revealing that DNA must be double-stranded, with two nucleotide strands wound helically about each other like a spiral staircase. Watson and Crick stated that the two strands of nucleotides were held together by hydrogen bonds linking purines (A and G) from one strand with pyrimidines (C and T) on the other strand and vice versa; this statement has become known as *base pairing*. Any purine-pyrimidine pair can follow any other in the DNA chain, so an extremely large number of code sequences can result from different arrangements of these two pairs of nitrogenous bases. In all species, there is a constancy in base pairing (adenine to thymine; guanine to cytosine) between the two strands of the DNA double helix. The DNA of different species shows variation in which base follows which in a strand.

Key Terms

Miescher	^{32}P	purines
deoxyribonucleic acid, DNA	nucleotides	X-ray diffraction
Griffith	base pairs	Franklin
Avery	cytosine	double helix
Hershey and Chase	thymine	Watson and Crick
bacteriophages	pyrimidines	principle of base pairing
^{35}S	adenine	base sequence
	guanine	

Objectives

1. State which two classes of molecules were (before 1952) suspected of housing the genetic code.
2. Identify the types of research carried out by Miescher, Griffith, Avery and colleagues, and Hershey and Chase, and state the specific advances in our understanding of genetic behavior made by each.
3. Explain how bacteriophages were used to study genetics.
4. State which piece of research demonstrated that DNA, not protein, governed inheritance.
5. Draw the basic shape of a deoxyribose molecule and show how a phosphate group is joined to it when forming a nucleotide.
6. Generally distinguish a *purine* from a *pyrimidine* on the basis of molecule structure; name the two purines and the two pyrimidines found in DNA.
7. Show how each nucleotide base would be joined to the sugar-phosphate combination drawn in Objective 5.
8. State the kinds of information about DNA structure that Rosalind Franklin discovered by using X-ray diffraction.
9. List the particular contributions of Watson and Crick to the modern understanding of DNA structure and behavior.
10. Explain what is meant by the pairing of nitrogen-containing bases (base pairing), and state the feature that causes bases of one strand to join with bases of the other strand.
11. Describe how the Watson-Crick model reflects the constancy in DNA observed from species to species, yet allows for variations from species to species.

True-False

If false, explain why.

___ (1) Miescher first discovered nucleic acids he obtained from cell nuclei.

___ (2) ^{35}S can be used to label DNA, and ^{32}P can be used to label almost any protein.

___ (3) Some viruses contain DNA as their hereditary material.

___ (4) Griffith demonstrated that DNA was the substance that had been permanently transformed when the rough strain of bacteria was transformed to the disease-causing smooth strain of bacteria.

___ (5) Bacteriophages are bacteria that eat viruses.

___ (6) DNA contains sulfur but not phosphorus.

___ (7) Hershey and Chase were the first to demonstrate that DNA was unquestionably the molecule that contained the hereditary instructions. Each nucleotide consists of a six-carbon sugar, a phosphate group, and a nitrogen-containing base.

___ (8) DNA is composed of only four different types of nucleotides, each of which contains adenine, thymine, cytosine, or guanine.

___ (9) In the DNA of every species, the amount of adenine present always equals the amount of thymine, and the amount of cytosine always equals the amount of guanine (A = T, and C = G).

___ (10) In a nucleotide, the phosphate group is attached to the nitrogen-containing base, which is attached to the five-carbon sugar.

___ (11) Watson and Crick built their model of DNA in the early 1950s.

___ (12) Guanine pairs with cytosine and adenine pairs with thymine by forming hydrogen bonds between them.

11-II
(pp. 147–148)

DNA REPLICATION
  Assembly of Nucleotide Strands
  Replication Enzymes

Summary

Once DNA's double-stranded nature had been deduced, Watson and Crick immediately saw how such a molecule could be duplicated prior to mitosis or meiosis. If the hydrogen bond(s) that connect a purine with a pyrimidine in a base pair are broken by the appropriate enzyme and separated, each base can attract and form hydrogen bonds with its complementary nucleotide already synthesized and present in the cell's stockpile of nucleotides. Each parent DNA strand remains intact, and a new companion strand is assembled on each one. The only sequence that could be attached to one parent strand (say, ATTCGC) would be an exact complement of the base sequence (TAAGCG) that occurs on the other parent strand. Each single parent strand ends up forming a new complementary strand, with which it remains linked by means of hydrogen bonds. In each of the two resulting DNA molecules, one old DNA strand has been *conserved* as a partner for the newly formed strand. A variety of enzymes participate in DNA replication: some serve to unwind the double helix of DNA and some wind up the replicated regions. The major replication enzymes are DNA polymerases, which control the assembly of nucleotides on parent strands and check for mismatched base pairs; incorrect bases are replaced with correct bases. This quality control inspection function by DNA polymerases ensures that DNA replication will be extremely accurate.

Key Terms

replication	double helix	new complementary strand
free nucleotides	conserved parent strand	DNA polymerases

Objectives

1. Assume that the two parent strands of DNA have been separated and that the base sequence on one parent strand is ATTCGC. State the sequence of new bases that will complement that parent strand.
2. Describe how double-stranded DNA replicates from stockpiles of nucleotides.
3. Explain what is meant by "each parent strand is conserved in each new DNA molecule."
4. List four jobs that enzymes perform during the process of replication.
5. Describe the functions of DNA polymerases during DNA replication.

Self-Quiz Questions

True-False

If false, explain why.

___ (1) The hydrogen bonding of adenine to guanine is an example of complementary base pairing.

___ (2) The replication of DNA is considered a conserving process because the same four nucleotides are used again and again during replication.

___ (3) Each parent strand remains intact during replication, and a new companion strand is assembled on each of those parent strands.

___ (4) Some of the enzymes associated with DNA assembly repair errors during the replication process.

Labeling

Identify each indicated part of the accompanying illustration.

(5) _____

(6) _____ _____

(7) _____

(8) _____

(9) _____

(10) _____

(11) _____

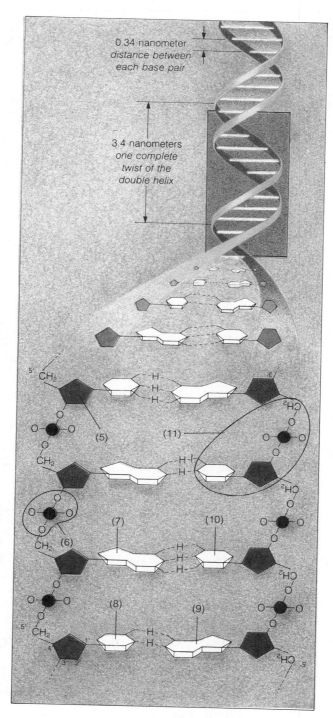

Concept Aid

The following helpful memory devices may be used:

1. Use pyrCUT to remember that the single-ring pyrimidines are cytosine, uracil, and thymine.

2. Use purAG to remember that the double-ring purines are adenine and guanine.

3. Pyrimidine is a *long* name for a *narrow* molecule; purine is a *short* name for a *wide* molecule.

4. To recall the number of hydrogen bonds between DNA bases, remember that AT = 2 and CG = 3.

**ORGANIZATION OF DNA IN CHROMOSOMES
SUMMARY**

Summary

Each chromosome has one DNA molecule coursing lengthwise through it. Most of the DNA and RNA in the nucleus is used to direct the activities of different genes. These molecules are long and thin; they avoid entanglement because they are highly organized. Proteins called *histones,* which make up about half of all proteins in the nucleus, are tightly linked to DNA and form a nucleoprotein fiber that resembles a chain of beads. Each bead of histone-DNA is a *nucleosome,* a complex coiling arrangement that winds up small lengths of DNA. This packing arrangement seems to be involved in controlling the gene activity required for protein synthesis.

Key Terms

histones   histone-DNA spool   nucleosome

Objectives

1. Describe how histone-DNA is organized to form nucleosomes.
2. List possible reasons for the highly organized packing of nucleoprotein into chromosomes.

**Self-Quiz
Questions**

Fill-in-the-Blanks

Each chromosome has one (1) _____ molecule coursing through it.

Eukaryotic DNA is complexed tightly with many (2) _____ . Some

(3) _____ proteins act as spools to wind up small pieces of (4) _____ .

A(n) (5) _____ is a histone-DNA spool. The way the chromosome is

packed is known to influence the activity of different (6) _____ .

CHAPTER TEST

UNDERSTANDING AND INTERPRETING KEY CONCEPTS

___ (1) Each DNA strand has a backbone that consists of alternating _____ .

   (a) purines and pyrimidines
   (b) nitrogen-containing bases
   (c) hydrogen bonds
   (d) sugar and phosphate molecules

___ (2) In DNA, complementary base pairing occurs between _____ .

   (a) cytosine and uracil
   (b) adenine and guanine
   (c) adenine and uracil
   (d) adenine and thymine

___ (3) Adenine and guanine are _____ .

    (a) double-ringed purines
    (b) single-ringed purines
    (c) double-ringed pyrimidines
    (d) single-ringed pyrimidines

___ (4) Franklin used the technique known as _____ to determine many of the physical characteristics of DNA.

    (a) transformation
    (b) transmission electron microscopy
    (c) density-gradient centrifugation
    (d) X-ray diffraction

___ (5) The significance of Griffith's experiment that used two strains of pneumonia-causing bacteria is that _____ .

    (a) the conserving nature of DNA replication was finally demonstrated
    (b) it demonstrated that harmless cells had become permanently transformed through a change in the bacterial hereditary system
    (c) it established that pure DNA extracted from disease-causing bacteria transformed harmless strains into "pathogenic strains"
    (d) it demonstrated that radioactively labeled bacteriophage transfer their DNA but not their protein coats to their host bacteria

___ (6) The significance of the experiments in which ^{32}P and ^{35}S were used is that _____ .

    (a) the semiconservative nature of DNA replication was finally demonstrated
    (b) they demonstrated that harmless cells had become permanently transformed through a change in the bacterial hereditary system
    (c) they established that pure DNA extracted from disease-causing bacteria transformed harmless strains into "killer strains"
    (d) they demonstrated that radioactively labeled bacteriophage transfer their DNA but not their protein coats to their host bacteria

___ (7) Franklin's research contribution was essential in _____ .

    (a) establishing the double-stranded nature of DNA
    (b) establishing the principle of base pairing
    (c) establishing most of the principal structural features of DNA
    (d) all of the above

___ (8) When Griffith injected mice—with a mixture of dead (pathogenic), encapsulated S cells and living, unencapsulated R cells of pneumonia bacteria—he discovered that _____ .

    (a) the previously harmless strain had permanently inherited the capacity to build protective capsules
    (b) the dead mice teemed with living pathogenic (R) cells
    (c) the killer strain R was encased in a protective capsule
    (d) all of the above

INTEGRATING AND APPLYING KEY CONCEPTS

Review the stages of mitosis and meiosis as well as the process of fertilization. Include what has now been learned about DNA replication and the relationship of DNA to a chromosome. As the stages are covered, be sure each cell receives the proper number of DNA threads.

12

FROM DNA TO PROTEINS

PROTEIN SYNTHESIS
    Transcription
    Classes of RNA
    Translation
    Mutation and Protein Synthesis

CONTROLS OVER GENE ACTIVITY
    Gene Control in Prokaryotes
    Gene Control in Eukaryotes
    Commentary: **Cancer—When Gene**
    Controls Break Down
SUMMARY

General Objectives

1. Know how the structure and behavior of DNA determine the structure and behavior of the three forms of RNA during transcription.
2. Know how the structure and behavior of the three forms of RNA determine the primary structure of polypeptide chains during translation.
3. Understand what is meant when reference is made to the "universal" nature of the genetic code.
4. Define *gene mutation;* explain types of mutations occurring in nucleotide sequences and cite an example of how protein synthesis may be adversely affected.
5. Understand how operon controls regulate gene expression in prokaryotes.
6. Understand how differentiation proceeds by selective gene expression during development.
7. Explain the relationship of cancer and a failure of gene controls.

12-I
(pp. 150–152)

PROTEIN SYNTHESIS
    Transcription
    Classes of RNA

Summary

We know that genes are portions of DNA that do not code directly for the production of proteins; that is, proteins are not assembled on the DNA. Rather, certain kinds of RNA are intermediaries constructed from ribose-containing nucleotides that are assembled in a complementary fashion on DNA, yet leave the DNA molecules intact in the nuclear region even as the RNA bears genetic messages to the sites of protein synthesis—the ribosomes. The linear sequence of nucleotide triplets becomes translated into a linear sequence of amino acids in a polypeptide chain, which is the basic structure of proteins. A protein consists of one or more chemically different polypeptide chains, and a single gene codes for each kind of polypeptide chain. RNA is usually a single strand of ribose-containing nucleotides in which the nitrogen-containing base *uracil* replaces thymine. Uracil base-pairs with adenine just as thymine does. Ribonu-cleotides are joined together by using DNA as a template.

Three kinds of RNA govern the assembly of amino acids into proteins. All three are assembled in a complementary fashion from the template of DNA by the process of *transcription*. Only one strand (the "sense" strand) of DNA is used as a template during transcription. The three differ in carbon-chain length and they have different shapes and functions. *Messenger RNA (mRNA)* molecules are linear, unbranched chains of at least sixty nucleotides. Messenger RNA alone carries the instructions for assembling a specific protein from the DNA in the nuclear region to the ribosomes. *Ribosomal RNA (rRNA)* becomes attached to proteins to form ribosomes, which are the actual protein assembly sites in the cytoplasm. rRNA is synthesized largely in the nucleolus of eukaryotic cells. *Transfer RNA (tRNA)*, the smallest of the three RNA forms, is a shuttle molecule that picks up whatever amino acid is dictated by the anticodon at one end of the RNA, attaches the amino acid to tRNA's opposite end, and moves over to the ribosome, where it awaits its chance to dump its load into the growing amino acid chain.

Transcription begins at a *promoter*, the base sequence signaling the start of a gene. Of the original DNA template (which guided the formation of RNA), the parts that actually dictate the sequence of amino acids in proteins are called *exons*. The intervening DNA sequences are called *introns* and will be snipped out as the exons are spliced together to form the mature mRNA transcript.

Key Terms

template	promoter	transfer RNA, tRNA
gene	transcript	introns
transcription	RNA polymerase	exons
RNA transcripts	ribosomal RNA, rRNA	nucleotide "cap"
translation	ribosome	nucleotide "tail"
uracil	messenger RNA, mRNA	mature mRNA transcript

Objectives

1. State how RNA differs from DNA in structure and function, and indicate what features RNA has in common with DNA.
2. Distinguish among the three types of RNA in terms of structure, function, where each is made, and where each carries out its function.
3. Describe the process of transcription and indicate how it differs from replication.
4. Tell what RNA code would be formed from the following DNA code: TAC-CTC-GTT-CCC-GAA.
5. Explain how terms such as *introns*, *exons*, *cap*, and *tails* are related to RNA transcript processing.

**Self-Quiz
Questions**

Fill-in-the-Blanks

(1) _____ is single-stranded and contains (2) _____ as its sugar instead of deoxyribose, and, in place of the base thymine, it has (3) _____ , which can form hydrogen bonds with the purine (4) _____ . During the process of (5) _____ , the two strands of the (6) _____ double helix

are separated; (7) _____ _____ , an enzyme, attaches to and moves along (8) [choose one] ( ) one of the strands, ( ) both strands, all the while speeding up the assembly of RNA nucleotide subunits onto (9) _____ regions of the exposed DNA. Then the (10) _____ strand detaches from the DNA and moves into the (11) _____ . Along the way, (12) _____ are snipped from future mRNA transcripts and (13) _____ are spliced together. At its ends, the transcript acquires a(n) (14) _____ _____ and a(n) (15) _____ _____ before it reaches the ribosome as a mature mRNA transcript.

12-II
(pp. 152–154)

PROTEIN SYNTHESIS (cont.)
   Translation

Summary

The DNA message is carried by RNA molecules to ribosomes in the cytoplasm, where amino acid sequences are assembled into polypeptide chains. The order of the various amino acids dictates the protein's final three-dimensional form. This form determines the substances the protein will interact with in helping to build and maintain the cell. There are only four different kinds of nucleotides in DNA and RNA, but there are twenty different amino acids commonly found in the proteins of living things. Crick, Brenner, and others deduced that the genetic code consists of nucleotide bases that are read linearly, three at a time, with the sequence of each triplet specifying an amino acid. Any of sixty-one such *triplets* specifies an amino acid; some different triplets specify the same amino acid, which makes the code redundant. Three other triplets serve to terminate the protein chain. Any nucleotide triplet in messenger RNA (RNA that carries protein assembly instructions from the nuclear region to the ribosomes in the cytoplasm) is now known as a *codon.* In all living organisms, essentially the same genetic code is the basic language of protein synthesis—compelling evidence for the fundamental unity of life at the molecular level.

A *codon* (base triplet on mRNA) is a hydrogen-bonding site for an anticodon. An *anticodon* is a complementary base triplet on tRNA; at the opposite end of the same tRNA molecule is a specific amino acid. The ribosome is composed of two subunits that come together when translation occurs. When the slot for its particular amino acid load clicks into place on the ribosome, the complementary *anticodon* hydrogen bonds to the *codon* of mRNA, and this temporary bonding positions the amino acid perfectly in order for the appropriate enzymes to form peptide bonds by dehydration synthesis and to incorporate the amino acid into the growing protein. That accomplished, the tRNA detaches from the codon and chugs off, searching for another amino acid like the first to transport to a ribosome. The mRNA strand continues to advance its codons through the groove in the ribosome until protein assembly is complete. In this way, protein chains of as many as 3,000 amino acid subunits are assembled from a collection of twenty different amino acids. Three stages of translation occur on the surface of ribosomes: *initiation, chain elongation,* and *chain termination.*

In summary, DNA transcribes its message into three forms of RNA, which cooperate at ribosomal sites to translate the message into a three-dimensional protein.

There are few exceptions to the universality of the genetic code. Most organisms use the same codons to select specific amino acids for protein synthesis.

Key Terms

Crick and Brenner
genetic code
"reading frame"
base triplet

codon
anticodon
ribosome
initiation

first tRNA binding site
second tRNA binding site
chain elongation
chain termination

Objectives

1. State the relationship between the DNA genetic code and the order of amino acids in a protein chain.
2. Explain the nature of the genetic code and describe the kinds of nucleotide sequences that code for particular amino acids.
3. Scrutinize Figure 12.4 in the main text and decide whether the genetic code in this instance applies to DNA, to messenger RNA, or to some other type of RNA.
4. Describe how the three types of RNA participate in the process of translation.
5. Determine which mRNA codons would be formed from the following DNA code: TAC-CTC-GTT-CCC-GAA.
6. Explain how the DNA message TAC-CTC-GTT-CCC-GAA would be used to code for a segment of protein and state what its amino acid sequence would be.
7. Summarize the steps involved in the transformation of genetic messages into proteins, using a diagram that shows the relationships among DNA, RNA, and proteins.

Self-Quiz Questions

Fill-in-the-Blanks

The order of (1) _____-_____ in a protein is specified by a sequence of nucleotide bases. The genetic code is read in units of (2) _____ nucleotides; each unit of three codes for (3) _____ amino acid(s). In the table that showed which triplet specified a particular amino acid, the triplet code was in (4) _____ molecules. Each of these triplets is referred to as a(n) (5) _____ . (6) _____ alone carries the instructions for assembling a particular sequence of amino acids from the DNA to the ribosomes, where (7) _____ of the polypeptide occurs. (8) _____ RNA acts as a shuttle molecule as each type brings its particular (9) _____-_____ to the ribosome where it is to be incorporated into the growing (10) _____ . A(n) (11) _____ is a triplet on mRNA that forms hydrogen bonds with a(n) (12) _____ , which is a triplet on tRNA.

(13) Given the following DNA sequence, deduce the composition of the mRNA transcript.
TAC-AAG-ATA-ACA-TTA-TTT-CCT-ACC-GTC-ATC

(14) From the mRNA transcript above, use Figure 12.4 of the main text to deduce the composition of the polypeptide sequence.

12-III
(pp. 155–157)

PROTEIN SYNTHESIS (cont.)
  Mutation and Protein Synthesis

Summary

Occasional errors occur during replication or during the separation of chromosomes during mitosis. In addition, high-energy radiation and mutagenic chemicals that enter cells can damage chromosomal strands. Repair enzymes are limited in just how well they can fix damage to DNA, so *deletions* (in which one or more base pairs are lost), *insertions* (in which one or more extra base pairs are incorporated into the original sequence), and base *substitutions* occasionally occur. The overall precision of DNA replication and repair is the source of life's fundamental unity; the rare "mistakes" account for much of its diversity. Although most mutations are harmful, some may confer upon their bearers a selective advantage to survive in changing environments and to leave more offspring even as others perish.

Key Terms

gene mutation   mutation rate   transposable elements
mutagens

Objectives

1. Explain why the base sequences in the DNA of a species must be kept constant from one generation to the next.
2. List some of the agents that can cause mutations.
3. Describe the types of gene mutations that occur in DNA from time to time.
4. Cite an example of a change in one DNA base pair that has profound effects on the human phenotype.

**Self-Quiz
Questions**

Fill-in-the-Blanks

(1) _____-_____ _____ and (2) _____ _____ that enter

cells can damage strands of DNA. (3) _____ occur when one or more

extra base pairs are incorporated into the original sequence; (4) _____

occur when one or more base pairs are left out of the original sequence.

(5) _____ _____ _____ is a genetic disease whose cause has

been traced to a change in a single DNA base pair; the result is that one

(6) _____ _____ is substituted for another in the beta chain of

(7) _____ .

12-IV
(pp. 157–158)

CONTROLS OVER GENE ACTIVITY
Gene Control in Prokaryotes

Summary

Cells require internal controls over which proteins are assembled at any given moment of the life cycle. These controls govern transcription, translation, and enzyme activity. Genetic information encoded in DNA first becomes transcribed as mRNA and then is translated (with the help of tRNA and rRNA) into polypeptide chains. In prokaryotes, controls over gene expression are basically short-range responses to changing environmental conditions (as in producing enzymes that will degrade or process particular food molecules that may appear in the environment).

Prokaryotes can respond to changes in the environment by activating or inactivating an *operon*—a set of genes that transcribes the appropriate pieces of mRNA, which in turn translates the code into a series of enzymes constituting a particular metabolic pathway. The term *operon* was coined to signify any set of protein-coding genes that operates as a coordinated unit under the direction of DNA control elements. The lactose operon of *Escherichia coli* is an example. The lactose operon permits *E. coli* to metabolize lactose as an energy source when it is present in the bacterial environment by *repressing* transcription; a repressor protein bound to the operator portion of the DNA prevents RNA polymerase from binding to the promoter portion of the DNA. This interferes with transcription and is a type of negative control mechanism.

Key Terms

operator	operon	lactose operon
Escherichia coli	repressor protein	RNA polymerase
promoter	negative control	regulator gene

Objectives

1. Describe the studies of *E. coli* nutrition and explain how the findings led scientists to develop the operon concept.
2. Diagram an operon; locate the regulator, the promoter, the operator, and the structural genes. Indicate where the repressor protein and RNA polymerase bind to the operon.
3. Describe the sequence of events that occurs on the chromosome of *E. coli* after you drink a glass of milk; use the function of each of the components listed in Objective 2.
4. Explain how the cells of *E. coli* manage to produce enzymes to degrade lactose when those molecules are present and to stop production of lactose-degrading enzymes when lactose is absent.
5. Explain why it is advantageous to a cell to be able to activate and inactivate its enzymes in response to changes in the environment.

Self-Quiz Questions

Fill-in-the-Blanks

(1) _____ _____ is a species of bacterium that lives in mammalian digestive tracts that provided some of the first clues about gene control. A(n) (2) _____ is any group of genes together with its promoter and operator sequence. Promoter and operator provide (3) _____ _____ . The (4) _____ _____ codes for the formation of mRNA, which assembles a repressor protein. The affinity of the (5) _____ for RNA polymerase dictates the rate at which a particular operon will be transcribed. Repressor protein allows (6) _____ _____ over the lactose operon. Repressor binds with operator and overlaps promoter when lactose concentrations are (7) _____ . This blocks (8) _____ _____ from the genes that will process lactose. This (9) [choose one] ( ) blocks, ( ) promotes, production of lactose processing enzymes. When lactose is present, lactose molecules bind with the (10) _____ _____ . Thus, repressor cannot bind to (11) _____ and RNA polymerase has access to the lactose processing genes. This gene control works well because lactose-degrading enzymes are not produced unless they are (12) _____ .

12-V
(pp. 158–161)

CONTROLS OVER GENE ACTIVITY (cont.)
Gene Control in Eukaryotes
Commentary: Cancer—When Gene Controls Break Down
SUMMARY

Summary

In eukaryotes, both short-range and long-range controls over gene expression govern development and differentiation. Genetic controls direct the unfolding of a single-celled zygote into all the specialized cells and structures of the adult form; in eukaryotes, the controls span longer periods of time and involve more intricate cellular arrangements in space than the controls in prokaryotes do. During development of all complex eukaryotes, cells come to differ in their positions in the body, in their developmental potential, and in their appearance, composition, and function. These differences are brought about by mitosis and *differentiation,* during which cells of the same genetic makeup become more or less abundant and become structurally and functionally different from one another according to the prescribed program for the species. Although all the cells in the body inherit the same genes, they activate or suppress some fraction of these genes in different ways to produce pronounced differences in their structure or functioning. *Selective gene expression* is controlled by agents that act within cells, between cells, and between cells and the environment.

In complex eukaryotes, differentiation of cells arises through selective gene expression; in a given cell type, some genes are activated and others are repressed.

In mammalian females, one of each pair of X chromosomes is inactivated in an early embryonic stage and always remains condensed (a Barr body) so that, of any pair of X chromosomes, only one X chromosome (either paternal or maternal) is being transcribed in a given cell. The process of differential transcription ensures that any adult XX female is a mosaic of X-linked traits. Human adult females with anhidrotic ectodermal dysplasia, an X-linked genetic disease, exhibit this mosaicism with patches of skin that lack sweat glands.

Throughout the life cycle, genes control a complex series of changes in cell growth and division. No one yet understands how.

Sometimes a single cell loses the controls that indicate when to stop cell division; it divides endlessly, crowding certain cells and interfering with vital cell functions. The mass that results will either be *benign*, with only its controls over cell division and growth gone awry, or *malignant*, with the recognition factors on its cell surfaces also disturbed. Benign tumors generally can be removed surgically. Malignant growth can invade and destroy surrounding tissues. Some malignancies do not stay as one solid tumor, but *metastasize*, or disperse to other sites in the body. Such cancers are generally treated with chemical agents, radiation, or both to destroy uncontrollably dividing cells selectively wherever they are in the body. Such traumatic changes in cellular controls seem to have a variety of causes. Specific viruses, chemicals, continual prolonged exposure to sunlight, and constant physical irritation of some tissue may increase susceptibility to some forms of cancer.

It now appears that a small number of altered regulatory genes may give rise to at least some kinds of cancer. *Proto-oncogenes* are inherent parts of vertebrate DNA, and they encode for proteins necessary in normal cell functioning. They become cancer-causing genes *(oncogenes)* only on those rare occasions when specific mutations alter their structure or their expression.

Key Terms

differentiate	tumor	cancer
selective gene expression	benign	oncogenes
anhidrotic ectodermal dysplasia	malignant	proto-oncogenes
	metastasis	carcinogens
Barr body		

Objectives

1. Define and distinguish between the terms *development* and *differentiation*. Cite examples from your body.
2. Define *selective gene expression* and explain how this concept relates to cell differentiation in multicelled eukaryotes.
3. Explain how X-chromosome inactivation provides evidence for selective gene expression; use the example of anhidrotic ectodermal dysplasia.
4. Explain, using examples from your own body, the gene control of cell division.
5. Give an example of cells that normally cease dividing. Then state how one such cell could be made to begin dividing again.
6. Define *tumor* and distinguish between *benign* and *malignant* tumors.
7. Explain what happens when a tumor *metastasizes*.
8. Describe the relationship of proto-oncogenes, environmental irritants, and oncogenes.

Self-Quiz Questions

Fill-in-the-Blanks

All cells in our bodies contain copies of the same (1) _____ . These genetically identical cells become structurally and functionally distinct from one another through a process called (2) _____ . (3) _____ _____ arises through (4) _____ _____ expression in different cells. X-chromosome inactivation produces adult females who are (5) _____ of X-linked traits. This effect is shown in human females affected by (6) _____ _____ _____ and provides evidence for (7) _____ _____ expression. No one knows exactly how, but genes control cell (8) _____ and (9) _____ .

True-False

If false, explain why.

___ (10) When cells become cancerous, cell populations decrease to very low densities and stop dividing.

___ (11) All abnormal growths and massings of new tissue in any region of the body are called *tumors*.

___ (12) Malignant tumors have cells that migrate and divide in other organs.

___ (13) The term *metastasis* means the acquiring of the shape or form typical of that species.

___ (14) Oncogenes are genes that combat cancerous transformations.

___ (15) Proto-oncogenes rarely trigger cancer.

CHAPTER TEST

UNDERSTANDING AND INTERPRETING KEY CONCEPTS

___ (1) Transcription _____ .

    (a) occurs on the surface of the ribosome
    (b) is the final process in the assembly of a protein
    (c) occurs during the synthesis of any type of RNA by use of a DNA template
    (d) is catalyzed by DNA polymerase

___ (2) _____ carries amino acids to ribosomes, where amino acids are linked into the primary structure of a polypeptide.

    (a) mRNA
    (b) tRNA
    (c) Introns
    (d) rRNA

___ (3) Transfer RNA differs from other types of RNA because it _____ .

    (a) transfers genetic instructions from cell nucleus to cytoplasm
    (b) specifies the amino acid sequence of a particular protein
    (c) carries an amino acid at one end
    (d) contains codons

___ (4) _____ dominates the process of transcription.

    (a) RNA polymerase
    (b) DNA polymerase
    (c) Phenylketonuria
    (d) Transfer RNA

___ (5) _____ and _____ are found in RNA but not in DNA.

    (a) Deoxyribose, thymine
    (b) Deoxyribose, uracil
    (c) Uracil, ribose
    (d) Thymine, ribose

___ (6) Each "word" in the mRNA language consists of _____ letters.

    (a) three
    (b) four
    (c) five
    (d) more than five

___ (7) If each nucleotide is coded for only one amino acid, how many different types of amino acids could be selected?

    (a) Four
    (b) Sixteen
    (c) Twenty
    (d) Sixty-four

___ (8) The genetic code is composed of _____ codons.

    (a) three
    (b) twenty
    (c) sixteen
    (d) sixty-four

___ (9) The cause of sickle-cell anemia has been traced to _____ .

    (a) a mosquito-transmitted virus
    (b) two DNA mutations that result in two incorrect amino acids in a hemoglobin chain
    (c) three DNA mutations that result in three incorrect amino acids in a hemoglobin chain
    (d) one DNA mutation that results in one incorrect amino acid in a hemoglobin chain

___ (10) Deletion, addition, or substitution of one to several bases in the nucleotide sequence of a gene is referred to as _____ .

    (a) chromosomal mutation
    (b) gene mutation
    (c) heterozygous DNA
    (d) code upset

___ (11) An example of a mutagen would be _____ .

   (a) a virus
   (b) ultraviolet radiation
   (c) certain chemicals
   (d) all of the above

___ (12) _____ refers to the process by which cells with identical genotypes become structurally and functionally distinct from one another according to the genetically controlled developmental program of the species.

   (a) Metamorphosis
   (b) Metastasis
   (c) Cleavage
   (d) Differentiation

___ (13) _____ binds to operator whenever lactose concentrations are low.

   (a) Operon
   (b) Repressor
   (c) Promoter
   (d) Operator

___ (14) Any gene or group of genes, together with its promoter and operator sequence, is a(n) _____ .

   (a) repressor
   (b) operator
   (c) promoter
   (d) operon

___ (15) The operon model explains the regulation of _____ in prokaryotes.

   (a) replication
   (b) transcription
   (c) induction
   (d) Lyonization

___ (16) In multicelled eukaryotes, cell differentiation occurs as a result of _____ .

   (a) growth
   (b) selective gene expression
   (c) repressor molecules
   (d) the death of certain cells

___ (17) One type of gene control discovered in female mammals is _____ .

   (a) a conflict in maternal and paternal alleles
   (b) slow embryo development
   (c) X-chromosome inactivation
   (d) operons

___ (18) Due to X-inactivation of either the paternal or maternal X chromosome, human females having anhidrotic ectodermal dysplasia _____ .

   (a) completely lack sweat glands
   (b) develop benign growths
   (c) are mosaics with patches of skin that lack sweat glands
   (d) develop malignant growths

___ (19) Which of the following characteristics seems to be most uniquely correlated with metastasis?

    (a) Loss of nuclear-cytoplasmic controls governing cell growth and
division
    (b) Changes in recognition proteins on membrane surfaces
    (c) "Puffing" in the chromosomes
    (d) The massive production of benign tumors

___ (20) Genes having the potential to induce cancerous formations are known
as _____ .

    (a) proto-oncogenes
    (b) oncogenes
    (c) carcinogens
    (d) malignant genes

INTEGRATING AND APPLYING KEY CONCEPTS

Genes code for specific polypeptide sequences. Not every substance in living
cells is a polypeptide. Explain how genes might be involved in the production
of a storage starch (such as glycogen) that is constructed from simple sugars.

    Suppose that you have been restricting yourself to a completely vegetarian
diet for the past six months. Quite unexpectedly, you find yourself in a social
situation that requires you to eat a half-pound sirloin steak. Would you expect
to digest the steak as easily as you digest soybean burgers? Explain your yes or
no answer in terms of transcriptional controls or feedback inhibition.

13
RECOMBINANT DNA AND GENETIC ENGINEERING

RECOMBINANT DNA TECHNOLOGY
    Plasmids
    Producing Restriction Fragments
    DNA Amplification
    Commentary: RFLPs and Genetic
    Fingerprinting
    Expressing a Cloned Gene
    Mapping the Human Genome

GENETIC ENGINEERING: RISKS AND PROSPECTS
    Genetically Engineered Bacteria
    Genetically Engineered Plants
    Genetically Engineered Animals
    Commentary: Human Gene Therapy
SUMMARY

General Objectives

1. Know how genetic recombination occurs naturally.
2. Define *recombinant DNA technology*.
3. Understand what *plasmids* are and how they may be used to insert new genes into recombinant DNA molecules.
4. Know how DNA can be cleaved, spliced, and cloned.
5. Explain the importance of *RFLPs*.
6. Be aware of several limits and possibilities for future research in *genetic engineering*.

13-I
(pp. 163–166)

RECOMBINANT DNA TECHNOLOGY
    Plasmids
    Producing Restriction Fragments
    DNA Amplification

Summary

Genetic experiments have been occurring in nature for billions of years. Mutations, recombination between homologues at meiosis, the novel assortments of alleles brought together at fertilization, and hybridization between species have all contributed to the current diversity among organisms. Prokaryotes ordinarily reproduce by prokaryotic fission, an asexual process in which the partitioning of a parental cell leads to two separate cells having basically similar parts. During *bacterial conjugation,* a donor cell transfers DNA through a tiny tube to a recipient cell. Conjugation results in genetic recombination—the production of individuals having new assortments of genes. This is regarded by many as a possible method of exchanging genetic information that helped diversify early prokaryotes.

Recombinant DNA technology involves procedures by which DNA molecules can be cut into fragments, then joined with similarly cut fragments from any organism to form hybrid recombinant DNA molecules that can be propagated in a line of dividing cells. *Plasmids* are small circles of DNA that are found in some bacteria in addition to the main circular DNA molecule. Genetic information carried on the larger, circular DNA molecule contains genes concerned with regular growth and development. Plasmids have few genes and replicate separately. Some plasmids contain genes that allow conjugation or confer resistance to some harmful substance in the environment. Scientists have determined how to cut and bundle DNA fragments into plasmids for incorporation into host cells. *Restriction enzymes* can be used to cut open plasmids at a particular site, and *DNA ligase,* an enzyme, can be used to attach some foreign DNA to the opened plasmid; then the plasmid is closed again in its circle. Particular DNA fragments can now be identified as specific nucleotide sequences of individual genes. A collection of DNA fragments produced by restriction enzymes and incorporated into DNA is called a *DNA library.* Under certain conditions, the hybrid DNA plasmid can be introduced into new host bacterial cells, and it will then be reproduced every time the bacterium undergoes cell division.

Cloned DNA is any DNA library that has been amplified in a line of dividing cells or in a cell-free system. The bacterium's metabolic machinery can churn out many multiple copies, or *clones,* of the genes carried on the hybrid plasmid ring; with repeated doublings of a bacterial population, millions of these genes can be manufactured overnight. This is known as *DNA amplification.* The most common method of DNA amplification is the *polymerase chain reaction.* DNA containing a gene is split and copied many times under enzyme control. Separated DNA fragments can be sorted by size and identified.

DNA amplification has also been achieved with a viral enzyme, *reverse transcriptase.* Single-stranded mRNA serves as a template to construct DNA strands with nucleotide sequences of a desired gene. Following assembly of the hybrid DNA/RNA molecule, the RNA template strand is degraded. Other enzymes convert the new single-stranded DNA to a double-stranded form. Such DNA molecules copied from mRNA are known as *cDNA,* not to be confused with cloned DNA. These techniques are used to map the genome of the species.

Recombinant DNA techniques hold great promise for transferring genes from one organism to another for manufacturing proteins such as insulin and other substances in a pure form for organisms that cannot make the required substances themselves. Entirely new species with entirely new assemblages of characteristics are now being synthesized in specially equipped laboratories following very strict safety rules.

Key Terms

recombinant DNA
  technology
plasmids
bacterial conjugation

genome
restriction enzymes
DNA ligase
DNA library

cloned DNA
reverse transcriptase
cDNA
polymerase chain reaction

Objectives

1. Define the term *recombinant DNA technology.*
2. Distinguish the genetic information carried on the large circular bacterial chromosome from that carried on plasmids.
3. Explain the function of *bacterial conjugation.*
4. Cite the relationships between bacterial enzymes used to fragment DNA and identifying nucleotide sequences for mapping *genomes* of species.
5. Explain the function of *restriction enzymes* and explain how *DNA ligase* is used to create a *DNA library.*

6. Tell how *cloned DNA* results from use of a DNA library.
7. Define *cDNA* and account for its production from *reverse transcriptase*.
8. Describe the *polymerase chain reaction* method of gene amplification.

Self-Quiz Questions

True-False

If false, explain why.

___ (1) Plasmids are organelles on the surfaces of which amino acids are assembled into polypeptides.

___ (2) Bacterial conjugation is a process by which a donor cell transfers DNA to a recipient cell.

___ (3) Restriction enzymes are produced by viruses to cut apart foreign DNA molecules.

___ (4) A DNA library is a collection of DNA fragments produced by an enzyme, reverse transcriptase.

___ (5) The most commonly used method of DNA amplification is known as the polymerase chain reaction.

Matching

Match the most appropriate letter with its numbered partner.

(6) ___ polymerase chain reaction

(7) ___ DNA ligase

(8) ___ cDNA

(9) ___ DNA library

(10) ___ cloned DNA

(11) ___ bacterial conjugation

(12) ___ recombinant DNA technology

(13) ___ plasmids

(14) ___ restriction enzymes

(15) ___ genome

A. All the DNA in a haploid set of chromosomes
B. A gene is split into two strands and then copied over and over by enzymes
C. A process by which a donor cell transfers DNA to a recipient cell
D. Connects DNA fragments
E. Small circular DNA molecules that carry only a few genes
F. DNA assembled through use of reverse transcriptase and coding on mRNA
G. Cuts DNA molecules
H. A collection of DNA fragments produced by restriction enzymes and incorporated into plasmids
I. Multiple, identical copies of DNA fragments from an original chromosome
J. Method of genetic engineering

13-II
(pp. 166–167)

RECOMBINANT DNA TECHNOLOGY (cont.)
    Commentary: **RFLPs and Genetic Fingerprinting**
    Expressing a Cloned Gene
    Mapping the Human Genome

Summary

Work on recombinant DNA is proceeding with *restriction fragment length polymorphisms* (RFLPs). Restriction enzymes are used to cut DNA into fragments of specific lengths. Except for identical twins, each person's DNA presents a unique pattern of fragment lengths produced by enzyme cuts. Radioactive probes may reveal abnormal restriction sites that represent mutant alleles causing genetic disease. The unique pattern of RFLPs in the DNA of each person provides a *genetic fingerprint* and may also be used in investigations of disputed paternity and crime.

Bacterial host cells lack the proper splicing enzymes and thus have difficulty expressing cloned genes unless introns have previously been cut out. For successful translation by such host cells, only exons must remain in a mature mRNA transcript. Research on human genes utilizes mature mRNA transcripts produced from cDNA.

The human genome is currently being mapped with "gene machines" that can determine the identity and order of nucleotides in cloned DNA. This procedure is known as *sequencing*.

Key Terms

restriction fragment length polymorphisms (RFLPs)   genetic fingerprint

Objectives

1. Define *restriction fragment length polymorphisms* and describe uses of the unique patterns of DNA fragments possessed by each person.
2. Explain why the term *genetic fingerprint* is applied to the use of one type of RFLP.
3. Explain the reason that host cells have difficulties expressing cloned genes.
4. Tell what is meant by *sequencing genes* and explain plans for mapping the human genome.

Self-Quiz Questions

Fill-in-the-Blanks

(1) _____ _____ _____ refers to the use of restriction enzymes to cut human DNA into fragments of specific lengths. (2) _____ refers to the fact that each person has a slightly unique pattern of sites on DNA where enzymes make their cuts. Each person has a(n) (3) _____ _____ or a unique pattern of RFLPs. Bacterial host cells lack the proper (4) _____ _____ to translate cloned genes unless the (5) _____ have been spliced out. Identification of the order and identity of nucleotides in DNA is called (6) _____ .

GENETIC ENGINEERING: RISKS AND PROSPECTS
  Genetically Engineered Bacteria
  Genetically Engineered Plants
  Genetically Engineered Animals
  Commentary: **Human Gene Therapy**
SUMMARY

Summary

Genetic engineering involves the transfer of genes between different species. Scientists debate the dangers of such work to organisms and the environment. One example is the removal of a gene from the *ice-minus bacteria* living on the leaves and stems of crop plants. The gene is one that enhances frost damage to plants. The genetically engineered cells cannot synthesize the ice-forming protein. This improves the genetics of the crop plant by deleting a harmful gene, but raises fears about possible dangers such new genes present. Potential problems of such research must be measured against the public good.

Genetic engineering of plants may allow an increase in the total food supply. Introduction of new genes to plants may allow crop plants to grow in environments formerly intolerable to them. For example, salt-resistant, high-yield plant strains may be developed.

Gene modification in animals has proven to be difficult because of failure of gene delivery; results have been mixed and largely unsuccessful.

Gene therapy refers to the process of inserting genes into body cells in an attempt to correct a genetic defect. Inserting genes into normal persons for the purpose of changing or improving is called *eugenic engineering*. Genetic manipulation raises tremendously important social and ethical issues for the human community. Who has the wisdom to make the crucial decisions on what changes will be beneficial or harmful?

Key Terms

ice-minus bacteria   gene therapy   eugenic engineering

Objectives

1. Explain the risks and positive outcomes from genetic engineering; state your own opinion as to the desirability of genetic engineering.
2. Cite examples of success and failure in the field of genetic engineering.
3. Define *gene therapy* and *eugenic engineering* and be able to discuss the possible social and ethical human questions that are raised by such procedures.

Self-Quiz Questions

Fill-in-the-Blanks

Bacterial strains used in (1) _____-_____ may be initially harmless,

but there is concern about danger to humans and the environment. A strain

of bacteria, *Pseudomonas syringae*, lives on stems and leaves of crop plants

and they become susceptible to (2) _____ _____ . The harmful

gene is called the ice-forming gene and the bacteria are known as

(3) _____-_____ bacteria; genetic engineers were able to remove the

harmful gene and test the modified bacterium on strawberry plants with no adverse effects. Inserting one or more genes into the (4) _____ _____ of an organism for the purpose of correcting genetic defects is known as (5) _____ _____ . Attempting to modify a human trait by inserting genes into sperms or eggs is a field called (6) _____ _____ .

CHAPTER TEST

UNDERSTANDING AND INTERPRETING KEY CONCEPTS

___ (1) Small circular molecules of DNA in bacteria are called _____ .

    (a) plasmids
    (b) desmids
    (c) pili
    (d) F particles
    (e) transferins

___ (2) Bacteria reproduce sexually by _____ .

    (a) fission
    (b) gametic fusion
    (c) conjugation
    (d) lysis
    (e) none of the above because bacteria only reproduce asexually

___ (3) Enzymes used to cut genes in recombinant DNA research are _____ .

    (a) ligases
    (b) restriction enzymes
    (c) transcriptases
    (d) DNA polymerases
    (e) replicases

___ (4) The total DNA in a haploid set of chromosomes of a species is its _____ .

    (a) plasmid
    (b) enzyme potential
    (c) genome
    (d) DNA library
    (e) none of the above

___ (5) An enzyme that heals random base-pairing of chromosomal fragments and plasmids is _____ .

    (a) reverse transcriptase
    (b) DNA polymerase
    (c) cDNA
    (d) DNA ligase

___ (6) A DNA library is _____ .

    (a) a collection of DNA fragments produced by restriction enzymes and incorporated into plasmids
    (b) cDNA plus the required restriction enzymes
    (c) mRNA-cDNA
    (d) composed of mature mRNA transcripts

___ (7) Amplification results in _____ .

    (a) plasmid integration
    (b) bacterial conjugation
    (c) cloned DNA
    (d) production of DNA ligase

___ (8) Any DNA molecule that is copied from mRNA is known as _____

    (a) cloned DNA
    (b) cDNA
    (c) DNA ligase
    (d) hybrid DNA

___ (9) The most commonly used method of DNA amplification is _____ .

    (a) polymerase chain reaction
    (b) gene expression
    (c) genome mapping
    (d) RFLPs

___ (10) Restriction fragment length polymorphisms are valuable because _____ .

    (a) they reduce the risks of genetic engineering
    (b) they provide an easy way to sequence the human genome
    (c) they allow fragmenting DNA without enzymes
    (d) they provide DNA fragment sizes unique to each person

---

INTEGRATING AND APPLYING KEY CONCEPTS

How could scientists guarantee that *Escherichia coli*, the human intestinal bacterium, will not be transformed into a severely pathogenic form and released into the environment if researchers use the bacterium in recombinant DNA experiments?

Crossword
Number Three

Across

2. reverse _____ is a special viral enzyme that assembles a DNA strand from a single-stranded mRNA molecule
7. mitotic cell divisions that convert a zygote or unfertilized egg into the early multicelled embryo
9. deviant; strangely irregular
12. a mold upon which a product that is complementary in shape is assembled
14. both alleles of a pair of genes are expressed in the heterozygote
15. cancer-causing agent
19. eased into place; populated
21. an altered regulatory gene that can induce cancerous transformations in cells
23. egg formation

24. one of the three wise men who followed the star to Bethlehem; a sorcerer or magician
25. a condition in which (as a result of nondisjunction) an individual inherits three homologous chromosomes
27. one piece of DNA spiralled around a histone spool
28. each person has as a genetic fingerprint a unique array of _____s
31. _____ group: a collection of genes that tend to be inherited together because they are part of the same chromosome
32. unwanted, unpleasant sound
33. a single-ringed nitrogen-containing base associated with RNA, but not DNA
35. the molecule that bears the genetic code from the nucleus to a ribosome and guides protein assembly
38. appearing in a phenotype
41. a genotype produced by nondisjunction of the Y chromatids in males during meiosis II
42. fusion of sperm and egg yielding a zygote
45. of or relating to inheritance
46. a procedure during the third or fourth month of pregnancy that samples the fluid surrounding the fetus in the uterus to detect a genetic disorder
49. a stage of nuclear division in which chromosomes are lined up at the spindle equator
50. possess
51. a circular extra DNA molecule in the cells of many bacterial species
52. charts of the genetic relationships among individuals
53. any gene (or group of genes) together with its promoter and operator sequences
55. gene not expressed in the heterozygote
57. the _____-square method predicts the genotypes that can result from a specific genetic cross
58. male gamete
60. all the DNA in a haploid set of chromosomes
61. a type of dwarfism caused by an autosomal dominant allele
64. a genetic disorder that arises when milk sugar cannot be metabolized; an example of autosomal recessive inheritance
67. a DNA segment that binds or releases a repressor protein
69. to walk lamely
71. examining a small portion of the total population
73. having two sets of chromosomes (one set from each parent)
76. a triplet in mRNA
78. DNA with gene arrays not present in the cells from which it was derived is _____ DNA
80. chromatids on different replicated homologous chromosomes
82. a double-ringed purine in both DNA and RNA
84. enzymes of special bacterial nucleases capable of cutting DNA strands during normal repair and degradation reactions
85. _____ chromatids are two lengthwise halves of a replicated chromosome
86. _____-breeding = homozygous for a particular trait
88. a process in which cancer cells in a multicelled organism travel to other places where they can form a new growth
89. any DNA assembled from an mRNA transcript
90. nitrogen-containing _____ ; examples are adenine, guanine, thymine, cytosine, and uracil
91. gene expressed in the heterozygote
95. maiden beloved by Zeus transformed into a heifer

97. true-_____ means to have consistent and recognizable in herited characteristics
98. not malignant
99. transferal of part of one chromosome to a nonhomologous chromosome
101. variation uninterrupted over its range of values
102. derived from a membrane associated with a developing embryo
103. single-ringed nitrogen-containing base

Down

1. relating to nonsex chromosomes
3. the alternative forms of a gene that appear at the same locus
4. chronic drunkard
5. _____-oncogene; potentially, a cancer-causing gene
6. J.R., the patriarch of "Dallas"
7. linear entity composed of links or monomers
8. a double-ringed nitrogen-containing base
9. a functional group of nitrogen and hydrogen atoms
10. the end of a forelimb in vertebrates; the hand, hoof, or claw (Latin)
11. place on a chromosome where one finds a particular gene (allele)
13. an agent that alters DNA, causing permanent change in an allele
16. a length of DNA that codes for a particular polypeptide sequence
17. simple, water-soluble proteins primarily associated with eukaryotic nucleic acids
18. arm or wrist jewelry
20. the process by which different cells become specialized in structure and function to do different tasks
22. a group of genetically identical organisms descended from one ancestor
26. a _____ cell has reproduction as its principal function
27. composed of a 5-carbon sugar, a phosphate group, and a nitrogen-containing base
29. sequence of an operon that serves as a binding site for RNA polymerase
30. in _____ dominance, a dominant allele cannot completely mask expression of the allele occupying its homologous locus
34. proteins that catalyze specific chemical reactions
35. a heritable change in a DNA molecule
36. division or apportioning a cell's cytoplasm following a nuclear division
37. a situation in which a single gene affects more than just one aspect of an organism's phenotype
38. human improvement by genetic control
39. photo selected for sex appeal; displayed on a wall
40. eggs (pl.)
43. intervening sequence in mRNA
44. a brewing container for a caffeinated beverage
47. during gamete formation, two alleles of a homologous pair are separated and end up in different gametes
48. the process of forming a mature sperm from a germ cell
54. a special region of the chromosome to which spindle microtubules attach during nuclear division
56. mistake
59. recently developed procedure that detects genetic disorders eight weeks into a pregnancy by using cells from a membranous sac around the embryo (abbrev.)
62. a long-legged wading bird

63. an enzyme that links together a series of monomers
65. the copying, often by a millionfold, of an isolated, cloned gene
66. a representation of the positions of genes on a given chromosome
68. the process that finishes transcribing a piece of RNA
70. a reduction divisional process that yields offspring nuclei with the haploid number of chromosomes
72. links up with a triplet of mRNA
73. liquid or moisture that falls in drops
74. implements used to row a boat
75. two that belong together
76. any of various malignant neoplasms that tend to invade and metastasize
77. the major molecule of heredity; the genetic blueprint for life
79. gap; _____ lip
81. the act, process, or result of change
83. one of ninety two naturally occurring substances composed of atoms having the same number of protons in each nucleus
87. an enzyme used to create a DNA library by making base pairing between chromosomal fragments and cut plasmids permanent
89. a cylindrical tobacco product
92. a nucleic acid that shuttles a specific amino acid to mRNA and the growing protein chain
93. group of three musicians or singers
94. a unit of measurement; 2.54 cm = one _____
96. having the color of light cinders
100. an insect of the genus *Formica*

14
MICROEVOLUTION

EMERGENCE OF EVOLUTIONARY
THOUGHT
MICROEVOLUTIONARY PROCESSES
    Variation in Populations
    Mutation
    Genetic Drift
    Gene Flow
    Natural Selection
EVIDENCE OF NATURAL SELECTION
    Stabilizing Selection

Commentary: Sickle-Cell Anemia
    —Lesser of Two Evils?
Directional Selection
Disruptive Selection
Sexual Selection
SPECIATION
    Defining the Species
    Divergence and Isolation
SUMMARY

General Objectives

1. Trace the beginnings of evolutionary thought.
2. Understand how variation occurs in populations and how changes in allele frequencies can be measured.
3. List and define the major forces of microevolution.
4. Describe four kinds of selection mechanisms that help shape populations.
5. Define speciation and explain the role of divergence and isolation in that process.

14-I
(pp. 172–176)

EMERGENCE OF EVOLUTIONARY THOUGHT

Summary

Early western scholars held that each kind of living thing had remained the same since creation. Sixteenth-century global exploration discovered an overwhelming complexity of new species. Eighteenth-century anatomists found puzzling similarities in the body plans of animals. For example, why do embryonic humans have bones that match tail bones of other animals? By mid-eighteenth century, geologists noted a layering of the earth that could only be explained by a slow process of change. The younger layers contained *fossils* that seemed to become more complex in form when compared with those found in deeper layers. Buffon, trying to reconcile new information with the ideas of creation, argued that species must have been created in several places and that they were imperfect. Available data suggested that *evolution* had occurred and that characteristics of populations had changed over long periods of time.

Spurred by the ideas of Lyell (who said Earth was far older than previously thought) and Malthus (who said that competition results when populations outgrow their resources) and by his experiences on the voyage of the *Beagle,*

Darwin developed the theory of evolution by natural selection; he was later joined by Wallace in suggesting that species do change over time. Heritable variations occur among members of a species. Each population produces more offspring than can survive to reproduce. Bearers of the traits that improve chances for surviving and reproducing under prevailing environmental conditions tend to produce more descendants than do other members of the population; this *differential reproduction* ensures that the most adaptive ("fit") traits will show up more frequently in the next generation.

Key Terms

fossils	Malthus	Wallace
Darwin	evolution	*Archaeopteryx*
Lyell	natural selection	"missing links"

Objectives

1. Relate information gathered in the sixteenth to eighteenth centuries that stimulated questions about the origins of species.
2. Indicate how the writings of Lyell and Malthus helped shape Darwin's early formulation of his theory.
3. Summarize the Darwin-Wallace theory of evolution by natural selection.
4. Describe Wallace's role in the formulation of the theory of evolution.
5. State the significance of a fossil named *Archaeopteryx*.

Self-Quiz Questions

True-False

If false, explain why.

___ (1) There is a great deal of evidence indicating that species have remained unchanged since the time of their creation.

___ (2) One expects to find the most recent fossil organisms in the upper layers of Earth's sedimentary rocks.

___ (3) Malthus believed that the food supply of a population increases faster than the population increases.

___ (4) Lyell's ideas on overpopulation and Malthus ideas on geologic change provoked Darwin's thinking on how organisms change with time.

___ (5) When resources become scarce, competition for similar resources promotes greater specialization among the competitors.

___ (6) Darwin first learned about the ideas of evolution when he received a letter from Wallace.

___ (7) *Archaeopteryx* was a unique fossil that supplied a missing link for the idea that a group of reptiles evolved into birds.

14-II
(pp. 177–181)

MICROEVOLUTIONARY PROCESSES
   Variation in Populations
   Mutation
   Genetic Drift
   Gene Flow
   Natural Selection

Summary

Individuals do not evolve; the unit of evolution is the population. The population is the interbreeding group. The *population* concept holds that variation in populations not only encourages evolutionary change but also is the product of evolutionary change. For any one locus on the chromosome, some alleles occur more often than others; thus, variation can be thought of in terms of *allele frequencies*.

When allele frequencies for a given gene locus remain constant throughout succeeding generations, the population is said to be at *equilibrium*, which is maintained only when all of a *set of specific conditions* occur. These conditions (no mutation, very large population, isolation from other populations of the same species, all members survive, mate, and reproduce, and random mating) are never met in nature. If all individuals in a population have an equal chance of surviving and reproducing, then the frequency of each allele in the population should remain constant from generation to generation (that is, be in *genetic equilibrium*). In nature, allele frequencies may change all the time. Some disturbances that cause changes are *mutation, genetic drift, gene flow,* and the *selection* pressures that bring about differential survival and reproduction of genotypes within a population. These processes of change in allele frequency are known as *microevolution*.

Key Terms

population	genetic equilibrium	bottleneck
allele frequencies	microevolution	gene flow
Hardy-Weinberg rule	mutation	adaptive trait
$p + q = 1$	genetic drift	differential reproduction
$p^2 + 2pq + q^2 = 1$	founder effect	natural selection

Objectives

1. Distinguish between the terms *population* and *species*.
2. Distinguish among morphological, physiological, and behavioral traits.
3. List five events that determine variation in traits.
4. Describe the relationship between *genetic equilibrium* and *allele frequencies*.
5. Explain the Hardy-Weinberg rule and list the five conditions that must be met before genetic equilibrium is possible.
6. List the major processes of *microevolution*.
7. Define *mutation* and explain how mutations cause allele frequencies to change in natural populations.
8. Define *genetic drift* and explain how this process causes allele frequencies to change in natural populations.
9. List two of the causes of *gene flow* and explain how gene flow can cause allele frequencies to change in natural populations.
10. Define *natural selection* and use the term *differential reproduction* in the definition.

Fill-in-the-Blanks

A(n) (1) _____ is a group of individuals of the same species that occupies a given area at a specific time. The (2) _____-_____ rule allows researchers to establish a theoretical reference point (baseline) against which changes in allele frequency can be measured. Variation can be expressed in terms of (3) _____ _____ , the relative abundance of different alleles carried by the individuals in that population. The stability of allele ratios that would occur if all individuals had equal probability of surviving and reproducing is called (4) _____ _____ . Over time, allele frequencies tend to change through infrequent, but inevitable (5) _____ , which are the original source of genetic variation. Random fluctuation in allele frequencies over time due to chance occurrence alone is called (6) _____ _____ ; it is more pronounced in small populations than in large ones. (7) _____ _____ associated with immigration and/or emigration also changes allele frequencies. (8) _____ _____ is the differential survival and reproduction of individuals of a population that differ in one or more traits. (9) _____ _____ is the most important microevolutionary process.

Problems

(10) In a population, eighty-one percent of the organisms are homozygous dominant and one percent are homozygous recessive. Find (a) the percentage of heterozygotes, (b) the frequency of the dominant allele, and (c) the frequency of the recessive allele.

(11) In a population of 200 individuals, determine how many individuals are (a) homozygous dominant, (b) homozygous recessive, and (c) heterozygous for a particular locus if $p = 0.8$.

14-III
(pp. 181–185)

EVIDENCE OF NATURAL SELECTION
    Stabilizing Selection
    Commentary: **Sickle-Cell Anemia—Lesser of Two Evils?**
    Directional Selection
    Disruptive Selection
    Sexual Selection

Summary

Natural selection is no longer in the realm of pure theory; examples of natural selection have been demonstrated in different populations.

    A process in which a form already well adapted to a given environment is selected for and maintained, even as extreme variants are selected against, is called *stabilizing selection.* Humans weighing about 7-1/2 to 8 pounds at birth are favored by stabilizing selection. Newborns weighing significantly more or less do not survive as well. Over hundreds of millions of years the body plan of the chambered nautilus has changed little; a single body plan has been conserved by stabilizing selection.

The transient polymorphism observed among peppered moths appears to be a case of *directional selection.* Because of a specific change in the environment, a heritable trait occurs with increasing frequency, and the whole population tends to shift in a parallel direction. The development of pesticide-resistant pests is another example of directional selection.

Disruptive selection, in which two or more distinct polymorphic varieties are favored and become increasingly represented in a population, tends to split up a population. An example of disruptive selection is provided by beak characteristics in a small population of Galápagos finches living in severe arid conditions. Survivors were those with long beaks that could open cactus fruits and those with deep, wide beaks that could crack cactus seeds and strip bark for insects.

Natural selection has favored the evolution of a different appearance of males and females in many species. Males are often more showy, larger, and more aggressive. *Sexual selection* is based on characteristics that provide individuals with a competitive edge in mating and producing offspring.

The survival value of variant alleles must be weighed in the context of the environment in which they are being expressed; they are not advantageous or disadvantageous in themselves. The HbA (normal hemoglobin) and HbS (sickle-cell hemoglobin) form a balanced polymorphism that has been maintained in populations throughout West and Central Africa for as long as malaria has been prevalent. Heterozygotes are said to have sickle-cell trait (HbS/HbA) and have a much higher resistance to malaria. Such individuals are more likely to survive severe malarial infections. Thus, malaria is a *stabilizing selection* agent that maintains the HbS allele in African populations.

Key Terms

adaptive	directional selection	differential fertility
differential reproduction	disruptive selection	balanced polymorphism
stabilizing selection	sexual selection	HbS and HbA alleles

Objectives

1. Define *stabilizing selection* and give two examples.
2. Define *directional selection* and explain the example of the peppered moth.
3. Explain how nautiloids have managed to remain essentially the same for hundreds of millions of years.
4. Explain why the maintenance of the sickle-cell anemia gene (HbS) is a special case of stabilizing selection called *balanced polymorphism.* Explain why inheriting a gene for HbS (sickle-cell hemoglobin) might be advantageous for some humans living under certain situations.
5. Define *disruptive* selection and give one example.
6. Define *sexual selection* and state the role of natural selection in this process.

Self-Quiz Questions

Fill-in-the-Blanks

When individuals of different phenotypes in a population differ in their ability to survive and reproduce, their alleles are subject to (1) _____

_____ . (2) _____ _____ favors the most common phenotypes in the population. (3) _____ _____ occurs when a specific change in the environment causes a heritable trait to occur with increasing frequency and the whole population tends to shift in a parallel direction. (4) _____ _____ favors the development of two or more distinct polymorphic varieties such that they become increasingly represented in a population and the population splits into different phenotypic variations. (5) _____ provide an excellent example of stabilizing selection because they have existed essentially unchanged for hundreds of millions of years. A shift in color of (6) _____ _____ in England provides a good example of (7) _____ selection. When populations of disease-carrying flies become increasingly resistant to an insecticide known as DDT, this is an example of (8) _____ selection. Two different but successful adaptations of beak types in a small population of Galápagos finches represents an example of (9) _____ selection. In Central Africa persons with the sickle-cell trait (HbS/HbA) have a greater chance of surviving (10) _____ infection than those lacking the sickle-cell gene. This is a special case of (11) _____ selection known as balanced (12) _____ . The more colorful, showy, and larger appearance of male pheasants when compared with females is the result of (13) _____ selection.

14-IV
(pp. 185–188)

SPECIATION
    Defining the Species
    Divergence and Isolation
SUMMARY

Summary

The process whereby species originate is known as _speciation_. A _species_ is composed of one or more populations of individuals who can interbreed under natural conditions and produce fertile offspring and who are reproductively _isolated_ from other populations.

    Evolution may be thought of as successive changes in allele frequencies brought about by such occurrences as mutation, genetic drift, gene flow, and selection pressure. Usually studies of evolution focus on some local breeding unit: a population of localized extent within a larger population system. The usual barriers to allele exchange between local breeding units of a large population are those that create _geographic isolation:_ severe storms, major floods, earthquakes, the uplift of mountain ranges, and the separation of continents. Following geographic isolation, mutation, genetic drift, and selection pressures may operate on the different local breeding units. Over time, these forces may lead to _divergence,_ a buildup of differences in allele frequencies between isolated populations of the same species.

Speciation has occurred when some restriction on interbreeding between populations is followed by enough genetic differentiation that, even if individuals from those populations do not make contact, they cannot or will not interbreed. *Reproductive isolating mechanisms* are aspects of structure, function, or behavior that prevent interbreeding between populations that are undergoing or have undergone speciation. Differences in reproductive structure or physiology, in timing of reproduction, in behavior, and in ecological factors may serve as isolating mechanisms. Gene flow between populations is prevented.

Interbreeding between two populations may be prevented by mechanical isolation in cases when differences in reproductive organs exist. Hybrid inviability and infertility, as in the case of mules, may maintain isolation between two populations. Complex courtship rituals that differ between species provide examples of behavioral isolation. Barriers to gene flow may exist because of geographic isolation; a large river or mountain ranges are examples.

A *polyploid* organism has three or more copies of each type of chromosome characteristic of its species. Polyploidy may arise through complete meiotic nondisjunction followed by abnormal gamete formation or when germ cells duplicate DNA but do not divide. Viable gametes may form when the original chromosomes have the "extra" identical chromosomes to pair with in meiosis. Especially among plants, polyploidy and/or hybridization are two other speciation routes; about one-half of all flowering plant species are polyploid. Wheat is a successful polyploid hybrid.

Key Terms

speciation	mechanical isolation	geographic barriers
species	hybrid inviability and	polyploidy
divergence	infertility	hybridization
reproductive isolating	behavioral isolation	
mechanism		

Objectives

1. Define the terms *species* and *speciation*.
2. Describe the role of *divergence* in populations.
3. State the meaning of *reproductive isolating mechanism* and give examples of how the following promote divergence: mechanical isolation; hybrid inviability and infertility; behavior isolation; geographic barriers.
4. Define *polyploidy* and indicate how speciation can occur by polyploidy and hybridization.

Self-Quiz Questions

Fill-in-the-Blanks

(1) _____ is the process whereby species are formed. A(n) (2) _____ is composed of one or more populations of individuals who can interbreed under natural conditions and produce fertile, reproductively isolated offspring. (3) _____ may be described as a process where differences in alleles accumulate between populations. Any aspect of structure, function, or behavior that prevents interbreeding is a(n) (4) _____ _____ mechanism. Physical barriers that prevent gene flow, as in the case of large

rivers changing course or forests giving way to grasslands, are known as
(5) _____ _____ . Plants or animals with complete extra sets of
chromosomes are known as (6) _____ . Wheat is an example of a species
formed by (7) _____ and (8) _____ .

CHAPTER TEST

UNDERSTANDING AND INTERPRETING KEY CONCEPTS

___ (1) An acceptable definition of evolution is changes in organisms _____ .

    (a) that are extinct
    (b) since the flood
    (c) over time
    (d) in only one place

___ (2) The two scientists most closely associated with the concept of evolution are _____ .

    (a) Lyell and Malthus
    (b) Henslow and Buffon
    (c) Henslow and Malthus
    (d) Darwin and Wallace

___ (3) The unit of evolution is the _____ .

    (a) population
    (b) individual
    (c) fossil
    (d) missing link

___ (4) Changes in allele frequencies brought about by mutation, genetic drift, gene flow, and natural selection are called _____ processes.

    (a) genetic equilibrium
    (b) microevolution
    (c) founder effects
    (d) independent assortment

For questions 5 to 7, choose from these answers:

    (a) disruptive selection
    (b) genetic drift
    (c) directional selection
    (d) stabilizing selection

___ (5) Cockroaches becoming increasingly resistant to pesticides is an example of _____ .

___ (6) The founder effect is a special case of _____ .

___ (7) Different alleles that code for different forms of hemoglobin have led to _____ in the midst of the malarial belt that extends through West and Central Africa.

For questions 8 to 10, choose from these answers:

(a) isolating mechanisms
(b) polyploidy
(c) genetic equilibrium
(d) sexual selection

___ (8) Hardy and Weinberg invented an ideal population that was in _____ and used it as a baseline against which to measure evolution in real populations.

___ (9) Plants that blossom at different times and chromosome sets that no longer can match up effectively during fertilization are examples of _____ .

___ (10) Many domestic varieties of fruits, vegetables, and grains have evolved larger vegetative and reproductive structures through _____ .

For questions 11 to 13, choose from these answers:

(a) speciation
(b) divergence
(c) reproductive isolating mechanism
(d) polyploidy

___ (11) Two species of sage plants with differently shaped floral parts that prevent pollination by the same pollinator is an example of a _____ .

___ (12) An ancestral species of pine tree that, through time, gives rise to several genetically isolated populations of pines is an example of _____ .

___ (13) As time passes, _____ occurs as differences in alleles accumulate between two populations.

INTEGRATING AND APPLYING KEY CONCEPTS

Can you imagine any way in which directional selection may have occurred or may be occurring in humans? Which factor(s) do you suppose are the driving force(s) that sustain(s) the trend? Do you think the trend could be reversed? If so, by what factor(s)?

15

MACROEVOLUTION

EVIDENCE OF MACROEVOLUTION
    The Fossil Record
    Dating Fossils
    Comparative Morphology
    Comparative Biochemistry
MACROEVOLUTION AND EARTH HISTORY
    Origin of Life
    Drifting Continents and Changing Seas
    Extinctions and Adaptive Radiations
    The Archean and Proterozoic Eras

The Paleozoic Era
The Mesozoic Era
Commentary: The Dinosaurs
—A Tale of Global Impacts, Radiations, and Extinctions
The Cenozoic Era
ORGANIZING THE EVIDENCE
—PHYLOGENETIC CLASSIFICATION
SUMMARY

General Objectives

1. Cite what biologists generally accept as evidence that supports their belief in macroevolution. Explain how observations from comparative morphology and comparative biochemistry are used to reconstruct the past.
2. Understand the factors that encourage increased rates of *speciation* and the formation of larger taxonomic groups. Know also the factors that bring about extinction and replacement of species.
3. Explain the ideas of mass extinctions, adaptive radiations, adaptive zones, and key innovations.
4. Explain the meaning of a phylogenetic classification system.

15-I
(pp. 189–194)

EVIDENCE OF MACROEVOLUTION
    The Fossil Record
    Dating Fossils
    Comparative Morphology
    Comparative Biochemistry

Summary

The evolution of groups of organisms larger than the species (groups of species) is referred to as *macroevolution*. Large-scale patterns, trends, and rates of change may be observed in the fossil record. The fossil is the most direct bit of evidence that evolution has occurred. Fossil formation requires rather quick burial in a medium that retards decomposition. Not all organisms become fossilized when they die. The quality and completeness of the fossil record varies as a function of the types of organisms (for example, shelled versus soft-bodied), the places where they died (eroding soils versus natural traps favoring fossilization), and the geologic stability of the region (tectonically active zones versus undisturbed sedimentary plains).

The geologic time scale initially was defined as a progression of four broad eras: *Proterozoic* ("first life"), *Paleozoic* ("ancient life"), *Mesozoic* ("middle life"), and *Cenozoic* ("modern life"). Within the past three decades, fairly firm boundaries have been assigned to these geologic intervals by using radioactive dating methods on an enormous number of rock samples taken from all over Earth. Research revealed that the origin and evolution of Earth itself occurred during an even more ancient era, the *Archean*. The earliest bacteria-like forms actually originated in the Archean, not the earlier-named Proterozoic. As we move from the most ancient fossilized cells that resemble bacteria, the fossil organisms preserved within younger and younger rocks become progressively more abundant, more diverse, and more like modern forms.

Comparative morphology is an area of study that compares the forms of organisms and any revealed structural patterns of major taxa. An example is furnished by the embryonic similarity observed in all vertebrate embryos—the presence of gill slits. Adult forms of organisms often differ markedly from their embryonic forms. Such differences are explained by mutations in regulatory genes that control the rate and growth of different body parts. Newborn chimpanzees and humans have nearly identical proportions of skull bones. These two organisms have very similar genes but mutations in the regulatory genes bring about the familiar differences in the proportions of their adult skulls.

Related organisms may show structural similarities or *homologous structures* due to descent from a common ancestor. The vertebrate forelimb with its numerous and diverse adaptations is an example of *morphological divergence* of homologous structures from an ancestral form. Organisms may be unrelated but have similar structures due to similar use. Such *analogous structures* illustrate *morphological convergence*. Convergence is illustrated by the streamlined body shape of sharks, penguins, and porpoises; fins or finlike structures stabilize these bodies in a similar environment.

Comparative biochemistry compares similarities in gene products to establish evolutionary relationships. *Neutral mutations* in different classes of proteins (amino acids substituted without loss of function) have accumulated through time at different rates and can be used to establish divergence of two species from a common ancestor. Immunological comparisons (which examine how foreign proteins react with antibodies) and DNA hybridization (which determines similarity in the nucleotide sequences between organisms of different species) are used to establish evolutionary relationships.

Key Terms

macroevolution	Cenozoic	morphological divergence
Archean	comparative morphology	analogous structures
Proterozoic	comparative biochemistry	morphological convergence
Paleozoic	homologous structures	neutral mutations
Mesozoic		

Objectives

1. Define macroevolution and explain the value of fossil evidence.
2. Explain why there are gaps in the fossil record.
3. Identify the five major eras of Earth's history and be familiar with their time spans.
4. Cite the method used to establish the age of fossil-containing rocks.
5. Define comparative morphology and distinguish between homologous and analogous structures; relate these terms to morphological divergence and morphological convergence.

6. Explain the role of regulatory genes in the development of differences in organisms with very similar gene sequences.
7. Understand the role of comparative biochemistry in establishing evolutionary relationships; cite examples.

Self-Quiz Questions

Fill-in-the-Blanks

The establishment of large-scale patterns, trends, and rates of change among groups of species is known as (1) _____ . Most of the information about the history of life on Earth comes from (2) _____ . Earth history can be divided into five great eras; beginning with the oldest, they are the (3) _____ , the (4) _____ , the (5) _____ , the (6) _____ , and the (7) _____ . These eras are based on abrupt (8) _____ in the (9) _____ record. Evidence for macroevolution comes from detailed comparisons of body form and structural patterns of major taxa; this is known as (10) _____ _____ . It is believed that variations among genetically similar adult vertebrates come about by mutations of (11) _____ _____ . The wings of birds and bats show genetic relationships and are termed (12) _____ structures. Such departure from a common ancestral form is known as (13) _____ _____ . Penguin wings and shark fins perform similar functions in a similar environment; these structures are said to be (14) _____ and demonstrate (15) _____ _____ . Divergence between two species can be dated using accumulation rates of neutral (16) _____ . Measurements of antigen-antibody reactions are used to make (17) _____ _____ . (18) _____ _____ is a method of converting DNA of different species to single-stranded forms and allowing the forms to recombine as a measure of similarity.

15-II
(pp. 194–202)

MACROEVOLUTION AND EARTH HISTORY
  Origin of Life
  Drifting Continents and Changing Seas
  Extinctions and Adaptive Radiations

Summary

A cohesive theory linking the flow of Earth's history and the history of life is being developed. Billions of years in the past, our solar system was born from dense clouds of dust left from exploding, dying stars. By 4.6 billion years ago, the cloud was a flat, slowly rotating disk with a young sun at the center. Our planet took form through the gravitational compression of dust and debris that swirled around the primordial sun. Five billion years ago, Earth was a cold, homogeneous mass, but through contraction and radioactive heating it developed a core that was dense and hot. By 3.8 billion years ago, a crust that periodically erupted volcanic outpourings from the inferno below had formed, and a dense atmosphere with almost no free oxygen was being retained by Earth's gravitational field as our planet settled into an orbit at a favorable distance around the sun. Lack of oxygen, the presence of water, and temperatures on the planet favored the development of life, and somewhere between 4 and 3.8 billion years ago, living cells appeared in the mineral-rich primeval seas. An experiment by Stanley Miller demonstrated that all the building blocks required for life can form under abiotic conditions, given the primitive Earth atmosphere of hydrogen, methane, ammonia, water, and an abundant supply of energy. The mechanism by which independently formed organic molecules combined into a permanent self-replicating living system is not yet known. Clay crystals in tidal flats may have served as the first templates for the synthesis of large chains of amino acids, the proteins. The first cells were probably membrane-bound sacs surrounding nucleic acids that served as patterns for protein synthesis. Sidney Fox demonstrated that heating a dry mixture of amino acids formed protein chains. When cooled in water, the protein chains assembled themselves into small, stable, water-filled spheres that possessed many characteristics of cell membranes. Thus, the work of Miller, Fox, and others showed that self-replicating systems surrounded by protective membranes could have arisen spontaneously under conditions that might have existed on the early Earth.

The first living cells arose in a very unstable environment and even today organisms contend with a changing and often violent environment. *Plate tectonics* describes the Earth's crust (lithosphere) as being comprised of slablike plates that float on a hot underlayer, the mantle. During the Paleozoic, an early continent named Gondwana drifted south from the tropics, then across the South Pole and then northward. Gondwana later fused with other land masses to form a single world continent, Pangea, surrounded by one huge ocean. Pangea began breakup during the Mesozoic; the drifting of its fragments continues today. Sporadic *mass extinctions* followed by recovery periods and *adaptive radiations* are two trends dominant in the evolutionary history of life. Mass extinctions are sudden rises in extinction rates that exceed the constant background extinction rate. Major groups of species with narrow distributions and specialized adaptations have been eradicated during mass extinctions. Adaptive radiations of organisms that fill adaptive zones commonly occur during the initial few millions of years following a mass extinction. The adaptive radiation of mammals following a mass extinction between the Mesozoic and Cenozoic eras is an example.

Key Terms

Stanley Miller
self-replicating systems
chemical competition
template
clay crystals

left-handed forms
Sidney Fox
plate tectonics
Gondwana
Pangea

mass extinction
adaptive radiation
adaptive zones
key innovation

Objectives

1. List in sequence the events and processes that cooperated to make Earth a planet suitable for supporting life. Indicate how long ago these processes were occurring.
2. Describe how life might have spontaneously arisen on Earth between 4 billion and 3.8 billion years ago. Present evidence that supports the evolution of the first primitive organism.
3. Understand how ancient prokaryotes are thought to have changed the early atmosphere of Earth.
4. Describe the movements of tectonic plates in the Paleozoic, Mesozoic, and Cenozoic eras; relate this to the evolution of life forms.

Self-Quiz Questions

Fill-in-the-Blanks

About (1) [choose one] ( ) 10 ( ) 4.6 ( ) 3.8 ( ) 3.2 billion years ago, our planet took form as a cooling dense cloud of dust and gas. By (2) [choose one] ( ) 4.6 ( ) 3.8 ( ) 3.2 ( ) 2 billion years ago, Earth had a hot, dense core and was hurtling through space as a thin-crusted inferno. Rain fell, minerals were stripped from rocks, and the oceans were formed. (3) _____ _____ probably were the absorbing agents that served to assemble (4) _____ _____ into proteins. Stanley Miller constructed a reaction chamber containing circulating hydrogen, methane, ammonia, and (5) _____ . Within one week, many (6) _____ _____ and other organic compounds had formed. Sidney Fox heated dry amino acids to form (7) _____ chains. These were placed in hot water and self-assembled into small, stable, membranous (8) _____ that showed properties of (9) _____ _____ . (10) _____ _____ refers to Earth's slablike plates floating on a hot, plastic (11) _____ . Gondwana and other land masses crunched to form a single world continent called (12) _____ that broke up; its movements continue today. These and other environmental changes of Earth's environment provoked two trends that occurred in the evolution of life, (13) _____ _____ and (14) _____ _____ . In (15) _____ _____ , a lineage fills the environment with new species through bursts of microevolution. This occurs when there are unfilled (16) _____ _____ . At times, evolutionary access results when a(n) (17) _____ _____ occurs in a species.

MACROEVOLUTION AND EARTH HISTORY (cont.)
The Archean and Proterozoic Eras
The Paleozoic Era
The Mesozoic Era
Commentary: The Dinosaurs—A Tale of Global Impacts,
Radiations, and Extinctions
The Cenozoic Era

Summary

Earth's unstable crust prevented formation of stable land masses until about 3.7 billion years ago. Many active volcanoes were present. Rocks 3.5 billion years old contain prokaryotic cells that resemble today's simple anaerobic bacteria. Fermentation probably evolved first but before the end of the Archean Era, photosynthetic bacteria had appeared and were releasing the first free oxygen into the atmosphere. Between 2.5 billion and 700 million years ago, photosynthetic bacteria ruled the warm, shallow seas; eukaryotic cells began to evolve. Stromatolites were constructed by mat-forming bacteria. Animals who grazed on photosynthesizers appeared.

Oxygen accumulated in the Proterozoic atmosphere, which prevented further chemical synthesis outside of cells. The presence of oxygen also opened an adaptive zone that allowed adaptation of a key innovation in some bacteria—aerobic respiration. The evolution of increasingly complex aerobic organisms would follow. The earliest known plants are green algae fossils over 1.2 billion years old. Green and red algae fossils, plant spores, and fungi have been found in Australian rock formations dated at 900 million years. At the Proterozoic-Paleozoic boundary 750 million years ago, animals with shells, spines, and armor plates suddenly appeared.

During the Paleozoic Era, continents were alternately submerged and drained by shallow seas. Most of the major animal phyla had evolved by the end of the Cambrian period. One of the dominant animals was the trilobite.

Gondwana drifted southward during the Ordovician; this opened up new shallow marine environments for evolutionary experimentation. Gondwana sat on the South Pole by the late Ordovician. Shallow seas drained, huge glaciers formed, and in this ice age reef life vanished everywhere in the first worldwide mass extinction.

During the Silurian period and into the Devonian approximately 400 million years ago, Gondwana began moving northward. Reef organisms recovered as a radiation of predatory armor-plated fishes occurred. Lobe-finned fishes that began to breathe air invaded the land, while muddy shores furnished a habitat for the evolution of the first small but erect land plants. At the Devonian-Carboniferous boundary, another worldwide mass extinction occurred.

During the Carboniferous period, plants and insect groups underwent major adaptive radiations. Repeated submerging and draining of the land created huge swamps in which large treelike plants lived; this vegetation eventually formed the coal deposits of Great Britain and North America.

The Permian period was a time with abundant insects, amphibians, and primitive reptiles. The greatest of all mass extinctions happened near the close of the Permian when many land and sea populations perished. At this time all the land masses collided to form a huge continent, Pangea, alone in one ocean; these collisions affected temperature and climate on a worldwide scale. Some of the remaining survivors were small reptilian ancestors of dinosaurs, birds, snakes, and lizards.

The Mesozoic Era lasted approximately 175 million years; its subdivisions are the Triassic, Jurassic, and Cretaceous periods. Pangea fragmented into three separating land masses while creating the Atlantic basin. Following the Permian extinction, new lines of organisms evolved through adaptive radiations. Small mammals originated in the Triassic period while magnificent reptiles lived in the air, oceans, and swamps. At the boundary of the Triassic-Jurassic, a mass extinction due to unknown causes took place. The dinosaurs dominated Earth during the late Jurassic and early Cretaceous and then suddenly disappeared at the end of the Cretaceous period. Incoming asteroids impacting the Earth's surface may have altered Earth's habitats so much that a cycle of extinctions began that terminated with the elimination of nearly all remaining dinosaurs. Flowering plants originated early in the Cretaceous and then underwent a major radiation that continues today. The gymnosperms had begun to decline and their former habitats were quickly taken by the ascending flowering plants.

By *Cenozoic* times North America, Europe, and Africa were moving their separate ways; brittle fragmentation of coastlines, severe volcanism, and massive uplifting of mountain ranges promoted a variety of cooler, drier environments. Sea levels changed, as did Earth's overall temperature. Grasslands were formed; plant-eating animals and their predators evolved and occupied them. With the land essentially cleared of major reptilian predators, the *mammals* and *birds* eventually diversified, along with the *flowering plants.* Tropical forests were fragmented into a patchwork of new environments, and many forest inhabitants were forced into new life styles in mountain highlands, deserts, and plains. One such evicted form gave rise to the human species.

Key Terms

Archean Era	armor-plated fishes	Permian
Proterozoic Era	vertebrate jaw	Mesozoic Era
Paleozoic Era	Devonian	Triassic
Cambrian	lobe-finned fishes	Jurassic
trilobites	Carboniferous	Cretaceous
Ordovician	reptiles	Cenozoic Era
Silurian		

Objectives

1. Describe the environment on Earth's surface during the Archean and Proterozoic eras.
2. Describe how prokaryotic metabolism changed the primitive atmosphere of Earth.
3. Describe the principal events that occurred in each of these periods of the Paleozoic Era: Cambrian, Ordovician, Silurian, Devonian, Carboniferous, Permian. State whether aquatic or terrestrial conditions prevailed and which organisms were most abundant and conspicuous during each period.
4. List the principal ideas of the plate tectonics theory and state which earthly phenomena are explained by the theory.
5. Name the environmental changes that permitted evolving plants and animals to shift from an aquatic environment to moist land and from moist land to dry land.
6. Name the animals and plants that were the pioneers in the transition to land.
7. Describe the tectonic events that were shaping land environments during Mesozoic and Cenozoic times, and state the types of organisms that evolved to exploit the different types of environments.
8. Speculate about factors that may have caused the extinction of the dinosaurs.
9. Explain how a knowledge of Earth's geologic history helps explain why there is such a great diversity of organisms on Earth today.

10. State whether you think humans are successfully adapted to their environments, and provide some evidence to support your statement.
11. Understand how classes, orders, and families come to thrive, become extinct, or undergo sustained trends in their general pattern of characteristics.
12. Describe some of the factors that have brought about extinction and replacement during the Ordovician, Devonian, Permian, Triassic, and Cretaceous
periods.

Self-Quiz
Questions

Fill-in-the-Blanks

Rocks 3.5 billion years old contain fossils of (1) _____ _____ that
probably existed in tidal mud flats. Little or no free (2) _____ was in the
atmosphere, so (3) _____ , an anaerobic pathway, must have been the
first metabolic pattern to develop. Between (4) [choose one] ( ) 3.2 ( ) 2.5 ( ) 1
billion and 700 million years ago, (5) _____ _____ dominated the
shallow, warm seas and the first (6) _____ appeared. Mound-shaped
bacterial cell mats found in rock formations more than 2 billion years ago are
called (7) _____ . Oxygen accumulated in the (8) _____ atmosphere.
This prevented further synthesis of (9) _____ _____ and opened up
new worldwide (10) _____ _____ in which organisms such as
multicelled plants, fungi, and animals evolved that carried on (11) _____
respiration. The Paleozoic Era is subdivided into the Cambrian, Ordovician,
Silurian, Devonian, Carboniferous, and Permian periods. Nearly all animal
phyla had evolved by the end of the (12) _____ period; (13) _____
were the dominant animal group. Gondwana drifted southward and seas
flooded land to open new, shallow marine environments for flourishing reef
organisms during the (14) _____ period. In the area of the South Pole,
glaciers formed on Gondwana and many forms of reef life became extinct.
Gondwana drifted northward during the Silurian and the Devonian. Reef
organisms recovered as a major radiation of (15) _____-_____ fishes
occurred. (16) _____-_____ fishes that were ancestral to amphibians
began to invade the land as small-stalked plants began to evolve in muddy
land-water margins. Another global (17) _____ _____ occurred at
the Devonian-Carboniferous boundary. During the (18) _____ period,
major adaptive radiations of insects and treelike coal-forming plants occurred
in swamp forests. When Pangea formed by collisions of all land masses,
nearly all land and sea species perished in the greatest mass extinction known
as the (19) _____ period drew to a close. Great changes in geology and

climate occurred but the small reptilian ancestors of dinosaurs, birds, snakes, and lizards survived. The Mesozoic Era is subdivided into Triassic, Jurassic, and Cretaceous periods and lasted about 175 million years. Mammals originated during the (20) _____ period. During a(n) (21) _____ _____ at the Triassic-Jurassic boundary, many marine organisms were lost. (22) _____ ruled Earth during the late Jurassic and early Cretaceous periods. There is physical evidence that Earth's collision with a(n) (23) _____ may have brought about dinosaur extinction. In the early Cretaceous period, (24) _____ _____ emerged and began a major radiation amid the decline of gymnosperms that continues today. When the (25) _____ era began, major mountain ranges were formed, the great land masses underwent major renovations, and climate changes took place that greatly altered the course of evolution. New environments promoted a great diversification of (26) _____ , (27) _____ , and (28) _____ _____ .

15-IV
(pp. 208–210)

ORGANIZING THE EVIDENCE—PHYLOGENETIC CLASSIFICATION SUMMARY

Summary

Phylogeny is the evolutionary history of a species, the study of which encompasses the most ancestral species through sequences leading to the most recent descendant. Change may occur within an established *lineage* (as, for example, in large, long-legged birds). Gradual changes in traits such as horn size occur as primitive genera evolve into advanced ones. Such an evolutionary pattern may take the shape of a branching tree with an ancestral trunk. Each twig or branch represents a single line of descent, or lineage. The branch points are where speciation events occur. According to the *gradualistic model*, this slow evolutionary pattern is the main trend in the history of life. However, not all transitions in the fossil record are gradual; some major groups of organisms appeared relatively abruptly and were already highly developed and diverse when they did appear. Several explanations have been proposed to account for such abrupt appearances. The rapid crossing of adaptive thresholds, the *punctuational model* states, occurs in times of intense pressure in small, isolated populations. This model also holds that rapid speciation has been the principal mode and that most morphological change has occurred during these times of intense pressure.

Sometimes important gaps occur in the fossil record. Classifying the millions of diverse species on the tree of life is an immense task. Carolus Linnaeus devised the two-name (genus and species) system to help catalog the immense diversity of life forms. Similar species are placed in the same genus, similar genera are placed in the same family, similar families are placed in the same order, similar orders are placed in the same class, similar classes are placed in the same phylum (or division), and these categories of inclusiveness are placed in one of the currently recognized five kingdoms of life.

Key Terms

phylogeny   gradualism   punctuation

lineage

Self-Quiz Questions

Fill-in-the-Blanks

The evolutionary history of a group that considers all ancestors and descendants is its (1) _____ . When evolutionary relationships are viewed as a branching tree, a branch represents a single line of descent or (2) _____ . Evolution can be viewed either by the (3) _____ model, which envisions speciation as accounting for only a small amount of large-scale change, or by the (4) _____ model, which sees higher taxa originating from the rapid crossing of adaptive thresholds. Linnaeus devised a two-name system for cataloging diverse life forms, the first name is the (5) _____ name and the second name is the (6) _____ name. Systems of classification begin with these two names and are based on several categories such as families and orders, which become increasingly more (7) _____ until all members are included in one of the five (8) _____ of life.

CHAPTER TEST

UNDERSTANDING AND INTERPRETING KEY CONCEPTS

For questions 1 to 8, choose from these answers:

     (a) Archean
     (b) Cenozoic
     (c) Mesozoic
     (d) Paleozoic
     (e) Proterozoic

___ (1) Dinosaurs and gymnosperms were the dominant forms of life during the _____ Era.

___ (2) The Alps, Andes, Himalayas, and Cascade Range were born during major reorganization of land masses early in the _____ Era.

___ (3) The composition of Earth's atmosphere changed during the _____ Era from one that was anaerobic to one that was aerobic.

___ (4) Invertebrates, primitive plants, and primitive vertebrates were the principal groups of organisms on Earth during the _____ Era.

___ (5) Before the close of the _____ Era, the first photosynthetic bacteria had evolved.

___ (6) The _____ Era ended with the greatest of all extinctions, the Permian extinction.

___ (7) Late in the _____ Era flowering plants arose and underwent a major radiation.

___ (8) The _____ Era included adaptive zones into which plant-eating mammals and their predators radiated.

For questions 9 to 11, choose from these answers. More than one letter may be chosen for one question.

      (a) gradualistic model
      (b) punctuational model
      (c) convergence
      (d) divergence
      (e) adaptive radiation

___ (9) Penguins and porpoises serve as examples of _____ .

___ (10) Kangaroos, koala bears, and opossums serve as examples of _____ .

___ (11) Evolution of the major groups of mammals following dinosaur extinction occurred by _____ .

For questions 12 to 15, choose from these answers:

      (a) key innovation
      (b) punctuation
      (c) gradualism
      (d) extinction

___ (12) An example of a(n) _____ is the modification of forelimbs into wings; this opened new adaptive zones.

___ (13) The vertebrate invasion of the land at the end of the Devonian was followed by rapid evolution; this is an example of _____ .

___ (14) The disappearance of dinosaurs from Earth is an example of _____ .

___ (15) Fossil foraminiferans (tiny shelled protists) show a gradual change of form (over lengthy time periods) within one lineage; this is an example of _____ .

For questions 16 to 20, choose from these answers:

      (a) Homologous
      (b) Analogous
      (c) Macroevolution
      (d) Plate tectonics
      (e) Phylogeny

___ (16) _____ structures resemble one another due to common descent.

___ (17) _____ is the study of the slabs floating on Earth's underlying mantle.

___ (18) _____ Structures include the shark fin and the penguin's wing.

___ (19) _____ of the horse includes consideration of all known horse fossils up to the living horse species.

___ (20) _____ refers to changes in groups of species.

UNDERSTANDING AND APPLYING KEY CONCEPTS

(1) Imagine that in the next decade three more Chernobyl disasters happen, the oceans acquire critical levels of carcinogenic pesticides that work their way up the food chains, and the ozone layer shrinks dramatically in the upper atmosphere. Describe the macroevolutionary events that you believe might happen.

(2) As Earth becomes increasingly loaded with carbon dioxide and various industrial waste products, how do you think living forms on Earth will evolve to cope with these changes?

16

HUMAN EVOLUTION: A CASE STUDY

THE MAMMALIAN HERITAGE
THE PRIMATES
    Trends in Primate Evolution
    Primate Origins
THE HOMINIDS
    Australopiths

Stone Tools and Early *Homo*
Homo erectus
Homo sapiens
SUMMARY

General Objectives

1. Understand the general physical features and behavioral patterns attributed to early primates. Know their relationship to other mammals.
2. Trace primate evolutionary development through the Cenozoic Era.
3. Understand the distinction between hominoid and hominid and distinguish between *Australopithecus* and *Homo*.

16-I
(pp. 212–216)

THE MAMMALIAN HERITAGE
THE PRIMATES
    Trends in Primate Evolution
    Primate Origins

Summary

When the world climate began to grow cooler and drier during the Cenozoic, birds and mammals, which have internal and external adaptations to help them maintain a relatively constant body temperature even when environmental temperatures rise and fall, rose to dominance. Humans are classified in the class Mammalia. *Dentition* and the extended period of *infant dependency* and *learning* are two mammalian features highlighted in the evolution of humans. The Primate family tree includes prosimians (lemurs and others) and the anthropoids (monkeys, apes, and humans). *Hominoids* include apes, humans, and recent human ancestors. The term *hominids* is applied only to members of the human lineage. The *primates* may be defined by several evolutionary trends: (1) modification of the skeleton for *bipedalism;* (2) modification of hands to accomplish prehensile and opposable movements; (3) less reliance on smell and enhanced daytime vision; (4) dentition changes toward smaller and less specialized teeth; (5) increase in brain mass and complexity; (6) behavioral evolution; and (7) coevolution of the human brain and human culture.

Primates apparently arose from mammals resembling night-foraging small rodents or tree shrews more than 60 million years ago. Between 54 million and 38 million years ago (Eocene), descendants of these animals had moved to the trees. Eyes adapted for sensing colors, shapes, and movements in a three-dimensional field and body and limbs adapted for climbing trees were the key characteristics of primate evolution. By 35 million years ago (Oligocene) the tree-dwelling ancestors of monkeys and apes had evolved. About 23 to 20 million years ago, as the moving continents neared their current positions, an adaptive radiation of apelike forms, the hominoids, occurred. Speciation may have been favored by cooler, drier climates, fewer forests, and new grasslands. The origin of Miocene apes known as *dryopiths* occurred; between 10 and 6 million years ago, some of their descendants gave rise to ancestors of modern gorillas, chimpanzees, and humans.

Key Terms

dentition	humans	opposable
infant dependency	hominids	hominoids
learning	bipedalism	culture
primates	prehensile	dryopiths

Objectives

1. List the two features that are central to the story of human evolution.
2. State when the earliest primates evolved and describe them.
3. Briefly describe changes in the characteristics of evolving primates that lived from 54 to 38 million years ago; 35 million years ago; 23 to 20 million years ago. Correlate the changes in primates with their changing environment.
4. Explain why the environmental changes that occurred from 23 to 20 million years ago might have encouraged an adaptive radiation of apelike forms—the first hominoids.
5. Name the group of Miocene apes whose descendants gave rise to the ancestors of modern gorillas, chimpanzees, and humans.

Self-Quiz Questions

Fill-in-the-Blanks

Two mammalian features are central to the story of human evolution, (1) _____ and infant (2) _____ and (3) _____ . Humans, apes, monkeys, and prosimians are all (4) _____ . Apes, humans, and recent human ancestors are classified as (5) _____ . The term (6) _____ refers only to members of the human lineage. (7) _____ refers to the ability of humans to sustain a two-legged gait for extended time periods. Several evolutionary trends in primate evolution have been identified. (8) _____ became modified to make prehensile and opposable movements. Changes in eyes provided enhanced (9) _____ _____ . Teeth became smaller and modified as (10) _____ changed from fruit, leaves, and insects to a mixed form. The (11) _____ increased in mass and complexity. Longer life spans, single births, increased offspring-parental bonding, and longer learning periods can be correlated with evolution of the brain and culture development; this trend in primate evolution is known as (12) _____ evolution. Primates evolved from ancestral (13) _____ more than 60 million years ago. The first known primates looked like small (14) _____ . Descendants of these early primates lived in the (15) _____ . By 35 million years ago, tree-dwelling ancestors of (16) _____ and (17) _____ had appeared. Between 23 and 20 million years ago, as continents moved near their current positions, climates changed, and forests gave way to grasslands, an adaptive radiation of apelike forms, the (18) _____ , appeared. At this time, Miocene apes known as (19) _____ appeared. Between 10 and 6 million years ago, their descendants gave rise to the ancestors of gorillas, chimpanzees, and (20) _____ .

16-II
(pp. 216–220)

THE HOMINIDS
    Australopiths
    Stone Tools and Early *Homo*
    Homo erectus
    Homo sapiens
    SUMMARY

Summary

The first hominids evolved between 8 and 4 million years ago and were identified by three common features: bipedalism, omnivorous feeding, and continued brain expansion and reorganization. These three features represent considerable *plasticity*, or adaptability, to a wide variety of demands that might be presented by new and unpredictable environments. As the climate became cooler and drier, some hominids adapted to mixed woodlands while others lived in the large and expanding grasslands.

Australopiths, the earliest hominids, are grouped into two wide groups: slightly built forms (*Australopithecus afarensis* and *A. africanus*) and more heavily built forms (*Australopithecus boisei* and *A. robustus*). Many divergent forms had arisen. The hipbone from a slightly built form, named Lucy, indicated upright walking or bipedalism. Bipedalism freed the hands for development of dexterity. Cranial capacity had increased to about 400 cubic centimeters and the jaw was slightly bow-shaped. The tooth structure of *A. africanus* indicated an omnivorous diet, while the teeth of *A. robustus* and *A. boisei* suggested a diet of plant materials.

The earliest member of the genus *Homo*, *Homo habilis*, lived about 2.5 million years ago. *H. habilis* had a larger brain and more generalized tooth structure and was the first toolmaker. These earliest rock-flake tools were discovered by Mary Leakey at Olduvai Gorge and were used to process foods of bone and flesh. Evidence at Olduvai also shows that by 1.9 million years ago, our ancestors had developed the concept of "home."

Between 1.5 million and 300 thousand years ago (Pleistocene), radical glacier-induced climate changes appear to have forced a larger-brained species, *Homo erectus*, to migrate from Africa to Southeast Asia, China, and Europe. The brain size of *H. erectus* was nearly the same as that of modern humans and correlates with the long travels, improved tools, and fire-making ability of this species.

Modern humans, *Homo sapiens*, probably evolved from *H. erectus* somewhere between 300 and 200 thousand years ago. These forms coexisted for a time before *H. erectus* became extinct. Face and skull structure more nearly resembled modern humans. One group belonging to the genus *Homo*, the Neandertals, lived in Europe, the Near East, and China. Neandertals, who were hunters and gatherers, had coarse facial features and slightly larger brains than modern humans have. They suddenly disappeared about 35 or 40 thousand years ago. Anatomically modern humans, *H. sapiens sapiens*, emerged about 40 thousand years ago; human evolution has been cultural since that time.

Key Terms

hominids	*A. africanus*	*Homo sapiens sapiens*
plasticity	*A. robustus*	Mary Leakey
australopiths	*A. boisei*	Pleistocene
Lucy	*Homo habilis*	Olduvai Gorge
Australopithecus afarensis	*Homo erectus*	Neandertal

Objectives

1. Describe three features that early hominids had in common and explain why *plasticity* of these features was important.
2. Describe the two broad categories of australopiths and their principal features.
3. Relate the characteristics of early *Homo* to tool making and using tools, diet, and the concept of home.
4. Correlate the characteristics of *Homo erectus* with adaptations to severe climate changes; list other accomplishments of this species.
5. Explain the origin of modern humans and compare the characteristics of this species with *Homo erectus* and Neandertals.

Self-Quiz Questions

Fill-in-the-Blanks

The first (1) _____ emerged between 8 million to 4 million years ago. All early hominids had three features in common: omnivorous feeding, further brain development, and (2) _____ . These features had a(n) (3) _____ that allowed adaptation to different environmental demands. The earliest known hominids are called the (4) _____ . Early *Homo* is associated with making (5) _____-_____ tools and the concept of (6) _____ . In response to extreme shifts in (7) _____ *Homo erectus* migrated out of Africa. There was little difference in the brain size of (8) _____ _____ when compared with modern humans, and this species apparently made use of improved (9) _____ and controlled (10) _____ . Somewhere between 300 and 200 thousand years ago, modern humans, (11) _____ _____ apparently arose from (12) _____ _____ stock. About 40 thousand years ago, *Homo sapiens sapiens* emerged; they were anatomically modern (13) _____ . Prior to this, groups of a coarse-faced species lived in Europe and Asia, the (14) _____ . Recent human evolution has been (15) _____ rather than biological.

CHAPTER TEST

UNDERSTANDING AND INTERPRETING KEY CONCEPTS

__ (1) Which of the following is *not* considered to have been a key character in early primate evolution?

    (a) Eyes adapted for discerning color and shape in a three-dimensional field
    (b) Body and limbs adapted for tree climbing
    (c) Bipedalism and increased cranial capacity
    (d) Eyes adapted for discerning movement in a three-dimensional field

__ (2) Primitive primates generally live _____ .

    (a) in tropical and subtropical forest canopies
    (b) in temperate savanna and grassland habitats
    (c) near rivers, lakes, and streams in East Africa
    (d) in caves where there are abundant supplies of insects

__ (3) The term *hominoids* includes _____ .

    (a) monkeys, apes, and humans
    (b) prosimians and lemurs
    (c) apes, humans, and recent human ancestors
    (d) only members of the human lineage

__ (4) The two mammalian features that, early on, became central to the story of human evolution are the _____ and the _____ .

    (a) thumb; brain
    (b) brain; extended period of infant dependency and learning
    (c) brain; ears
    (d) teeth; extended period of infant dependency and learning

__ (5) The hominid evolutionary line stems from a divergence (fork in a phylogenetic tree) from the age line that apparently occurred _____ .

    (a) somewhere between 8 million and 4 million years ago
    (b) about 3 million years ago
    (c) during the Eocene epoch
    (d) less than 2 million years ago

__ (6) The earliest known hominids are called _____ .

    (a) dryopiths
    (b) australopiths
    (c) *Homo erectus*
    (d) Miocene apes

___ (7) Johanson's Lucy was a(n) _____ .

    (a) dryopith
    (b) australopith
    (c) *Homo habilis*
    (d) Neandertal

___ (8) A hominid hunter-gatherer of Europe and Asia that became extinct nearly 35 to 40 thousand years ago was _____ .

    (a) a dryopith
    (b) an australopith
    (c) *Homo erectus*
    (d) Neandertal

Matching

Choose the one most appropriate answer for each.

(9) ___ anthropoids

(10) ___ australopiths

(11) ___ dryopiths

(12) ___ hominoids

(13) ___ Neandertals

(14) ___ prosimians

A. A group that includes apes, humans, and recent human ancestors

B. A population of archaic *Homo sapiens* with large browridges and cranial capacities larger than modern humans

C. A group of organisms that includes monkeys, apes, and humans

D. A group of organisms that includes lemurs and others

E. Bipedal organisms that lived about 4 million years ago; transitional between Miocene apes and later hominids

F. Transitional apelike forms that could climb about in trees and walk on the ground

---

INTEGRATING AND APPLYING KEY CONCEPTS

Suppose that someone told you that sometime between 12 million and 6 million years ago dryopiths were forced by predatory larger members of the cat family to flee the forests and take up residence in estuarine, riverine, and sea coastal habitats where they could take refuge in the nearby water to evade the tigers. Those that, through mutations, became naked, developed an upright stance, developed subcutaneous fat deposits as insulation, and developed a bridged nose had advantages in watery habitats that other dryopiths who remained inland never developed. As time went on, predation by the big cats and competition with other animals for available food caused most of the terrestrial dryopiths to become extinct, but the water-habitat varieties survived as scattered remnant populations, adapting to easily available shellfish and fish, wild rice and oats, various tubers, nuts, and fruits. It was in these aquatic habitats that the first food-getting tools (baskets, nets, and pebble tools) were developed, as well as the first words that signified different kinds of food.

How does such a story fit with current speculations and evidence of human origins? How could such a story be demonstrated to be true or false?

Crossword
Number Four

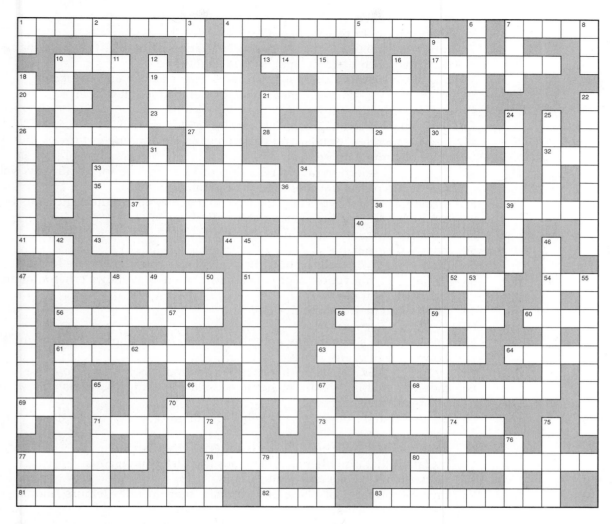

Across

1. walking on two feet
4. stability of allelic and genotypic ratios in a population over succeeding generations
7. _____ tectonics
10. central storage tissue of many dicot plants
13. _____ infertility: the inability of offspring of genetically dissimilar species to have offspring of their own
17. of or relating to the family that includes humans (abbrev.)
19. flap on the front of a suit jacket
20. _____ of the valley; a flower
21. slow morphological change within species from genetic drift, directional selection, and other processes by which allele frequencies change
23. interrogative pronoun
26. knowing (Latin)
27. sixty-one in roman numerals
28. heritable change in a DNA molecule
30. seed or fruit from a cereal grass

32. salt (Latin)
33. morphological _____ : selection leads to departures from ancestral form of species
34. _____ reproduction: tendency of bearers of adaptive traits to reproduce more successfully than bearers of less adaptive traits
35. abbreviation for road
37. rapid morphological change during speciation, with relatively little change occurring once a particular species is established
38. one of two or more alternative forms of a gene at a given gene locus
39. *en* _____ (French), meaning a food embedded in gelatin
41. at this time, now; however
43. dark bluish green to greenish blue; also, a kind of duck
44. large-scale rates, trends, and patterns of change among groups of species since beginning of life
47. _____ selection favors retention of intermediate forms already well adapted to prevailing conditions in a population
51. formation of new types of organisms that are similar to their ancestors but can no longer interbreed effectively with them
52. feminine personal pronoun
54. fifty-two in roman numerals
56. plate _____ tells us how the present-day continents were formed
58. _____ innovation: a principal new change that allows environmental exploitation in new ways
59. dress worn by woman in India
60. _____ *sapiens*
61. this selection shifts the phenotypic character of the population as a whole
63. how often an event occurs
64. _____ *Gay*: airplane that dropped an atomic bomb on Hiroshima
66. radiation that fills vacant niches in ecosystems with closely related species in a relatively short time
68. era that lasted from approximately 240 million years ago to 65 million years ago
69. an individual grain of edible cereal species *Avena sativa*
71. mutation that is neither advantageous nor disadvantageous
73. the termination of a taxonomic group
77. member of order of placental mammals that includes lemurs, monkeys, apes, and humans
78. this selection favors extreme forms of a trait and operates against intermediate forms
80. structures in different animals that are similar in construction and develop from the same embryonic tissues
81. two or more species from dissimilar and only distantly related lineages adopt a similar way of life and come to resemble one another as they live in similar environments
82. _____ extinction: abrupt termination of several large categories of organisms simultaneously
83. a group of organisms of the same species occupying a given area at a specific time

Down

1. along
2. a tiny round mark, spot, or point
3. pertaining to the study of the form and structure of organisms
4. theory of _____ by natural selection
5. color of light with a wavelength of approximately 700 nm

6. morphology that examines the similarities and differences in structure between two organisms
7. a type of evergreen tree
8. organ that is sensitive to light rays
9. fake, fraud; a spurious imitation
10. having several sets of chromosomes in the somatic cells of a species
11. a group that includes both apes and humans; superfamily
12. gene _____ : physical movement of alleles into and out of populations
14. Babi _____ : a site outside Kiev where Germans killed Jews in 1941
15. adaptive _____ fills vacant niches in ecosystems with closely related species in a relatively short time
16. edible vegetable of the genus *Allium*
18. flexibility and adaptability to a wide range of demands
21. sticky substance that is sometimes chewed
22. evolutionary relationships that begin with most ancestral species and include all branchings that lead to all of their descendants
24. similar in function or behavior
25. a remnant or trace of an organism of a past geological age
29. last letter of the Greek alphabet, Ω
31. selection that involves traits females use to select mates that give an individual an advantage in producing offspring
33. genetic _____ : the chance increase or decrease in relative abundance of different alleles
36. changes in allele frequencies within populations or species; small-scale evolution
40. a normally large population is drastically reduced by unfavorable conditions
42. characteristic
45. earliest known representatives of the human family
46. an era noted for vertebrate and land plant origins and formation of Pangea
47. choice; a kind of weeding out
48. a metric unit of volume equal to 1.056 liquid quarts; British spelling
49. _____ Buddhism: a belief that enlightenment can be attained through meditation, self-contemplation, and intuition
50. a vaporous substance of low density
53. _____ -Weinberg principle
55. mechanism that prevents interbreeding between populations that are undergoing or have undergone speciation
57. Greek goddess of the rainbow; also, the colored portion of the eye
59. to _____ : see, feel, taste, smell, hear
61. tooth structure and arrangement
62. the sum total of behavior patterns of a social group
65. line of descent
67. *Homo* _____ lived during Pleistocene Ice Ages
68. Big _____ : a popular hamburger
70. mythical fire-belching reptile
72. shelf
74. integrated *pest management*
75. fruit of oak tree
76. _____ Desert in Mongolia
79. male sheep
80. joint where pelvis meets upper leg bone

17

VIRUSES, MONERANS, AND PROTISTANS

VIRUSES
MONERANS
    Characteristics of Bacteria
    Major Groups of Bacteria
    About the "Simple" Bacteria
THE RISE OF EUKARYOTIC CELLS
    Commentary: Speculations on the Origin
    and Evolution of Eukaryotes

PROTISTANS
    Slime Molds
    Euglenids
    Chrysophytes
    Dinoflagellates
    Flagellated and Amoeboid Protozoans
    Sporozoans
    Ciliated Protozoans
SUMMARY

General Objectives

1. Explain what a virus is and describe how virus particles are reproduced.
2. Distinguish major structural differences between prokaryotic and eukaryotic cells.
3. List the modern descendants of three prokaryotic lineages that diverged from a common ancestor over 2.5 billion years ago.
4. Describe the principal moneran forms and ways of living.
5. Generally describe the eight categories of protistans. Describe the four categories of protistans. Tell how protistans differ from monerans, viruses, and multicellular eukaryotes. Give some common names of protistans.

17-I
(pp. 222–225)

VIRUSES

Summary

Viruses have a set of nucleic acids (DNA or RNA) sheathed in a protective protein coat, but they are not capable of metabolism and cannot reproduce themselves unless they take over the chemical machinery found within a living host cell. Viruses are an expression of parasitic simplification.

    The general method of cell attack in the *lytic pathway* occurs when the virus attaches to a cell surface; this is followed by entry of the entire virus or its nucleic acid core into the cell cytoplasm. The viral nucleic acids are then replicated; this is followed by transcription and translation of the viral gene to produce viral proteins. The new nucleic acids and proteins associate into new viral particles, which are released with cell lysis (membrane rupture). In a *lysogenic pathway*, infected host cells are not killed but viral genes are incorporated into host cell chromosomes. The viral nucleic acid is replicated with host cell DNA. RNA viruses utilize *reverse transcription* to produce viral DNA transcripts for insertion into a host chromosome. *Bacteriophages* are viruses that attack bacterial cells. Various *animal viruses* infect vertebrates and invertebrates. *Plant viruses* cause a multitude of plant diseases. Viruses are the causative agents of human diseases such as smallpox, influenza, AIDS, polio, and possibly cancer. They damage livestock and crops, and they play an important role in keeping certain bacterial populations within bounds. There are presently no cures for AIDS and herpes infections in humans.

Viroids are infectious pieces of single-stranded RNA that lack protein coats and are smaller than the smallest virus. They afflict citrus, avocado, potato crops, and other crops and may be implicated in the development of some forms of cancer.

Key Terms

microbes	bacteriophage	lytic pathway
virus	animal viruses	lysogenic pathway
host cell	plant viruses	reverse transcription
lysis	viroid	

Objectives

1. State the principal characteristics of viruses, indicate how they might have evolved, and list three human diseases caused by viral agents.
2. Define what is meant by *bacteriophage;* briefly distinguish between *lytic pathway* and *lysogenic pathway.*
3. Describe *viroids* and discuss how their existence might affect humans.

Self-Quiz Questions

Fill-in-the-Blanks

Each virus particle consists of (1) _____ or _____ surrounded by a(n) (2) _____ _____ . (3) _____ contain the blueprints for making more of themselves, but cannot carry on metabolic activities unless they can pirate the metabolic machinery of a(n) (4) _____ _____ . A group of viruses that infect bacteria are the (5) _____ . Herpes and Spanish influenza are caused by (6) _____ _____ . Viral diseases that affect crop yield are caused by (7) _____ _____ . Naked strands of RNA that cause disease and depend on host cells are called (8) _____ .

17-II
(pp. 225–229)

MONERANS
  Characteristics of Bacteria
  Major Groups of Bacteria
  About the "Simple" Bacteria

Summary

Some bacteria, chemosynthetic autotrophs, extract energy from inorganic molecules. Some are *photosynthetic autotrophs*, which use light energy, and others are *heterotrophs* (decomposers and parasites that extract energy by breaking down parts of the bodies of dead or living organisms). Bacteria generally reproduce by asexual *binary fission*, although many bacteria form endospores (which resist drying out, irradiation, disinfectants, and acids). Although each bacterium is a functionally independent unit, bacteria often remain linked together in pairs, chains, or clusters following division. The bacterial chromosome is a circular DNA molecule; additional DNA is carried in the form of smaller plasmids.

All eubacteria possess a semirigid cell wall composed of polysaccharide-protein molecules (peptidoglycan), and some species have a sticky capsule or slime layer outside the wall. Some have *bacterial flagella* anchored to wall and cytoplasm. Bacteria differ somewhat in shape: cocci are spherical, bacilli are rodlike, and spirilla are spiral. Two groups of bacteria have distinctly different molecules embedded in their walls and respond differently to the Gram-stain technique. Gram-positive bacteria stain deep purple while Gram-negative bacteria lose the purple stain during decolorization but retain a pink-to-reddish color.

There are two major groups of bacteria, the genetically and structurally distinct *archaebacteria* and the *eubacteria.* Archaebacteria includes methanogens, extreme halophiles, and thermoacidophiles; these groups tolerate low-oxygen environments. Eubacteria are grouped according to nutritional styles. *Photosynthetic eubacteria* use sunlight energy during photosynthesis to form ATP, which they use later to run cell processes. The cyanobacteria, with their enzyme-producing and carbohydrate-storing *heterocyst* cells, are examples. Some cyanobacteria, like *Anabaena,* can fix nitrogen (that is, extract it from air and attach it to the molecules they use metabolically). *Chemosynthetic eubacteria* include nitrifying bacteria that strip electrons from ammonia or nitrite to form ATP. *Heterotrophic eubacteria* include agents causing syphilis, gonorrhea, and Lyme disease. *Clostridium botulinum* forms resistant *endospores* and produces a deadly toxin that causes botulism. *Escherichia coli* lives in the human gut and benefits its host by producing vitamin K and compounds useful in digesting fats. All bacteria seem to be able to cause diseases under the right conditions, even *E. coli.* Actinomycetes produce many antibiotics. *Azotobacter* and *Rhizobium* can fix nitrogen to enrich soil and aid plant growth. *Lactobacillus* can digest milk sugar and aid milk digestion in newborn mammals. Other eubacteria species are helpful to humans in producing fermented milk products such as cheese and yogurt.

Even though bacteria are small and less structurally complex than eukaryotic cells, they are by no means simple. The several different biochemical means bacteria have for dealing with their environment represent a high level of complexity. Bacteria sense chemicals and often are able to move toward or away from chemical sources. A few bacteria, such as *Myxococcus xanthus,* move purposely as cell colonies to trap other microbes, which fall prey to their enzyme secretions.

Key Terms

monerans	spirillum, spirilla	chemosynthetic eubacteria
photosynthetic autotrophs	Gram-negative	nitrifying bacteria
chemosynthetic autotrophs	Gram-positive	heterotrophic eubacteria
heterotrophs	archaebacteria	Lyme disease
binary fission	methanogens	endospore
peptidoglycan	extreme halophiles	*Clostridium botulinum*
capsule	thermoacidophiles	botulism
slime layer	photosynthetic eubacteria	*Escherichia coli, E. coli*
bacterial flagella	cyanobacteria	*Lactobacillus*

coccus, cocci   heterocysts   *Myxococcus*
bacillus, bacilli

Objectives

1. Distinguish among the following terms: *photosynthetic autotroph, chemosynthetic autotroph,* and *heterotroph.*
2. Describe how bacteria reproduce and the nature of bacterial genetic material.
3. Describe the diversity of body plans that bacteria of different groups have.
4. Distinguish Gram-positive from Gram-negative bacteria.
5. List three essential ways that archaebacteria differ from eubacteria.
6. Briefly describe the three nutritional modes of the eubacteria.
7. Explain how endospore formation by bacteria can concern humans.
8. Tell which bacteria produce antibiotics.

**Self-Quiz
Questions**

Fill-in-the-Blanks

Bacteria that are (1) _____ autotrophs extract energy from inorganic molecules. Most bacteria are (2) _____ of one type or another. Some bacteria have (3) _____ _____ secured to both the cell wall and the plasma membrane. Bacteria whose walls stain (4) _____-_____ appear purple through the light microscope. The (5) _____ include three bacterial groups known as methanogens, extreme halophiles, and thermoacidophiles. The (6) _____ are photosynthetic eubacteria that produce cells in the chains; some cells in the chains, the (7) _____ , produce enzyme molecules. The (8) _____ _____ are chemosynthetic eubacteria that play a very important role in the (9) _____ _____ . *Clostridium botulinum* produces a deadly toxin that taints food and causes a form of poisoning called (10) _____ . *Clostridium* can produce (11) _____ that are very resistant and can germinate to form new bacterial cells. *Escherichia coli* is a heterotrophic eubacterium that lives in (12) _____ hosts and produces needed vitamin K and other compounds for that host.

17-III
(pp. 229–234)

THE RISE OF EUKARYOTIC CELLS
    Commentary: **Speculations on the Origin and Evolution of Eukaryotes**
PROTISTANS
    Slime Molds
    Euglenids
    Chrysophytes
    Dinoflagellates

Summary

When DNA and RNA nucleotide sequences from different bacteria are studied, the evidence still continues to suggest that archaebacteria, eubacteria, and the ancestors of modern eukaryotes once evolved from a common ancestor. The term *eukaryote* includes single-celled protistans, multicelled plants, fungi, and animals. These diverse groups all have a distinct nucleus and other membrane-bound organelles. According to some biologists, the nucleus of eukaryotes may have originated by the plasma membrane infoldings of bacteria. Further, organelles of eukaryotes may have had their origin as smaller but independent cells that were captured and continued to live symbiotically within one cell. For example, the mitochondrion may have once been a free living cell; mitochondria contain their own genetic instructions. Simple eukaryotic multicelled organisms, such as *Volvox* and *Trichoplax,* suggest how the first multicelled organisms appeared. Several cells living together and sharing the benefits of different roles may have been advantageous to all cells in the group.

Slime molds exhibit features of other kingdoms. Some slime molds have individual cells like bacteria that come together and then differentiate into fruiting bodies that produce fungal-like spores. Part of the slime mold life cycle resembles that of animals in that it shows movement. Plasmodial slime mold cells move as a single mass of cytoplasm; cellular slime molds mass together but retain their identity as separate cells.

Euglenids are flagellated, photosynthetic protistans with rather complex cells; they prefer rather stagnant waters. Euglenid cells possess an *eyespot*, which orients the cell to receive the most favorable light for photosynthesis.

Chrysophytes are photosynthetic protistans that live in freshwater and marine waters; the group includes yellow-green algae, golden algae, and diatoms. Other pigments in the cells mask the chlorophyll. Many have hard cellular parts made of silica. Diatoms, with their glassy shells, are mined as diatomaceous earth, which has many economic uses such as abrasives, filters, and insulation.

Dinoflagellates are primitive photosynthetic eukaryotes with stiff cellulose plates and are motile members of the aquatic and marine plankton. Many populations "bloom" suddenly in response to suddenly available nutrients. Red dinoflagellates sometimes show such *red tide* population explosions.

Dinoflagellates color the seas red or brown while producing a neurotoxin that can severely affect sea life and humans who eat seafoods, even causing death.

Key Terms

protistans, Protista	multicellularity	diatoms
Margulis	euglenids, *Euglena*	dinoflagellates
symbiosis	eyespot	plankton
cellular slime molds	chrysophytes	red tides
plasmodial slime molds	golden algae	

Objectives

1. Explain how eukaryotic cells might have evolved from ancestral prokaryotes.
2. Describe possible advantages that the evolution of multicellularity brought to organisms.
3. Define the term *protistan* and be able to describe important characteristics of slime molds, euglenids, chrysophytes, and dinoflagellates.

Fill-in-the-Blanks   Single-celled eukaryotic organisms are classified in the Kingdom (1)

_____ . The eukaryotic nucleus may have originated from infoldings of

the (2) _____ plasma membrane. Eukaryotic organelles may be captured

(3) _____ cells. Slime molds are either (4) _____ or (5) _____

and show animal-like movements but produce fungal-like (6) _____ . (7)

_____ are flagellated, photosynthetic protistans with a light-sensing

eyespot. Photosynthetic protistans that often have hard silica-based parts and

include yellow-green algae, golden algae, and diatoms are known as (8)

_____ . Drifting collections of marine or aquatic protistans and animal

larvae are generally called (9) _____ . Yellow-green, brown, or red

photosynthetic members of plankton that possess stiff cellulose plates with

flagella in grooves are the (10) _____ . Some forms of (11) _____

produce harmful neurotoxins and bloom in the sea in such numbers that the

phenomenon is called (12) _____ _____ .

17-IV
(pp. 234–237)

PROTISTANS (cont.)
    Flagellated and Amoeboid Protozoans
    Sporozoans
    Ciliated Protozoans
SUMMARY

Summary

Protozoans ("first animals") are grouped by their movement types, either amoeboid or with flagella. Their habits are predatory or parasitic.

Those flagellated protozoans known as *trypanosomes* live a portion of their life cycle in the salivary glands of insects as a means of ensuring transfer to new hosts; African sleeping sickness is caused by a trypanosome. Some forms of *Trichomonas* cause damage to the urinary and reproductive membranes in humans and are transmitted during sexual intercourse. *Cysts* are walled, resistant, resting forms of protozoans. *Giardia* cysts are found in waters contaminated with feces; the adult form causes intestinal disturbances and sometimes even death. Resistant cysts are formed that exit the body with feces.

Amoeboid protozoans include amoebas and foraminiferans. Movement is accomplished by temporary cell extensions, the pseudopods ("false feet"). Amoeba is commonly studied in biology laboratories. *Entamoeba histolytica* is parasitic in humans, causing severe intestinal disturbances, known as amoebic dysentery. Cysts leave the body with feces. Shelled foraminiferans dwell mostly in the seas; their shells have many holes through which the tiny pseudopods extend.

Sporozoans are parasitic. *Plasmodium* species are subtropical or tropical sporozoans that cause malaria. Mosquitoes transmit this long-infecting sporozoan to birds and humans. Cells of ciliated protozoans like *Paramecium* have a high degree of structural complexity. Many cilia on their surfaces propel these cells through the water and move food into a gullet. Digestion occurs in cytoplasmic vesicles and wastes leave the cell at an anal pore. Contractile vacuoles remove excess water.

Key Terms

protozoans	amoeboid protozoans	sporozoans
flagellated protozoans	amoeba	*Plasmodium*
cysts	*Entamoeba histolytica*	malaria
trypanosomes	amoebic dysentery	ciliated protozoans
Giardia	foraminiferans	*Paramecium*

Objectives

1. Cite two flagellated protozoans that cause disease; define *cyst*.
2. Explain how amoeboid protozoans move and feed; briefly give the life cycle of *Entamoeba histolytica*.
3. Tell about the nutritional mode of sporozoans; define *malaria*.
4. List the features common to most ciliated protozoans; cite an example.

Self-Quiz Questions

Fill-in-the-Blanks

Flagellated protozoans cause some well-known human diseases. (1) _____ infection is spread by sexual intercourse and damages membranes of the urinary and reproductive systems; (2) _____ causes mild intestinal disturbances and forms resting stages called (3) _____ . Amoebas move by sending out (4) _____ , which surround food and engulf it. (5) _____ secrete a hard exterior covering of calcareous material that is peppered with small holes through which tiny, food-trapping pseudopodia extend. *Entamoeba histolytica* causes a severe intestinal disorder known as (6) _____ _____ . (7) _____ is a famous (8) _____ that causes malaria. *Paramecium* is a ciliated protozoan that lives in (9) _____ environments and depends on (10) _____ _____ for eliminating the excess water constantly flowing into the cell. *Paramecium* has a(n) (11) _____ through which food enters the cell; digestion occurs in cytoplasmic (12) _____ .

Matching

Choose the most appropriate answer for each.

13. ___ trypanosomes
14. ___ *Plasmodium*
15. ___ *Entamoeba*
16. ___ *Paramecium*
17. ___ pseudopods
18. ___ foraminiferans
19. ___ amoeba
20. ___ *Trichomonas*

A. Amoeboid protozoan with perforated, calcareous "shells"
B. False feet
C. Spread by sexual contact
D. A ciliated protozoan
E. Parasitic protozoan that causes malaria
F. Commonly studied in school biology laboratories to demonstrate cytoplasmic flow
G. Amoebic dysentery
H. African sleeping sickness

CHAPTER TEST

UNDERSTANDING AND INTERPRETING KEY CONCEPTS

___ (1) The basic structure of a virus may be described as _____ .

   (a) DNA or RNA surrounded by a carbohydrate coat
   (b) carbohydrate surrounded by DNA or RNA
   (c) DNA or RNA surrounded by a protein coat
   (d) protein surrounded by DNA or RNA

___ (2) Bacteriophages are _____ .

   (a) viruses that parasitize bacteria
   (b) bacteria that parasitize viruses
   (c) bacteria that phagocytize viruses
   (d) composed of a protein core surrounded by a nucleic acid coat.

___ (3) The only members of Kingdom Monera are the _____ .

   (a) animals
   (b) plants
   (c) bacteria
   (d) fungi

___ (4) _____ use simple inorganic substances as their energy source.

   (a) Photosynthetic autotrophs
   (b) Chemosynthetic autotrophs
   (c) Heterotrophs
   (d) Photosynthetic heterotrophs

___ (5) The three major shapes of bacteria are _____ .

   (a) spherical, cylindrical, and helical
   (b) spherical, cylindrical, and square
   (c) oval, cylindrical, and helical
   (d) oval, cylindrical, and square

___ (6) A Gram-positive bacterial cell is colored _____ after being immersed in the Gram stain.

  (a) red
  (b) pink
  (c) purple
  (d) green

___ (7) The most ancient bacteria appear to be the _____ .

  (a) eubacteria
  (b) chemosynthetic eubacteria
  (c) photosynthetic eubacteria
  (d) archaebacteria

___ (8) Population blooms of _____ cause red tides and extensive fish kills.

  (a) *Euglena*
  (b) specific dinoflagellates
  (c) diatoms
  (d) *Plasmodium*

___ (9) For an organism to be considered truly multicellular, _____ .

  (a) its cells must be heterotrophic
  (b) the organisms cannot be parasitic
  (c) there must be division of labor and cellular specialization
  (d) the organisms must at least be motile

___ (10) Slime molds, euglenids, chrysophytes, dinoflagellates, flagellated and amoeboid protozoans, amoeboid protozoans, and ciliated protozoans are all classified as _____ .

  (a) animals
  (b) monerans
  (c) plants
  (d) protistans

___ (11) Which of the following play an important role in the cycling of nitrogen-containing substances?

  (a) cyanobacteria
  (b) prions
  (c) viroids
  (d) photosynthetic flagellates

Matching

Choose the most appropriate answers for each.

(12) ___ diatoms

(13) ___ *Clostridium*

(14) ___ bacteriophages

(15) ___ slime molds

(16) ___ dinoflagellates

(17) ___ amoeba

(18) ___ euglenids

(19) ___ viruses

(20) ___ *Giardia*

A. Cause red tides
B. Nonliving but cause disease
C. Creep about like animals and form spores
D. Possess light-sensitive eyespots
E. Obtain food by pseudopods
F. Flagellated protozoan, causes mild intestinal infections, forms cysts
G. Produces endospores
H. Viruses that infect bacteria
I. Live in overlapping silica shells

INTEGRATING AND APPLYING KEY CONCEPTS

Suppose that genetic engineers could successfully introduce chloroplasts into human zygotes and that this new combination established a symbiosis. Describe all of the changes that you can imagine would occur in the appearance and behavior of the resulting individuals.

18
FUNGI AND PLANTS

PART I. KINGDOM OF FUNGI
    General Characteristics of Fungi
    Major Groups of Fungi
    Lichens and Other Symbionts
    Commentary: A Few Fungi We Would
    Rather Do Without
PART II. KINGDOM OF PLANTS
    General Characteristics of Plants

Evolutionary Trends Among Plants
Red, Brown, and Green Algae
Bryophytes
Lycophytes, Horsetails, and Ferns
Existing Seed Plants
Gymnosperms
Angiosperms—The Flowering Plants
SUMMARY

General Objectives

1. Describe the nutritional mode of fungi and explain its importance to nature.
2. Describe the various types of fungal body plans and general patterns of reproduction.
3. Name at least one specific example of each of the five groups of true fungi.
4. Give a suitable definition of a *plant*.
5. Outline the evolutionary advances that converted marine algal ancestors into forms that could exist on wet land. Then state the structural advances that converted primitive marsh plants into dry-land flowering plants.

18-I
(pp. 240–244)

PART I. KINGDOM OF FUNGI
    General Characteristics of Fungi
    Major Groups of Fungi
    Lichens and Other Symbionts
    ***Commentary:* A Few Fungi We Would Rather Do Without**

Summary

Fungi are heterotrophic eukaryotes. Because their vegetative bodies are generally branched and filamentous, allowing them to come in contact with a large volume of their surroundings, they have access to raw materials that may be dilute or scarce. Some (for example, those that cause most plant diseases and athlete's foot) are *parasitic,* but most fungal species are *saprobes,* feeding on the remains of dead organisms or their waste products. By bringing about the decay of organic material, they help recycle vital substances through communities of organisms. Most fungi rely on enzyme secretions that promote digestion *outside* the fungal body, followed by nutrient absorption across the plasma membrane of individual cells. The vegetative body of most *true fungi* is a *mycelium*—a mesh of branched, tubular filaments called *hyphae* (singular, hypha). Many fungi have hyphae composed of elongated cells with chitin-reinforced walls. An important part of their life cycle is the production of *fungal spores,* which are walled, resistant cells. A spore germinates to form a new mycelium. Most true fungi can reproduce asexually through spore formation from a parent mycelium. Many also can reproduce meiotic spores as part of a sexual cycle.

About 60,000 species of fungi have been identified. *Chytrids* are single-celled saprobes or parasites of muddy or aquatic habitats that feed on decaying plant materials or living organisms. *Water molds* are major decomposers in aquatic habitats; they parasitize algae or aquatic animals. *Zygospore-forming fungi* are generally saprobes and rely mostly on asexual reproduction. The group includes bread molds and *Pilobolus.* Zygospores are thick-walled sexual spores that germinate to give rise to stalked, spore-producing structures. *Sac fungi* bear spores in saclike structures that are often concentrated on the surfaces of complex fruiting bodies shaped like globes or cups. The group contains single-celled yeasts (bread and alcohol production), edible species such as morels and truffles, and many disease-causing species. *Club fungi,* important decomposers, are structurally the most complex of all fungi. Spores are borne at the end of club-shaped cells. They include the familiar cultured edible mushrooms as well as extremely toxic forms. Other types are the shelf fungi, coral fungi, and puffballs, as well as those causing serious plant diseases: the rusts and smuts.

Imperfect fungi apparently lack a sexual stage and are placed in this informal category; they remain there unless a sexual phase is discovered. Many cause human and crop diseases. Some, like *Penicillium,* are commercially and medically important.

Many fungi have a negative effect on human existence. Various fungi spoil food supplies, one causes athlete's foot, while others attack ornamental and crop plants. One water mold, the cause of late blight of potato in Ireland (1845–1860), influenced history when its widespread destruction caused a cycle of death and disease that eliminated one-third of the Irish population. Ergotism is caused by a fungus parasitic on rye and other grain plants.

Lichens are composite organisms composed of an alga and a fungus living interdependently; the fungus generally obtains as much as eighty percent of the photosynthetically derived food produced by the alga. Some green algae thrive better as part of the lichen than when living independently. Lichens are found on tree bark, bare rocks, and soil. Animals in the arctic tundra feed on lichens. Lichens cannot thrive in heavy pollution; their absence indicates the presence of pollutants.

A *mycorrhiza,* or "fungus-root," is a symbiotic relationship described as fungal hyphae living densely wrapped around roots of forest trees and shrubs. The fungus obtains carbohydrate food from the plants and supplies mineral ions to the plant in times of soil shortage. Mycorrhizae are easily damaged by acid rain. The fossil record shows that mycorrhizal relationships are ancient and probably increased survival for both organisms as invasion of the new land habitat occurred.

Key Terms

fungi	fungal spore	club fungi
saprobes	chytrids	imperfect fungi
parasite	water molds	lichen
mycelium, -lia	zygospore-forming fungi	mycorrhiza
hypha, hyphae	sac fungi	

Objectives

1. Describe the general structure of a fungus and its relationship to the fungal method of obtaining nutrients.

2. Distinguish between fungi that are parasites and saprobes. Mention one way in which parasitic fungi harm humans and one way in which saprobic fungi benefit humans.
3. List the ways that fungi can reproduce.
4. List the six principal groups of fungi.
5. State the fundamental contribution of fungi to ecosystems.
6. Give two examples of parasitic fungi that have played havoc with the production of crop plants.
7. Describe the symbiotic relationship known as a *lichen*.
8. Explain the symbiotic relationships in a mycorrhiza.

Self-Quiz Questions

Fill-in-the-Blanks

As to their nutritional mode, fungi are (1) _____ . Fungi obtaining nutrients from nonliving organic matter are (2) _____ ; those that obtain nutrients directly from living host tissues are (3) _____ . The vegetative body of most true fungi is a(n) (4) _____ , which is a mesh of branched, tubular filaments called (5) _____ . The fungal (6) _____ is a walled, resistant cell that can be dormant and germinate to form a new mesh of hyphae, the (7) _____ . Fungal spores can be produced by meiosis in the (8) _____ cycle or by mitosis in a(n) (9) _____ cycle. (10) _____ are single-celled fungi that have rootlike absorptive structures that feed on decaying plants or live as parasites on plants, animals, and fungi. Black bread mold is an example of a(n) (11) _____-_____ fungus. Yeasts, morels, and truffles are examples of (12) _____ fungi. Rusts, smuts, puffballs, and shelf fungi are examples of (13) _____ fungi. The (14) _____ _____ are fungi whose sexual phase is absent or undetected. (15) _____ are composite organisms that comprise an alga and fungus living interdependently. (16) _____ help many complex land plants absorb certain vital nutrients from the soil.

18-II
(pp. 245–248)

PART II. KINGDOM OF PLANTS
General Characteristics of Plants
Evolutionary Trends Among Plants
Red, Brown, and Green Algae

Summary

Plants are multicellular, generally photosynthetic organisms. They range in size from microscopic algae to giant redwoods. Most plants are vascular with fluid-conducting systems in well-developed root and shoot systems. Red, brown, and green algae are included as plants. Bryophytes are nonvascular plants that lack true roots, stems, and leaves.

Ancestors of modern land plants presumably evolved from green algal types that lived in aquatic habitats 700 million years ago. Simple stalked plants had adapted to the land habitat by 300 million years later. After 55 million more years passed, forests of large, woody, fernlike plants had evolved. A variety of plant features have developed over a long time: (1) complex vascular tissues—the xylem and the phloem—transport substances throughout the plant and raise the photosynthetic areas up into the sunlight; (2) the development of a waxy cuticle that helps the plant retain water; (3) the development of stomata that enable gas exchange to be regulated; (4) the reduction of the haploid *gametophyte* generation, so that the diploid *sporophyte* (which has highly developed vascular tissues) dominates most of the life cycle; (5) the evolution of plants producing two spore types from plants that produced one spore type; (6) the evolution of gametes from motile into nonmotile types; and (7) the development of plants bearing seeds in fruits from earlier seedless plants.

Red, brown, and green *algae* (sing., alga) are still treated as members of the plant kingdom. The diverse body forms and colors (green, red, purple, or green-black) of the *red algae* are adapted to a wide range of communities, but they are found principally as producers in deep marine tropical waters. They resemble the blue-green algae with respect to their photosynthetic membranes and pigments including phycobilins. They have cell walls hardened by calcium carbonate and several species produce agar. The *brown algae,* with xanthophyll pigments, are almost all marine and include kelps—some of which are anchored by holdfasts, and some of which thrive in the open sea. Most of the complexly organized kelps show alternation of generations, and one species of giant kelp has developed phloemlike tissue. Some species produce commercially valuable algin. The diverse *green algae* generally dwell in freshwater communities, but there are also many species that live in saltwater or in moist soil. Green algae also show alternation of generations and variation in the type of gametes produced. Green algae, like complex land plants, have *chlorophyll a* and *b,* carotenoids, and xanthophylls in their plastids, store carbohydrates as starch, and have a cellulose framework within the cell wall. These features provide an evolutionary link between the green algae and complex higher plants.

Key Terms

vascular plant

red algae

brown algae

green algae

nonvascular plant

sporophyte

gametophyte

plant spore

Objectives

1. List the general characteristics of plants.
2. Summarize the key evolutionary trends among plants.
3. Distinguish the red, brown, and green algae from one another.

Self-Quiz Questions

Fill-in-the-Blanks

(1) _____ plants have specialized tissues that transport water and solutes through the plant organs. (2) _____ tissue transports water and dissolved

minerals throughout the plant body; (3) _____ tissue transports photosynthetic products through the plant body. The (4) _____ is the spore-producing body and the (5) _____ is the gamete-producing body in a plant life cycle. The (6) _____ phase is the dominant phase of complex land plants and it is this phase that produces (7) _____ spore types. The gametes in complex land plants are (8) _____ and fertilization does not require a watery medium. The most complex vascular plants produce (9) _____ , which contain the embryo (10) _____ phase of the plant life cycle. (11) Some _____ algae have stonelike cell walls, participate in coral reef building, and are major producers of marine communities. The (12) _____ algae contain xanthophylls, live offshore or in intertidal zones, and have many representatives with large sporophytes known as kelps; some species produce (13) _____ , a valuable thickening agent.

Green algae are thought to be ancestral to more complex plants because they have the same types and proportions of (14) _____ pigments, have (15) _____ in their cell walls, and store their carbohydrates as (16) _____ .

18-III
(pp. 248–250)

PART II. KINGDOM OF PLANTS (cont.)
  Bryophytes
  Lycophytes, Horsetails, and Ferns

Summary

Bryophytes are *nonvascular* transitional land plants that evolved adaptations for obtaining and conserving water, even though they do not possess vascular tissue, roots, or the ability to reproduce sexually in dry-land environments: (1) aboveground parts have a waxy cuticle with numerous stomata; (2) reproductive cells are surrounded by protective cell layers (the *sporangium* houses *spores,* the *archegonium* contains an egg, and the *antheridium* houses sperms, preventing water loss); and (3) embryo sporophytes begin development inside female gametophytes. Bryophytes include the generally small mosses, liverworts, and hornworts, which grow close to the soil surface of moist environments. They have leaflike, stemlike, and rootlike parts. The familiar moss plant is the photosynthetic gametophyte. Eggs and sperms develop at protective gametophyte plant tips. Sperms swim to eggs in a water film; following fertilization the zygote divides to form an attached sporophyte, which is dependent on the gametophyte for water and nutrients. The development of a *protected embryo sporophyte*—attached to and nourished by gametophyte tissue—must have been an important advance for life on land; it was first seen among the bryophytes.

The primitive seedless vascular plants (lycophytes, horsetails, and ferns) once dominated the ancient landscape in much larger forms than their present-day relatives. Although they are adapted for life on land, the sporophytes mostly prefer moist habitats but possess vascular tissues; these tissues are lacking in the gametophytes. Sperms require moisture to reach eggs.

Treelike *lycophytes* lived about 350 million years ago; their modern descendants are small forest-dwelling club mosses. Spores develop on surfaces of special leaves clustered tightly into cones. Spores develop into small, free living gametophytes.

Ancient ancestors of modern horsetails were also treelike forms; only species of the genus *Equisetum* have survived. Horsetails are found in moist habitats. Underground stems (rhizomes) produce photosynthetic aerial branches and scalelike leaves; true roots anchor the rhizome in the soil. Air-dispersed horsetail spores form in protective layers at shoot tips. Spores germinate to form small green gametophytes.

Most ferns live in tropical and temperate zones and vary greatly in size. Most have underground stems; leaves (fronds) are finely divided but occur in diverse forms. Rust-colored spots on the undersurface of fronds are clusters of sporangia, the *sori* (singular, *sorus*). Sporangia release spores into the air. Spores germinate into small, green, heart-shaped gametophytes.

Key Terms

bryophytes	hornworts	horsetails
sporangium, -ia	protected embryo	rhizomes
archegonium, -ia	sporophyte	ferns
antheridium, -ia	lycophytes	frond
mosses	club mosses	sorus, sori
liverworts		

Objectives

1. State three bryophyte features that were adaptive to life on land.
2. Define *sporangium, archegonium,* and *antheridium.*
3. Compare the life cycles of mosses and ferns.
4. Describe lycophytes, horsetails, and ferns.

Self-Quiz Questions

Fill-in-the-Blanks

The aboveground parts of bryophytes usually are covered by a(n) (1) _____ layer with numerous stomata. Protective tissue layers surround (2) _____ cells. The (3) _____ houses spores; the (4) _____ houses an egg; the (5) _____ houses sperms. Bryophyte sporophyte embryos begin early development (6) _____ the female gametophyte. (7) _____ are the most common bryophytes. The moss (8) _____ is attached and dependent on the moss (9) _____ . Seedless vascular plants include the (10) _____ , the (11) _____ , and the (12) _____ .

Labeling

Identify each indicated part of the accompanying illustration (p. 192).

(13) _____

(14) _____

(15) _____   (19) _____
(16) _____   (20) _____
(17) _____   (21) _____
(18) _____

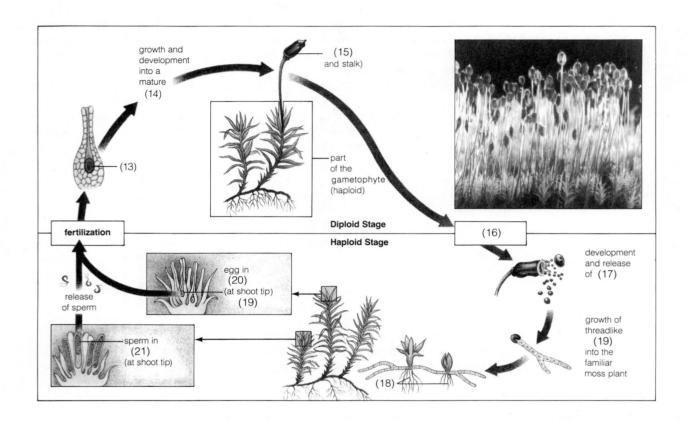

18-IV
(pp. 250–256)

PART II. KINGDOM OF PLANTS (cont.)
Existing Seed Plants
Gymnosperms
Angiosperms—The Flowering Plants
SUMMARY

Summary

Seed-bearing plants are the most successful vascular plants. A Devonian ancestor called a *progymnosperm* is like the type believed to be ancestral to gymnosperms and angiosperms. *Gymnosperm* seeds are unprotected on the surface of reproductive parts, but *angiosperm* seeds are borne within protective flower parts.

Gymnosperms include the familiar conifers, cycads, ginkgos, and gnetophytes. Familiar conifers are pines, spruces, firs, hemlocks, junipers, cypresses, and redwoods. Cone-shaped clusters of leaves bear sporangia. Pine is an example of a conifer life cycle. Male cone scales bear sporangia in which mother cells produce haploid *microspores* by meiosis. Each spore becomes a pollen grain containing a male gametophyte. Female cone scales bear *ovules* containing mother cells, which undergo meiosis to produce haploid spores; only one functions as a *megaspore* that develops into a female gametophyte. In spring, air-borne pollen is carried to female cones, a process called *pollination*. Sperm-bearing pollen tubes grow from pollen grains toward female gametophytes as eggs form in female gametophytes. Zygote formation and embryo development follow fertilization. A pine *seed* includes the embryo, female gametophyte tissue that serves as food reserve, and protective seed coats. Gymnosperms occupied many environments during the Mesozoic. Today, their numbers are reduced, but about 550 species live in northern latitudes and high altitudes. They supply lumber and many other products for humans.

For more than 100 million years, the angiosperms (flowering plants) have been the most successful plants. They have many diverse adaptations for life in different habitats, and in size range from tiny duckweeds to giant *Eucalyptus* trees. Most angiosperms are free living and photosynthetic. Angiosperms are grouped into two classes: the *monocots* (grasses, palms, lilies, orchids, and principal crop plants) and the *dicots* (shrubs, trees, herbs, cacti, and water lilies). As an adaptation for life on land, diploid sporophytes dominate the life cycle as they retain and nourish gametophytes. Endosperm tissue within seeds nourishes embryos; *seeds* are packaged in protective fruits for dispersal. Angiosperms alone have reproductive structures called *flowers*. Floral structures coevolved with pollinators such as insects, bats, birds, and rodents, a probable factor in the great success of angiosperms.

Key Terms

gymnosperms	megaspores	seed
angiosperms	pollen grain	monocots
cycad	pollen tube	dicots
ginkgos	pollination	flowers
gnetophytes	fertilization	herbaceous
conifers	microspore mother cell	endosperm
microspores	megaspore mother cell	embryo sac
ovule	female gametophyte	

Objectives

1. Define *gymnosperm;* describe general characteristics of gymnosperms and list several examples.
2. Outline the principal steps of the conifer life cycle. Tell in which plant part the spores and gametes are formed. Distinguish *pollination* from *fertilization*.
3. List three parts of a seed and explain how each is produced.
4. Cite characteristics of angiosperms; define *monocot* and *dicot*.
5. Discuss the placement of gymnosperms in the evolutionary time scale.
6. Compare the life cycle of an angiosperm (monocot) with that of a gymnosperm. See p. 253 and p. 255 of the main text for comparison.

Self-Quiz Questions

Fill-in-the-Blanks

The most successful vascular plants are seed plants, the (1) _____ ,

and the (2) _____ . (3) _____ are the most successful gymnosperms.

All conifers have (4) _____ -shaped leaf clusters, which bear the

(5) _____ . Pine trees produce (6) _____ kinds of spores in

(7) _____ kinds of cones. Mother cells in sporangia on male cones

produce haploid (8) _____ by meiosis. These develop into (9) _____ grains, each of which contains a(n) (10) _____ gametophyte. A mother cell inside an ovule undergoes meiosis to produce haploid (11) _____ . One (12) _____ survives and develops into a female (13) _____ . Spring air currents move pollen from (14) _____ pine cones to (15) _____ pine cones; this is called (16) _____ . Pollen grains develop tubes containing (17) _____ while eggs form in the female (18) _____ . Fertilization results in (19) _____ formation and (20) _____ development. Pine seeds include the (21) _____ , a female (22) _____ , and an outer wrapping, the (23) _____ coat. The flowering plants are called (24) _____ . The two classes of angiosperms are (25) _____ and (26) _____ . (27) _____ are the most successful plants on earth. Angiosperms have unique reproductive structures called (28) _____ . Many diverse floral structures coevolved with animal (29) _____ .

Labeling

Identify each indicated part of the accompanying illustration (p. 195).

(30) _____

(31) _____ _____

(32) _____ _____

(33) _____ _____

(34) _____

(35) _____

(36) _____

(37) _____ _____

CHAPTER TEST   **UNDERSTANDING AND INTERPRETING KEY CONCEPTS**

____ (1) Which of the following describes the fungal mode of nutrition?

   (a) heterotrophs
   (b) saprobes
   (c) parasites
   (d) all of the above

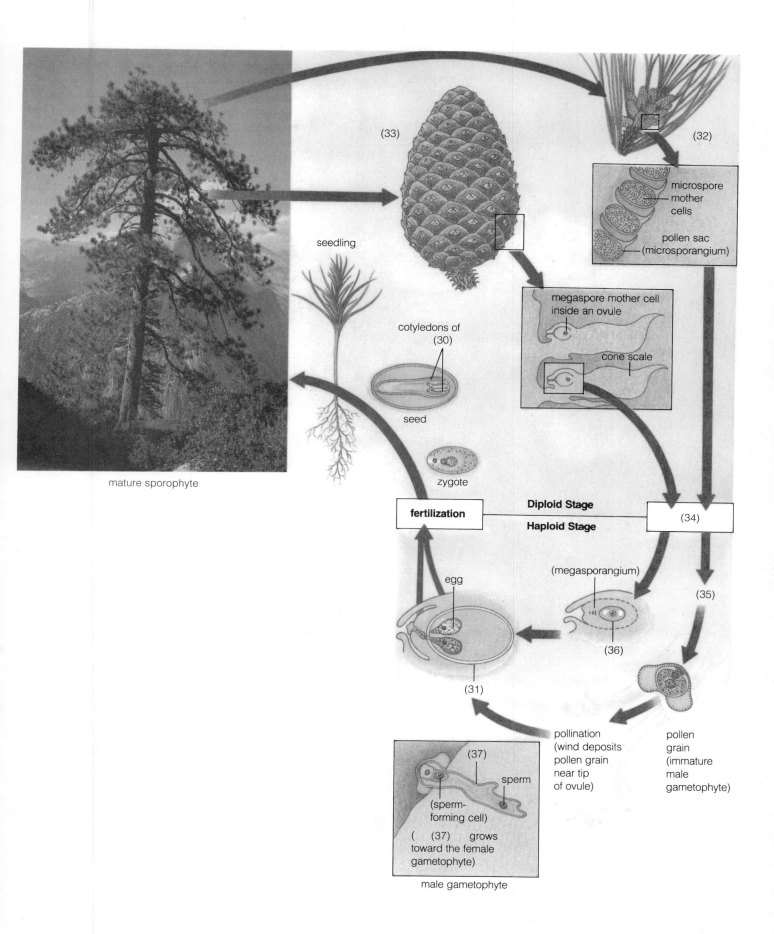

mature sporophyte

seedling

(33)

(32)

microspore
mother
cells

pollen sac
(microsporangium)

cotyledons of
(30)

megaspore mother cell
inside an ovule

cone scale

seed

zygote

fertilization | **Diploid Stage**

Haploid Stage | (34)

egg

(megasporangium)

(35)

(36)

(31)

pollen
grain
(immature
male
gametophyte)

pollination
(wind deposits
pollen grain
near tip
of ovule)

(37)

sperm

(sperm-
forming cell)

((37) grows
toward the female
gametophyte)

male gametophyte

___ (2) Most true fungi send out cellular filaments called _____ .

    (a) mycelia
    (b) hyphae
    (c) mycorrhiza
    (d) asci

___ (3) Fungal spores can be produced in a(n) _____ cycle by meiosis and in a(n) _____ cycle by mitosis.

    (a) sexual; sexual
    (b) sexual; asexual
    (c) asexual; asexual
    (d) asexual; sexual

___ (4) The _____ is a group of fungi that is lacking or has a yet undetected sexual stage.

    (a) water molds
    (b) zygospore-forming fungi
    (c) sac fungi
    (d) club fungi
    (e) imperfect fungi

___ (5) A _____ is a symbiotic relationship between an alga and a fungus.

    (a) morel
    (b) lichen
    (c) mycorrhiza
    (d) mushroom

___ (6) A _____ is a symbiotic relationship between fungal hyphae and roots of forest shrubs and trees.

    (a) morel
    (b) lichen
    (c) mycorrhiza
    (d) mushroom

___ (7) Plants possessing xylem and phloem are called _____ plants.

    (a) gametophytes
    (b) nonvascular
    (c) vascular
    (d) seedless

___ (8) A gametophyte is _____ .

    (a) a gamete-producing plant
    (b) haploid
    (c) both a and b
    (d) the plant produced by the fusion of gametes

___ (9) Red, brown, and green "algae" are found in the Kingdom _____ .

    (a) Plantae
    (b) Monera
    (c) Protista
    (d) all of the above

___ (10) Red algae _____ .

    (a) are primarily marine organisms
    (b) are thought to have developed from green algae
    (c) contain xanthophyll as their main accessory pigments
    (d) all of the above

___ (11) Stemlike structure, leaflike blades, and gas-filled floats are found in the species of _____ .

    (a) red algae
    (b) brown algae
    (c) bryophytes
    (d) green algae

___ (12) Because of pigmentation, cellulose walls, and starch storage similarities, the _____ algae are thought to be ancestral to more complex plants.

    (a) red
    (b) brown
    (c) blue-green
    (d) green

___ (13) Bryophytes _____ .

    (a) have vascular systems that enable them to live on land
    (b) include lycopods, horsetails, and ferns
    (c) include mosses, liverworts, and hornworts
    (d) have true roots but not stems

___ (14) In horsetails, lycopods, and ferns _____ .

    (a) spores give rise to gametophytes
    (b) the main plant body is a gametophyte
    (c) the sporophyte bears sperm- and egg-producing structures
    (d) all of the above

___ (15) _____ are seed plants.

    (a) Cycads and ginkgos
    (b) Conifers
    (c) Angiosperms
    (d) all of the above

___ (16) An ovule of a gymnosperm develops into the _____ after fertilization.

    (a) flower
    (b) fruit
    (c) seed
    (d) sporangium

___ (17) In complex land plants the diploid stage is resistant to adverse environmental conditions such as dwindling water supplies and cold weather.

The diploid stage progresses through this sequence: _____ .

(a) gametophyte → male and female gametes
(b) spores → sporophyte
(c) zygote → sporophyte
(d) zygote → gametophyte

___ (18) The rapid expansion of angiosperms late in the Mesozoic Era appears to be related to their coevolution with _____ .

(a) dinosaurs
(b) gymnosperms
(c) insects
(d) mammals

___ (19) Monocots and dicots are groups of _____ .

(a) gymnosperms
(b) club mosses
(c) angiosperms
(d) horsetails

INTEGRATING AND APPLYING KEY CONCEPTS

Explain why totally submerged aquatic plants that live in deep water never develop heterosporous life cycles.

19
ANIMALS

OVERVIEW OF THE ANIMAL KINGDOM
    General Characteristics of Animals
    Body Plans
SPONGES
CNIDARIANS
FLATWORMS
    Turbellarians
    Flukes
    Tapeworms
ROUNDWORMS (NEMATODES)
    Commentary: A Rogues' Gallery of
    Parasitic Worms
ROTIFERS
TWO MAIN EVOLUTIONARY ROADS
MOLLUSKS
    Gastropods
    Bivalves
    Cephalopods

ANNELIDS
ARTHROPODS
    Arthropod Adaptations
    Chelicerates
    Crustaceans
    Insects and Their Relatives
ECHINODERMS
CHORDATES
    Invertebrate Chordates
    Evolutionary Trends Among the
    Vertebrates
    Fishes
    Amphibians
    Reptiles
    Birds
    Mammals
SUMMARY

General Objectives

1. Describe the major advances in body structure and function that made invertebrates and vertebrates increasingly large and complex.
2. Discuss the relationship between segmentation and the development of paired organs and paired appendages.
3. Explain why the insects and their relatives are the most successful animals.
4. List the several trends that appear in the evolution of fishes, amphibians, reptiles, birds, and mammals.
5. Reproduce from memory a phylogenetic tree that expresses the relationships among the major groups of animals.

19-I
(pp. 259–264)

OVERVIEW OF THE ANIMAL KINGDOM
    General Characteristics of Animals
        Body Plans
SPONGES
CNIDARIANS

Summary

The animals we are most with familiar with are *vertebrates*, but most of the world's animals are *invertebrates*, lacking a backbone.

Animals are diploid heterotrophs that usually reproduce sexually, but some can reproduce asexually. As occurs in plants, a period of embryonic development is part of the animal life cycle as a zygote becomes a multicelled embryo composed of three formative ("germ") types of tissue; in animals, these tissues are ectoderm, mesoderm, and endoderm. These formative tissues eventually become the adult tissues and organs. Most animals are motile to some extent—at least in some portion of the life cycle.

Increasing complexity in animals is judged by evaluating five body features: type of body symmetry, degree of cephalization, gut type, body cavity type, and segmentation. By careful study, animals may be arranged by the major trends in their evolution.

Sponges are simple animals that have been very successful. Between 2 billion and 1 billion years ago, the first animals appeared on Earth; by the dawn of the Cambrian, there existed already well-developed multicellular animals. *Sponges*, which are relatively simple vaselike animals that inhabit shallow seas, evolved from one of the early animal forms. Sponges lack symmetry and have no organs. Even their tissues differ from those in other animals. The sponge body is composed of a layer of semifluid material between inside and outside layers of flattened cells. Needles or fibers are embedded in the semifluid layer to lend structural support. Water flows in through pores in the body wall, then out through the opening at the top. The interior is lined with microvilli-adorned *collar cells*, which trap and ingest microscopic organisms carried in the water flowing through the many pores. Some food may be passed to amoebalike cells within the semifluid layer. Sponges reproduce sexually and young sponges pass through a *larval stage*. Some sponges also may reproduce asexually by fragmentation.

Cnidarians (jellyfish, corals, *Hydra,* and sea anemones) have simple organs distributed in radial symmetry. Both *medusa* and *polyp* body forms are common. A saclike gut with one opening makes up the main body cavity of both body forms. Cnidarians have true tissues: an outer being epithelium, and an inner jellylike layer, the mesoglea. Responses to environmental stimuli occur through a system of nerve cells, the nerve net. Stinging cells (nematocysts) aid in food capture and defense. Corals have calcium-reinforced external skeletons that contribute to building reefs. Reproduction may include asexual and sexual methods. The zygote formed in sexual reproduction develops into a planula larva: a roving, ciliated embryonic form.

Key Terms

vertebrates	coelom, coelomate	cnidarians
invertebrates	peritoneum	hydrozoans
multicellular	false coelom	sea anemones
radial symmetry	segmentation	medusa
bilateral symmetry	sponges	polyp
cephalization	collar cells	nerve net
gut	amoeboid cells	mesoglea
saclike gut	larval stage	nematocysts
"complete" gut	epithelia	planula

Objectives

1. Define vertebrate and invertebrate.
2. List the three formative tissue types found in the embryos of most multi-celled animals.

3. Explain how body symmetry, cephalization, type of gut, body cavity type, and segmentation are used in establishing evolutionary relationships.
4. Distinguish *radial symmetry* from *bilateral symmetry*.
5. List two characteristics that distinguish sponges from other animal groups.
6. Describe the two cnidarian body types.
7. Explain how radial symmetry might be more advantageous to floating or sedentary animals than would bilateral symmetry.
8. Design an illustration that summarizes a generalized cnidarian life cycle. Include these terms: *egg, sperm, zygote, planula, polyp, medusa.*
9. Tell how cnidarians obtain food, and describe what they do with food they cannot digest.
10. Name examples of several cnidarians.

Self-Quiz Questions

Fill-in-the-Blanks

Animals that lack backbones are known as (1) _____ . Animals are (2) _____ , which means "constructed of many cells." Animals cannot produce their own food and so are (3) _____ . Animals are diploid organisms that reproduce (4) _____ or in some cases, (5) _____ . Animal life cycles include a period of (6) _____ development. During at least a portion of their life cycle, animals move or are (7) _____ . (8) _____ symmetry is the symmetry of a wheel. (9) _____ symmetry is the symmetry of a crayfish or a human. A(n) (10) _____ is where food is digested and absorbed. Some guts are (11) _____-like, with only one opening; other guts are (12) _____-like with specialized regions for food processing. The most common body cavity is the (13) _____ , which is a body cavity lined with (14) _____ and occurring between the gut and body wall of most bilateral animals. In some animal lines of evolution, a "false" (15) _____ evolved which was not lined with (16) _____ . A coelomate animal composed of a series of body units is (17) _____ . (18) _____ form the most primitive major group of multicellular animals. They are nourished by microscopic organisms extracted from the water that flows in through pores in the body wall and captured by (19) _____ _____ . Sponges may reproduce sexually and young sponges pass through a(n) (20) _____ stage. Some sponges are able to reproduce asexually through (21) _____ . Cnidarians have (22) _____ symmetry and stinging cells called (23) _____ , which aid in defense and food capture. Most cnidarian life cycles have a(n) (24) _____ larval stage. The two cnidarian body types are the (25) _____ and the (26) _____ .

Labeling   Identify each indicated part of the accompanying illustration.

(27) _____ _____   (29) _____ _____

(28) _____ _____   (30) _____ _____

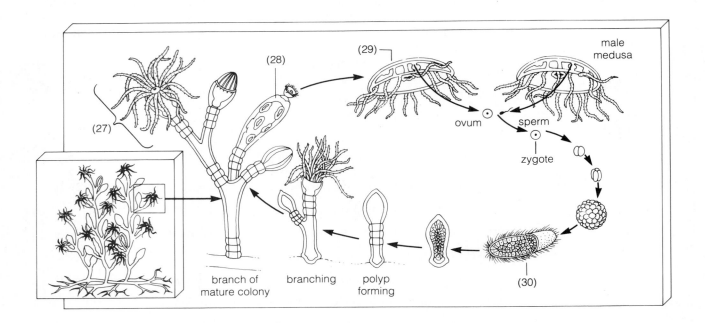

19-II
(pp. 265–268)

FLATWORMS
  Turbellarians
  Flukes
  Tapeworms
ROUNDWORMS (NEMATODES)
  Commentary: **A Rogues' Gallery of Parasitic Worms**
ROTIFERS

Summary

The group known as *flatworms* contains turbellarians, flukes, and tapeworms. Except for tapeworms these animals take in food through a muscular pharynx (tube) before entering a saclike gut. Flatworms are bilateral, cephalized, possess an organ system, and lack a coelom. Three germ layers are found in the embryo; muscles and reproductive structures originate from the mesoderm. In more complex animals, the embryonic mesoderm gives rise to muscles, blood, and bone.

Most turbellarians live in the sea; planaria is an example of a freshwater species. They feed on small living animals or cells and tissues from dead and wounded ones. Body fluids are regulated by an organ system of tubes, called the *proto-nephridia,* which extend to *flame cells* that drive away excess body fluid. Planarian reproduction often occurs by the body dividing in half, then each half regenerates the missing half. Planarians are hermaphroditic; sperms are exchanged in sexual reproduction.

Flukes (trematodes) are chiefly *parasites* of vertebrates that suck in cells, tissue fluids, or blood from the inside or outside surfaces of the host's body.

Fluke life cycles are often complex, involving definitive hosts (where maturation and reproduction occur) and intermediate hosts (some development, no reproduction). Humans serve as definitive hosts for several troublesome flukes such as the Southeast Asian blood fluke *(Schistosoma)*.

Adult tapeworms (cestodes) live as parasites in vertebrate digestive tracts. They lack a gut and absorb nutrients from the host's digested food. Attachment to the intestinal wall is by hooks and suckers, or both, on the *scolex*. New body segments, each a hermaphroditic *proglottid*, bud from the scolex. Fertilized eggs accumulate in older, larger proglottids, which exit the body with feces; the eggs become available for new hosts. Humans can be infected by pork, beef, and fish tapeworms if that flesh is eaten after inadequate cooking.

Parasitic and free living roundworms (nematodes) are found almost everywhere. Many parasitize plants and animals, including several human parasites such as hookworm, pinworm, *Trichinella* (pork and game animals), and the roundworm that causes elephantiasis. Roundworms are bilateral, circular in cross section and tapered at both ends. The body plan is a tube (the gut) within another tube (the body wall). Reproductive organs are in a false coelom between the gut and body wall; nutrients are moved through the body cavity by fluids. A tough cuticle covers the body.

Rotifers are tiny aquatic animals that feed on bacteria and single-celled algae with their crowns of cilia; they have a false coelom.

Key Terms

flatworms	parasites	roundworms (nematodes)
turbellarians	host	false coelom
pharynx	*Schistosoma*	cuticle
proto-nephridium, -dia	schistosomiasis	hookworm
flame cells	tapeworms (cestodes)	Trichinella
hermaphroditic	scolex	rotifers
flukes (trematodes)	proglottids	

Objectives

1. List the three main types of flatworms and briefly describe the body plan of each.
2. Describe the structural adaptations of flukes and tapeworms that allow them to be successful parasites.
3. Describe the body plan of roundworms, comparing their various organ systems with those of the flatworm body plan.
4. List some of the common flatworm and roundworm parasites that affect humans (see commentary, pp. 267–268).
5. Describe a rotifer and tell where it usually lives and what it eats.

Self-Quiz Questions

Fill-in-the-Blanks

Turbellarians, flukes, and tapeworms belong to an animal group called the (1) _____ . Planarians can reproduce asexually by dividing and (2) _____ missing halves. Parasitic (3) _____ suck in cells, tissue juices, or blood at internal or external host surfaces. The host in which the internal parasite matures and reproduces is the (4) _____ host. Schistosomiasis is

a discomforting human condition caused by a(n) (5) _____ _____ .
Tapeworms attach to the intestinal wall by hooks and suckers on the (6)
_____ , which also buds new hermaphroditic body units, the (7)
_____ . (8) _____ are found almost everywhere as free living and
parasitic forms. Pinworms and hookworms are types of parasitic (9)
_____ . Between the gut and body wall of a roundworm is a(n) (10)
_____ _____ , which contains (11) _____ organs. (12) _____
are tiny abundant aquatic animals with false coeloms that move about feeding
on bacteria.

<table>
<tr><td>19-III
(pp. 269–271)</td><td>**TWO MAIN EVOLUTIONARY ROADS**
MOLLUSKS
    Gastropods
    Bivalves
    Cephalopods</td></tr>
</table>

Summary

Two groups of animals descended from ancestral flatworms during Cambrian times. In *protostomes* (annelids, arthropods, and mollusks), the first opening into the gut that arises during embryonic development becomes the mouth; the anus forms later. In *deuterostomes* (echinoderms and chordates), the first opening to the gut becomes the anus, and the second becomes the mouth. The distinction between protostomes and deuterostomes seems trivial, but it helps us to identify two major lines of animal evolution that diverged long ago.

The term *mollusk* refers to a soft-bodied animal; familiar representatives are snails, clams, and octopuses. The molluscan body plan presents a head, foot, a *mantle*, usually a calcified shell, and eyes and tentacles on those with a well-developed head. The radula is a toothed tonguelike organ that shreds food before entry into the gut. Mollusks also have a heart and respiratory *gills*, as well as reproductive and excretory organs.

The largest molluscan group, the *gastropods*, are snails and slugs. The coiling of a snail encloses and balances the stomach, digestive gland, and gonad above the rest of the body. In some gastropods such as sea slugs and sea hares, the shell is greatly reduced or completely lost. Due to torsion during development, the gills, anus, and kidney openings are found above the head; wastes do not enter the mouth due to beating cilia that sweep them away. Examples of mollusks that are *bivalves* are clams, scallops, oysters, and mussels. Clams range in size from a few millimeters to more than a meter across. The bivalve head is reduced, but the foot is enlarged and used in burrowing. Ciliated gills function in respiration and suspension feeding. Gill mucus traps bits of suspended food in water. Bivalves buried in mud or sand use tubes to obtain water and suspended food.

Cephalopods include swift predators like squids, octopuses, nautiluses, and cuttlefish. The largest invertebrate known is the giant squid, but some cephalopods are only a few centimeters long. Cephalopods lack a foot but have tentacles with suction pads that are used with beaklike jaws to capture and crush their prey. Forcible ejection of water from the mantle cavity through a siphon propels (a type of jet propulsion) cephalopods through the water with control. Highly active cephalopods have a closed circulation system (blood flows from the heart to the gills) to supply a high oxygen demand. Quick response and some learning is possible because of a well-developed nervous system and large brain. The structure of the cephalopod eye is similar to the human eye. Cephalopods are the most complex invertebrates known.

Key Terms

protostomes	radula	cephalopods
deuterostomes	gastropods	squids
mollusks	torsion	jet propulsion
mantle	bivalves	nautilus
foot	siphons	cuttlefish
gills	suspension feeding	

Objectives

1. Define *protostomes* and *deuterostomes* and give examples of each group.
2. Explain why zoologists care to make what seems to be such a trivial distinction between the two groups.
3. Give the general characteristics of the following animal groups and cite examples of each: gastropods, bivalves, and cephalopods.
4. List the rather advanced structural features possessed by the cephalopods.

Self-Quiz Questions

Fill-in-the-Blanks

(1) _____ include echinoderms and chordates; in this group, the first opening to the gut becomes the (2) _____ , and the second one to appear becomes the (3) _____ . The situation is reversed in the (4) _____ , which include mollusks, annelids, and arthropods. (5) _____ include snails, clams, and octopuses. The presence of a head, foot, and calcified shell is typical of many (6) _____ . The (7) _____ include snails and slugs and make up the largest molluscan group. Some mollusks preshred their food with a tonguelike organ, the (8) _____ . Due to torsion in development, some (9) _____ end up with their gills, anus, and kidney openings above their head. (10) _____ include the clams, scallops, oysters, and mussels. Most bivalves use their gills for (11) _____ _____ and (12) _____ . In the cephalopods, movement is accomplished by a type of (13) _____ _____ . The most highly evolved invertebrate nervous systems are generally considered to be those of the (14) _____ , which include squids and octopuses.

ANNELIDS

Summary

Annelids, the true segmented worms, include leeches, polychaetes, and earthworms. *Leeches* are found in freshwater, seawater, and the moist tropics. They eat small animals or kill them to suck their body juices. Familiar leeches feed on vertebrate blood. Leeches are equipped with a sucker at each end and the capacity to store large amounts of food. *Polychaetes* are mostly marine worms with different adaptations for attaching, crawling, or swimming. Some construct tubes for homes; diet includes smaller animals, algae, or organic debris. Crawling and grasping is possible by bristles, the *setae.* An example of an *oligochaete* is the familiar burrowing scavenger, the earthworm. Diet is mostly plant material taken in with dirt, an action that aerates soil and moves subsoil to the surface. Earthworms also have setae. In general, annelids are segmented, have a complete digestive system, a closed circulatory system, and kidneylike organs, the *nephridia,* that control body fluids. A thin, moist, flexible cuticle covers the body and allows gas exchange. Nerve cell clusters and a rudimentary brain provide neural control. Annelids probably evolved early from the protostome line.

Key Terms

annelids	polychaetes	segmentation
segmented worms	oligochaetes	nephridium, -dia
leeches	seta, setae	rudimentary brain

Objectives

1. Name the three groups of annelids and give a specific example from each.
2. Describe how the different structural and functional adaptations found in the three annelid groups fit them for their different environments.
3. Define *segmentation* and explain what possible advantages this body plan might have.
4. Explain how, without lungs, annelids exchange gases with the environment.

Self-Quiz Questions

Fill-in-the-Blanks

(1) _____ , which is a repeating series of body parts, is well-developed in annelids. (2) _____ include truly segmented worms such as earthworms, (3) _____ , and leeches; they differ from flatworms in having (first) a complete digestive system with a mouth and (4) _____ and (second) a(n) (5) _____ , which is a fluid-filled space between the gut and body wall. In a circulatory system, (6) _____ provides a means for transporting materials between internal and external environments. In each segment, there are clusters of (7) _____ _____ that control local activity and a pair of (8) _____ that act as kidneys. Most leeches (9) _____ small animals or kill them and suck out their juices. Some leeches feed on

vertebrate (10) _____ . (11) _____ are mostly marine worms that burrow, crawl, swim, or attach to substrates. Earthworms belong to an annelid group called (12) _____ . As to feeding habit, earthworms are said to be (13) _____ . Earthworms have bristles embedded in the outer body wall called (14) _____ .

Labeling   Identify each indicated part of the accompanying illustration.

(15) _____   (19) _____ _____

(16) _____   (20) _____

(17) _____ _____   (21) _____

(18) _____

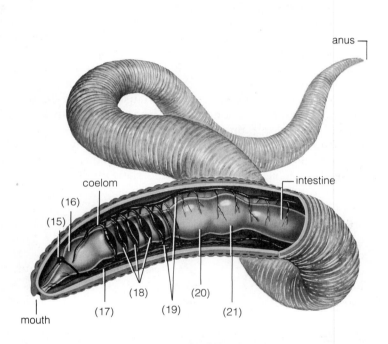

anus

coelom

(16)

(15)

intestine

mouth

(17)   (18)   (19)   (20)   (21)

19-V
(pp. 273–278)

ARTHROPODS
Arthropod Adaptations
Chelicerates
Crustaceans
Insects and Their Relatives

Summary

Arthropods as a group include the now extinct trilobites, chelicerates (horseshoe crabs, spiders, scorpions, ticks, and mites), crustaceans (copepods, crabs, lobsters, shrimps, and barnacles), and uniramians (centipedes, millipedes, and insects). Arthropods display more diversity and occupy more varied habitats than any other animal group. Five adaptations have contributed to the success of arthropods: (1) a hardened, light, often flexible, chitinous *exoskeleton;* (2) a life cycle in which sexually immature individuals feed, grow, and molt, sexually mature individuals mate, reproduce, and dispense their populations, and other individuals experience *metamorphosis;* (3) highly specialized jointed appendages adapted for different functions; (4) respiratory systems constructed with *tracheas* in terrestrial individuals; and (5) specialized sensory structures, which include the arthropod eye with its wide vision angle.

The much avoided chelicerates includes some venomous, eight-legged, many eyed spiders, scorpions, parasitic blood-sucking ticks that sometimes serve as disease vectors (Rocky Mountain spotted fever), mites that are free living scavengers, and parasitic mites.

Crustaceans live in the sea and freshwater but a few live on land. They are important parts of food webs including that of humans. Crustaceans such as crabs and lobsters have many segments, are covered by a carapace, and have large claws for defense and food processing. Paired eyes, jawlike mandibles, and antennae are found on the head. Tiny copepods have a single eye located in the center of the head.

Slow-moving *millipedes* and rapid-moving *centipedes* are close relatives of the insects. Nonaggressive millipedes scavenge decaying plant material while centipedes are aggressive carnivores.

The most successful invertebrates are the diverse insects. Their success has been supported by having small size, a life cycle that includes metamorphosis, and wings. These same characteristics position the insect to be a formidable competitor with humans for available resources, as well as to spread diseases that affect humans. Insects are beneficial in pollinating crop plants and attacking or parasitizing other insects detrimental to humans.

The three major body parts of an insect are the head, thorax, and abdomen. The head has paired sensory antennae and mouth parts adapted for different purposes (sucking, biting, chewing, or puncturing). Insects have three pairs of legs and usually two pairs of wings. Digestion occurs mostly in the midgut; water reabsorption and collection of nitrogen wastes are accomplished by small tubes, the *Malpighian tubules.* Insects have an impressive reproductive capacity.

Key Terms

arthropod	metamorphosis	copepod
trilobites	jointed appendages	millipedes
chelicerates	tracheas	centipedes
crustaceans	sensory structures	thorax
uniramians	carapace	abdomen
exoskeleton	mandibles	insect wings
chitin	antennae	Malpighian tubules
molting		

Objectives

1. Explain how the development of a thickened cuticle and a hardened exoskeleton affected the ways that arthropods lived.
2. State what ancestors are thought to have given rise to the arthropods and think of some structural features that might support this idea.
3. List and explain fully the five major arthropod adaptations that have enabled their great success.
4. Explain what is meant by division of labor in the life cycle and relate this to the concept of molting and metamorphosis.
5. Explain how insect wings differ from those of birds.
6. List the four major groups of arthropods and cite several examples of each (one is extinct).

Fill-in-the-Blanks

Arthropods developed a thickened (1) _____ and a hardened (2) _____ . Arthropod (3) _____ can occur with or without a sudden change in body form. Transformation of a larva into an adult form is called (4) _____ . As arthropods evolved, body segments became more (5) _____ including various (6) _____ , which became adapted for different functions. Tubes called (7) _____ are respiratory structures that contributed to arthropod diversity. Arthropod eyes have a wide angle of (8) _____ . Spiders, scorpions, ticks, and mites belong to the arthropod group called (9) _____ . (10) _____ include shrimps, crayfishes, lobsters, crabs, copepods, and pill bugs. (11) _____ and (12) _____ are close relatives of the insects. (13) _____ have bodies divided into a head, thorax, and abdomen with three pairs of legs and usually two pairs of wings. In insects, (14) _____ _____ absorb nitrogen wastes and reabsorb water.

19-VI
(pp. 278–289)

ECHINODERMS
CHORDATES
    Invertebrate Chordates
    Evolutionary Trends Among the Vertebrates
    Fishes
    Amphibians
    Reptiles
    Birds
    Mammals
SUMMARY

Summary

Of the small number of *deuterostomes* that have survived to the present, only the echinoderms and the chordates are prominent. *Echinoderm* means *spiny-skinned*, in reference to the bristling spines possessed by some members of the group. Echinoderms include crinoids, sea stars, sea urchins, sand dollars, sea cucumbers, and brittle stars. In echinoderms, radial symmetry has been superimposed on an earlier bilateral heritage (echinoderm larvae are bilateral). Most echinoderms still go through a free-swimming, bilaterally symmetrical larval stage, but adults generally show the radial symmetry typical of bottom-dwelling organisms that cannot move quickly. The echinoderm way of moving about is based on a *water-vascular system* of canals and *tube feet*. Each tube foot is equipped with a muscular, fluid-filled ampula. Coordinated extension and retraction of many suckered tube feet enable echinoderms to move body parts in specific directions, to burrow, cling to rocks, or grip food such as clams or snails. Some sea stars can push their stomach outside of their mouth to begin digestion of prey.

Chordates are bilateral organisms that are distinguished from other animals by a *notochord*, a tubular *nerve cord*, a *pharynx* with slits in its wall, and a *tail* that extends past the anus at some time in its life history.

Tunicates and lancelets are invertebrate chordates. *Tunicates* are sea animals that secrete a gelatinous or leathery tunic around themselves. They lack a coelom and have a bilateral larval stage with a notochord. After attachment, the larva undergoes a drastic metamorphosis in which tail and notochord disappear and the tunic thickens. The pharynx enlarges, develops slits, and then acts as a respiratory organ and a sieve for collecting food. *Lancelets* are small (5 centimeters or less) fishlike animals that taper at both ends and bury themselves in sand or mud with mouth up. Cilia bring food and water through gill slits in the pharynx.

The seven groups of living vertebrates are the jawless fishes, cartilaginous fishes, bony fishes, amphibians, reptiles, birds, and mammals. Several key evolutionary trends occurred in vertebrate evolution: (1) the notochord was replaced by a column of hard, bony units (vertebrae) strengthened by muscles; some bony units evolved into powerful jaws; (2) the nerve cord expanded regionally to form the spinal cord and complex brain; this was followed by the evolution of more complex sense organs; (3) outpockets from the pharyngeal wall evolved into lungs and heart; the heart became a more complex pump; and (4) the fins of certain fishes (lobe-finned fishes) represented the starting point for the evolution of legs, arms, and wings on present-day amphibians, reptiles, birds, and mammals.

Filter-feeding chordates were perhaps the ancestors of the fishes first seen in the fossil record more than 450 million years ago. One such group was agnathan fishes with a food-straining pharynx, no jaws, and a supporting notochord. Placoderms were another group of early jawed fishes with coverings of bony plates that became extinct about 325 million years ago. They were bottom-dwelling scavengers or predators that fed on agnathans. Today, fishes with streamlined, finned, and scaled bodies are the most numerous and diverse vertebrates.

Living *jawless fishes* (agnathans) are the lampreys and hagfishes. Their bodies are long and cylindrical with a notochord and cartilaginous skeleton. Lampreys rasp flesh from their prey with a sucking oral disk lined with hard, toothlike plates; juices are then sucked from prey such as salmon or trout. Hagfishes are large and wormlike with "feelers" around their mouths; they tear flesh from dead fishes.

Sharks, skates, and rays belong to the *cartilaginous fishes*. Most are streamlined predators with cartilaginous endoskeletons, hard scales, and pharyngeal gill slits. Teeth are modified scales that are continually shed and replaced. Predatory sharks are among the largest vertebrates. Skates and rays, with distinctive large fins mounted on the side of the head, have flattened teeth and feed along the bottom sediments for invertebrates.

The *bony fishes* evolved along four lines: lungfishes, crossopterygians (coelocanths), bichirs, and ray-finned fishes. The teleosts are the most successful ray-finned fishes; this group includes salmon, tuna, catfish, perch, minnows, and eels. Teleosts are capable of rapid, complex movements due to highly maneuverable fins, light, flexible scales, and efficent respiratory systems. The seahorse is placed in this category but does not possess the general characteristics of bony fishes.

Amphibians evolved from lobe-finned fishes. The land environment required brain development in regions related to vision and balance. Living amphibians include salamanders, frogs, toads, and wormlike apodans. Amphibians have mostly bony endoskeletons and usually four legs. The first four-legged animals probably walked like a present-day salamander, from side to side, as a fish moves through water. Adult amphibians eat most anything; eggs are laid in moist areas or water. Frogs and toads have long hindlimbs and can propel themselves through air or water. Their sticky-tipped tongue is attached at the front of the mouth; it flips out and captures prey. In water dwellers, respiration occurs through gills (if present) and skin. Land dwellers respire with lungs, skin, and pharynx. Some amphibians have toxin-producing skin glands. Skin of South African clawed frogs produces chemicals that defend against microbes.

Insects underwent a major adaptive radiation on land in the late Carboniferous that may have begun *reptilian* evolution from certain carnivorous amphibians. Jaws became adapted for feeding on insects, and limb bones became more efficient for moving on land. Life on land required a type of internal fertilization and an egg for development: the *amniotic egg.* Within this egg the embryo is protected within special metabolic membranes and a hard shell, while exchanging oxygen, carbon dioxide, and water. In the Mesozoic, ancestral reptilian populations underwent an adaptive radiation that established the ruling dinosaurs and their relatives. Most of them suddenly disappeared as the Cretaceous ended. Today's reptiles include turtles, crocodilians, snakes, and lizards. The most familiar reptiles are snakes and lizards. The large Komodo lizard hunts deer and water buffalo. Living reptiles have a bony endoskeleton, a skin that resists drying, lungs, and internal fertilization; they lay eggs. Internal body temperature is controlled by behavior and physiology. Today's reptiles exhibit complex social behavior.

Birds appear to have evolved from crocodilian-like reptiles that moved on two legs during Jurassic times about 160 million years ago. The oldest known fossil bird is *Archaeopteryx;* it resembles those Jurassic reptiles. The internal structures, scaly legs, and egg-laying habits of modern birds resemble those of reptiles. Structural and behavioral characteristics of the various birds vary tremendously. The ostrich and hummingbird represent extremes in size. Light, hollow bones and powerful breast muscles enable many birds to fly. Some primitive and domesticated breeds are too heavy to fly for long periods. All birds have large, strong, four-chambered hearts and efficient oxygen delivery.

Early in the Cenozoic Era, ancestral mammalian populations that had been lying low during the dinosaur dominated Mesozoic Era diverged, forming lineages that led to the modern mammals. Brain expansion was especially conspicuous in primate evolution as encoding and processing of information improved dramatically. The three major groups of mammals are egg-laying, pouched, and placental mammals. Mammalian embryos mostly develop internally, are liveborn, and are nourished by a mother's milk glands. Most mammals are covered with hair; adults have a permanent set of teeth, lungs, a four-chambered heart, and a well-developed cerebral cortex.

Key Terms

deuterostome lineage	lancelets	bony fishes
echinoderms	jaws	scales
tube feet	lungs	amphibians
water-vascular system	paired fins	reptiles
ampulla	coelocanths	amniotic egg
chordates	cerebral cortex	birds
notochord	agnathan	bird wing
nerve cord	placoderms	mammals
pharynx	jawless fishes	egg-laying mammal
gill slits	lamprey	pouched mammal
tunicates	hagfish	placental mammal
metamorphosis	cartilaginous fishes	

Objectives

1. Define *echinoderm* and list four groups of echinoderms.
2. List the essential characteristics of echinoderms.
3. Describe how locomotion occurs in echinoderms.
4. List three characteristics found only in chordates.
5. Describe the adaptations that sustain the sessile or sedentary life style seen in primitive chordates such as tunicates and lancelets.
6. State what sort of changes occurred in the primitive chordate body plan that could have promoted the emergence of vertebrates.
7. Describe the differences between primitive and advanced fishes, in terms of skeleton, jaws, special senses, and brain.
8. Describe the changes that enabled aquatic fishes to give rise to land dwellers.
9. Draw an amniotic egg, tell which animals produce them, and state how it enables a life on land.
10. Describe, from fishes to mammals, the basic changes that occurred in brain development and circulatory system development.

Self-Quiz Questions

Fill-in-the-Blanks

Only two prominent groups of (1) _____ have survived to the present—the (2) _____ and the chordates. In echinoderms, (3) _____ symmetry has been overlaid on an earlier bilateral heritage; most echinoderms still go through a free-swimming (4) _____ larval stage. Echinoderm locomotion is based on constant circulation of seawater through a(n) (5) _____-_____ system of canals and (6) _____ _____ . (7) _____ and (8) _____ are invertebrate chordates. The first fishes were (9) _____ ; examples of these are the lampreys and hagfishes, which (10) ( ) have ( ) lack jaws. Sharks, skates, and rays belong to a group called the (11) _____ fishes. Lungfishes, crossopterygians, bichirs, and ray-finned fishes are all (12) _____ fishes. Amphibians evolved from (13) _____-_____ fishes. Land-dwelling amphibians depend on lungs, skin, and (14) _____ for respiration. In the late Carboniferous, reptiles probably evolved from certain (15) _____ . Internal (16) _____ and the development of the (17) _____ egg were required for animals to live on land. Reptiles breathe entirely with their (18) _____ . Birds apparently evolved from crocodilian-like (19) _____ . The bird wing is constructed of feathers, powerful muscles, and lightweight (20) _____ . The evolution of modern (21) _____ accelerated greatly at the beginning of the Cenozoic Era. The three major groups of mammals are egg-laying, pouched, and (22) _____ mammals.

UNDERSTANDING AND INTERPRETING KEY CONCEPTS

___ (1) Sponges do not have _____ .

    (a) distinct cell types
    (b) nerve cells
    (c) muscles
    (d) a gut

___ (2) Which of the following is not a protostome?

    (a) earthworm
    (b) crayfish or lobster
    (c) starfish
    (d) squid

___ (3) Bilateral symmetry is characteristic of _____ .

    (a) cnidarians
    (b) sponges
    (c) jellyfish
    (d) flatworms

___ (4) Flukes and tapeworms are parasitic _____ .

    (a) leeches
    (b) flatworms
    (c) nematodes
    (d) annelids

___ (5) Insects include _____ .

    (a) spiders, mites, and ticks
    (b) centipedes and millipedes
    (c) termites, aphids, and beetles
    (d) all of the above

___ (6) Which of the following is *not* a characteristic of roundworms?

    (a) some are parasites
    (b) radial symmetry
    (c) cylindrical body tapered at both ends
    (d) covered with a tough flexible cuticle
    (e) all of the above

___ (7) Mollusks include _____ .

    (a) gastropods and bivalves
    (b) octopuses and slugs
    (c) snails and clams
    (d) all of the above

___ (8) Mollusks have a head, foot, and usually _____ .

    (a) setae
    (b) jointed appendages
    (c) paired antennae
    (d) a mantle

___ (9) True segmentation occurs in _____ .

    (a) mollusks
    (b) cephalopods
    (c) annelids
    (d) gastropods

___ (10) Lobsters, shrimps, spiders, centipedes, and ticks are all examples of _____ .

    (a) annelids
    (b) arthropods
    (c) mollusks
    (d) echinoderms

___ (11) _____ are kidneylike organs that control the volume and composition of fluids in _____ .

    (a) Tracheas; earthworms
    (b) Nephridia; crayfish
    (c) Tracheas; crayfish
    (d) Nephridia; earthworms

___ (12) The transformation of any larval arthropod into an adult form requires _____ .

    (a) differentiation
    (b) reproduction
    (c) metamorphosis
    (d) molting
    (e) both c and d

___ (13) Chelicerates include _____ .

    (a) spiders, scorpions, ticks, and mites
    (b) insects
    (c) crayfishes, lobsters, and crabs
    (d) millipedes and centipedes

___ (14) Malpighian tubules reabsorb water and take in nitrogen wastes in _____ .

    (a) starfishes
    (b) true fishes
    (c) earthworms
    (d) insects
    (e) flukes

___ (15) Tube feet and a water vascular-system are _____ adaptations.

    (a) arthropod
    (b) echinoderm
    (c) chordate
    (d) cnidarian

___ (16) A notochord, a nerve cord, a pharynx with gill slits, and a tail that extends past the anus are characteristics that distinguish _____ from all other animals.

(a) arthropods
(b) echinoderms
(c) chordates
(d) cnidarians
(e) mollusks

___ (17) Tunicates and lancelets are invertebrate _____ .

(a) flatworms
(b) arthropods
(c) annelids
(d) chordates
(e) mollusks

___ (18) An arrangement of fish groups from oldest to more recent would be _____ .

(a) bony fishes ⟶ cartilaginous fishes ⟶ jawless fishes
(b) cartilaginous fishes ⟶ jawless fishes ⟶ bony fishes
(c) jawless fishes ⟶ cartilaginous fishes ⟶ bony fishes
(d) bony fishes ⟶ jawless fishes ⟶ cartilaginous fishes

___ (19) An arrangement of land animal groups from oldest to more recent would be _____ .

(a) reptiles ⟶ amphibians ⟶ mammals
(b) mammals ⟶ amphibians ⟶ reptiles
(c) amphibians ⟶ birds ⟶ mammals
(d) amphibians ⟶ reptiles ⟶ mammals

___ (20) Which of following is mismatched?

(a) amphibians = frogs and toads
(b) reptiles = dinosaurs and lizards
(c) birds = ostrich and hummingbird
(d) mammals = platypus and kangaroo
(e) none of the above is mismatched

INTEGRATING AND APPLYING KEY CONCEPTS

Scan Figure 19.4 (main text) to verify that most highly evolved animals have bilateral symmetry, a complete digestive system, a true coelom, and segmented bodies. Why do you suppose that these features are considered to be more highly evolved than conditions shown lower in the figure?

Crossword
Number Five

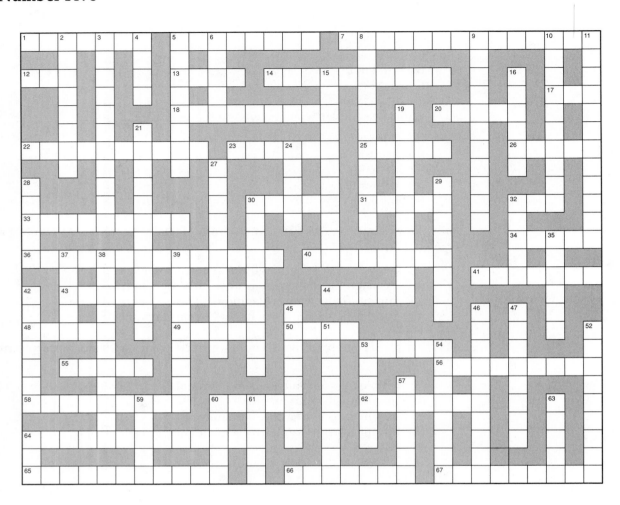

Across

1. region sensitive to light on some organisms
5. embryonic tissue from which muscles, bones, and the circulatory system are derived
7. a process some bacteria use to extract energy from chemical reactions involving inorganic substances (such as sulfur) and to synthesize their own food molecules
12. the digestive tract, or a portion thereof; especially, the intestines or stomach
13. nerve _____ ; in vertebrates, it is single, hollow, and dorsal
14. a subkingdom of kingdom Monera; organisms in this subkingdom have peptidoglycan in their cell walls
17. an alcoholic liquor distilled from fermented molasses or sugar cane
18. pathway in which viral DNA is transferred in a latent state to progeny cells
20. cells in some flatworms that regulate volume and salt concentrations of their body fluid
22. a membrane lining walls of abdominal cavity and enclosing viscera
23. a fold of tissue that secretes shell(s) of a mollusk
25. a unit of weight for precious stones, equal to 200 milligrams
26. should

30. fruit of oak tree
31. prohibited activity
32. female parent of a quadruped (such as a horse)
33. organisms that grow and feed on or in a different organism while contributing nothing to the survival of their host(s)
34. large intestine
36. organisms in this subkingdom of Monera lack peptidoglycan in their cell walls
40. _____ sac: contains a watery fluid in which the embryo of a mammal, bird, or reptile is suspended
41. a noncellular waxy covering of a plant's epidermis; also, in some invertebrates, the tough, inflexible chitinous or protein covering secreted by the epidermis
43. chemosynthetic anaerobic bacteria that produce methane; used in sewage treatment facilities
44. a noncelluar infectious agent than can replicate itself only while inside a host cell
48. in plants, a cellular product of meiosis that gives rise to one or more gametophytes
49. a male bee, especially a honeybee, generally without a sting, that does no work and produces no honey
50. _____ and rave
53. _____ leaf: symbol of Canada
55. gill _____ in the pharyngeal wall: a chordate characteristic occurring sometime in a chordate's life history
56. a flexible rodlike structure that may function as a primitive backbone in all chordates
58. a relationship involving two or more different organisms in a close association that may be (or may not be) mutually beneficial
60. organs of gas exchange in many aquatic organisms
62. tiny organism visible using a microscope
64. a change in structure and habits of an animal during normal growth, usually in postembryonic stage
65. plants with "naked seeds" not enclosed in a fleshy fruit
66. part of digestive tract that extends from nasal cavities to larynx where it joins with esophagus
67. an arthropod with two pairs of legs per segment

Down

2. in animals, embryonic tissue from which nervous system and skin develop
3. animals whose first embryonic opening becomes their mouth
4. in echinoderms, _____ feet are part of the water vascular system
5. mass of hyphae that compose a fungal body
6. in ferns, a cluster of sporangia
8. in cyanobacterium *Anabaena*, cells in which nitrogenase is synthesized
9. not having xylem and phloem
10. spore case
11. produces an animal body with a series of body units that may be externally similar to or quite different from one another
15. female sex organ of mosses and ferns, among others
16. a type of body cavity lined with peritoneum
19. animals with backbones
21. animal without a spine composed of individual bones
24. a population of cells that are dividing to an abnormal extent
27. an external supportive body covering, as in arthropods

28. an individual fungal filament
29. flowering plants with a single cotyledon (lilies, corn)
30. flowering plant
32. flowering plant with two cotyledons
35. a composite organism consisting of an alga and fungus living interdependently
37. a special, small role in a movie
38. male sex organ of mosses and ferns, among others
39. animals with a notochord and pharyngeal gill slits existing sometime in their life histories
42. animals with an exoskeleton and six jointed legs
45. kingdom of single-celled eukaryotes
46. relating to sheets of cells that line internal and/or external surfaces of animal bodies
47. relating to vessels that transmit plant or animal body fluids
51. excretory organs in annelids and some other invertebrates
52. a small asexual spore, such as that formed by some bacteria
53. vertebrate with hair
54. embryonic tissue of animals from which develops inner lining of gut and organs derived from it
57. an attachment structure of tapeworms
59. thick, sweet liquid
60. sudden bursts of wind
61. bursting open
63. slimy primitive autotrophs that dwell in watery environments
64. stout, thick-walled container of beverages or shaving soap

20
PLANT TISSUES

THE PLANT BODY: AN OVERVIEW
    Shoot and Root Systems
    Plant Tissues
    How Plant Tissues Arise: The Meristems
    Monocots and Dicots Compared
SHOOT SYSTEM
    Arrangement of Vascular Bundles
    Arrangement of Leaves and Buds
    Leaf Structure

ROOT SYSTEM
    Taproot and Fibrous Root Systems
    Root Structure
WOODY PLANTS
    Herbaceous and Woody Plants Compared
    Tissue Formation During Secondary
    Growth
    Early and Late Wood
SUMMARY

General Objectives

1. Describe the generalized body plan of a flowering plant.
2. Define and distinguish among the various types of ground tissues, vascular tissues, and dermal tissues.
3. Explain how plant tissues develop from meristems.
4. Know the functions of stems, leaves, and roots.
5. Explain what is meant by *primary* and *secondary growth;* describe how they occur in woody dicot roots and stems.

20-I
(pp. 293–294)

THE PLANT BODY: AN OVERVIEW
    Shoot and Root Systems
    Plant Tissues

Summary

There are more than 275,000 species of plants. Flowering plants are represented by plants as diverse as roses, cherry trees, corn, and dandelions.

Shoot systems are composed of stems and leaves with internal conduits to transport water, minerals, and organic substances among roots, leaves, and other plant parts. Usually stems are upright and function to expose leaves to light and flowers to pollinators. *Root systems* most often grow below ground and function to absorb water and dissolved minerals from the soil and conduct water and solutes to plant parts above ground. Roots also store food, anchor the plant, and provide structural support. *Primary growth* causes roots, stems, and leaves to increase in length; *secondary growth* causes roots and stems to increase in thickness.

Three kinds of tissues—dermal, vascular, and ground—extend continuously throughout the plant body. Tubes and fibers of *vascular* tissue are embedded in *ground* tissue, which in turn is covered and protected by *dermal* tissue.

Ground tissue comprises most of the plant body. It consists mainly of three types of cells: *parenchyma,* which participates in photosynthesis, food storage, wound healing, and regeneration of plant parts; *collenchyma,* thin-walled flexible cells that help support young plant parts; and *sclerenchyma,* rigid-walled strengthening cells that consist mainly of dead cells (fibers and sclereids).

Vascular tissues contain food-conducting tissues (phloem) and water-conducting tissues (xylem) that form a network throughout the entire plant. *Xylem* contains parenchyma cells that store water and food, sclereids, and fibers that provide mechanical support, and tracheids and vessel members that passively conduct water and dissolved salts. Like xylem, *phloem* also contains parenchyma, sclereids, and fibers. Unlike the tracheids and vessel members of xylem, which die as soon as they reach maturity, the food-conducting cells of phloem (called *sieve-tube members* or *sieve cells*) are alive when mature, as are their adjacent parenchyma cells (called companion cells), which help load and unload the sieve element pipelines.

Dermal tissues provide a protective cover for the plant body. In general, the *epidermis,* the tightly packed outermost layer of cells, covers the plant as a single, compact, and continuous layer of cells. The *cuticle* is the surface coat of waxes and fats on epidermal cell walls; it protects plant parts such as leaves and young stems, restricts water loss, and discourages microbial attack. Stomata are openings between pairs of specialized epidermal cells that allow movement of water vapor, carbon dioxide, and oxygen across epidermal cells. Periderm is a protective covering that replaces the epidermis of gymnosperms and angiosperms when they are growing.

Key Terms

shoot system	fibers	companion cells
root system	sclereids	dermal tissues
ground tissues	vascular tissues	epidermis
parenchyma	xylem	cuticle
collenchyma	phloem	periderm
sclerenchyma	sieve-tube members	

Objectives

1. Define *shoot system* and *root system.*
2. Name the three tissue systems that extend throughout the plant body.
3. Define and distinguish from each other the three principal tissue types that compose most of the ground system of the plant body.
4. State the function(s) of the three types of ground tissue and tell where each is located in the plant body.
5. Distinguish *xylem* from *phloem* in structure and function.
6. Distinguish *cuticle, epidermis,* and *periderm* from each other in structure, plant location, and function.

Self-Quiz Questions

Fill-in-the-Blanks

Flowering plants have well-developed (1) _____ systems, which include stems and leaves, and (2) _____ systems, which usually grow underground. (3) _____ is the most abundant cell type in ground tissue. (4) _____ and (5) _____ are ground tissues that provide mechanical support for plant parts. Vascular tissues consist of (6) _____ , which conducts water and ions from the roots to the photosynthetic areas, and (7) _____ , which conducts the products of photosynthesis away to storage areas; both help support the plant. The specialized cells of phloem that conduct sugars and solutes rapidly through the plant are called (8) _____ _____ members. (9) _____ _____ in the phloem assist sieve-tube members in moving sugars from photosynthetic regions to other plant parts. Dermal tissues include the outer cuticle-covered (10) _____ cells of the primary plant body and the thicker (11) _____ , which forms as roots and stems grow in diameter and become woody.

20-II
(pp. 294–295)

THE PLANT BODY: AN OVERVIEW (cont.)
How Plant Tissues Arise: The Meristems
Monocots and Dicots Compared

Summary

When a new embryo plant begins growth, dome-shaped cell masses at the tips of the root and shoot called *apical meristems* divide and elongate. Some cells in the meristems remain to divide perpetually rather than become stem or root tissues. Growth occurring by cell divisions in the root and shoot tips (meristems) is termed *primary growth.* Woody plants show *secondary growth,* or growth in diameter of roots and stems, that occurs through cell divisions in *lateral meristems.* Each spring the tips of roots and shoots of woody plants divide to provide primary growth while cells of the lateral meristem divide to add diameter to those organs.

The two classes of flowering plants are *monocots* and *dicots.* Examples of monocots are grasses, lilies, orchids, irises, cattails, and palms. Many familiar herbs, shrubs, and trees are dicots. Monocots have one embryonic seed leaf, the cotyledon, and parallel veined leaves; dicots have two cotyledons and net-veined leaves. The arrangement of tissues in cross section differs between monocots and dicots as well as the number of flower parts.

Key Terms

apical meristem	lateral meristems	dicot
primary growth	monocot	cotyledon
secondary growth		

Objectives

1. Define, locate, and cite the function of apical meristems.
2. Distinguish *primary growth* from *secondary growth.*
3. Explain the concept of the lateral meristem.
4. List the principal characteristics that distinguish monocots from dicots.

Self-Quiz
Questions

Fill-in-the-Blanks   Root and shoot tips have dome-shaped (1) _____ meristems where new cells form rapidly through mitotic divisions. (2) _____ growth originates at root and shoot tips. (3) _____ growth occurs primarily in (4) _____ plants and originates at areas of dividing cells, the (5) _____ meristems. The two classes of flowering plants are the (6) _____ and the (7) _____ . The presence of one (8) _____ on the flowering-plant embryo identifies (9) _____ plants; the presence of two (10) _____ on the flowering plant embryo identifies (11) _____ plants.

20-III
(pp. 295–299)

SHOOT SYSTEM
    Arrangement of Vascular Bundles
    Arrangement of Leaves and Buds
    Leaf Structure

Summary

Flowering plants have two general patterns of stem structure. Most *monocot* stems usually are uniformly thick along their length; they generally do not undergo secondary growth, and cross sections of their stems reveal that strands of vascular tissue are arranged in many *vascular bundles* scattered throughout the ground tissue; a sheath of fibers and parenchyma is usually found around each bundle. Most *dicot* stems are tapered; they often undergo secondary growth, and their vascular bundles are arranged in a ring that separates the ground tissue into pith and *cortex.*

For most vascular plants, leaves are the site for photosynthesis. Leaves develop on the sides of main stems or their branches. Leaves develop as small bulges from apical meristems and enlarge to become small leaves. As stem and leaves develop, elongation separates the new leaves. The *node* is the point on a stem where one or more leaves attach. The area of the stem between two successive nodes is the *internode*. A typical winter twig has a terminal *bud,* a mass of meristematic tissue protected by modified leaves, at its shoot tip. Lateral buds also occur on the sides of the stem in the upper angles where leaves attach to stems. Such buds can develop into leaves, flowers, or both. The number of buds found per node depends on the particular plant species.

Generally, dicot leaves have a stalklike petiole attached to the stem; joined to the petiole may be one leaf or several leaflets. Most monocot leaves lack a petiole; the base of the blade forms a sheath around the stem, as in corn. Leaves can have various shapes; some are lobed or divided into leaflets. Different types of hairs and scales may be found on leaf surfaces. As to duration, leaves may be deciduous or evergreen.

Generally, leaves have a broad surface for exposure to light and atmospheric gases. *Veins* that contain vascular bundles connect the stem's vascular tissue with photosynthetic parenchyma, which is loosely (air spaces) sandwiched between the upper and lower epidermis. *Stomata,* usually found on lower leaf surfaces, are gates to the interior of leaves that may be opened or closed to regulate the inward movement of carbon dioxide and the outward movement of water vapor and oxygen.

<table>
<tr><td>**Key Terms**</td><td>vascular bundle</td><td>bud</td><td>leaflet</td></tr>
<tr><td></td><td>cortex</td><td>terminal bud</td><td>deciduous</td></tr>
<tr><td></td><td>pith</td><td>lateral bud</td><td>evergreen</td></tr>
<tr><td></td><td>leaf, leaves</td><td>leaf blade</td><td>veins</td></tr>
<tr><td></td><td>node</td><td>petiole</td><td>stoma, stomata</td></tr>
<tr><td></td><td>internode</td><td></td><td></td></tr>
</table>

Objectives

1. Describe the primary structure of a generalized stem. Distinguish *nodes* from *internodes,* tell where buds are located, and indicate how cross sections of monocot stems differ from those of dicot stems.
2. Briefly describe how leaves develop and how they are arranged on stems.
3. Define the different types of buds on a stem and state their functions.
4. Describe how leaves might vary from one species to another; define *deciduous* and *evergreen.*
5. Generally describe the internal structure of a leaf; give the function of stomata.

Self-Quiz Questions

Fill-in-the-Blanks

A(n) (1) _____ bundle is a strandlike arrangement of primary xylem and phloem. Such bundles commonly are surrounded by a sheath of (2) _____ and (3) _____ cells. Cross sections of (4) _____ stems generally show (5) _____ _____ scattered throughout the ground tissue. The stems of most (6) _____ and conifers have vascular bundles arranged in a ring that divides the ground tissue into cortex and pith. A(n) (7) _____ is the point on a stem where one or more leaves attach. The bud at the stem tip is the (8) _____ bud; buds at the sides of the stem are called (9) _____ buds. Buds can develop into (10) _____ or (11) _____ or both. Plants that drop their leaves each fall are (12) _____ species. The principal function of a leaf is (13) _____ . Vascular bundles that form a network through leaves are known as (14) _____ . (15) _____ are tiny openings in lower leaf surfaces that function in (16) _____ exchange.

Labeling

Identify each indicated part of the accompanying illustration (p. 224).

(17) _____ _____

(18) _____ _____

(19) _____ _____
(20) _____
(21) _____

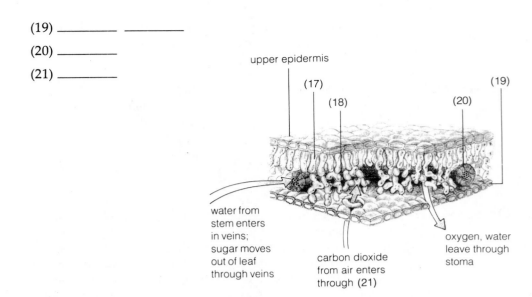

upper epidermis

(17)

(18)

(19)

(20)

water from
stem enters
in veins;
sugar moves
out of leaf
through veins

carbon dioxide
from air enters
through (21)

oxygen, water
leave through
stoma

20-IV
(pp. 299–301)

ROOT SYSTEM
    Taproot and Fibrous Root Systems
    Root Structure

Summary

A tremendous root surface area is required to absorb water and dissolved minerals for growth and maintenance of a plant. Roots anchor the plant, store food, and transport materials to other plant organs.

    The first root to grow from a dicot embryo is the primary root. Eventually numerous *lateral roots* branch from the primary root; the youngest are nearest the root tip. A *taproot system* is defined as the primary root and its lateral branchings. An example is the carrot. In monocots such as grasses, the primary root is short-lived and is replaced by many *adventitious roots* growing from the stem of the plant. This forms a *fibrous root system.*

    Division of root apical meristem cells produces cells that elongate and differentiate (mature) into primary tissues. A longitudinal slice through a root tip reveals a root cap, which protects the root apical meristem that produces it. Epidermis, ground tissue, and vascular tissues form behind the root cap. In the area where the undifferentiated cells mature, epidermal cells develop extensions, the *root hairs*, which increase root absorption. The central cylinder of vascular tissues is the *vascular column.*

    In cross section, the vascular column is surrounded by ground tissue, the *cortex*. These living cells communicate materials through cytoplasmic connections, the plasmodesmata. Water entering the root moves through cortex cells until the inner cell layer of the cortex, the *endodermis,* is encountered. Structural features of the endodermis control movement of water and dissolved minerals into the vascular column. The *pericycle* layer lies just inside the endodermis; cell divisions of the pericycle give rise to lateral roots.

Key Terms

primary root
lateral roots
taproot system
adventitious root
fibrous root system

root cap
root apical meristem
root epidermis
root hairs

vascular column
endodermis
cortex
pericycle

Objectives

1. State the principal functions of roots and explain how the structural components of roots allow those functions to be performed.
2. Distinguish between the taproot system and the fibrous root system in terms of structure and developmental origins.
3. Describe the zones of a root and their functions as seen in longitudinal section.
4. Tell the function of root hairs and the cell layer called *endodermis*.

Self-Quiz Questions

Fill-in-the-Blanks

(1) _____ systems are based on a primary root and its lateral roots; (2) _____ root systems are (3) _____ in that they arise from the young stem. The root (4) _____ protects the delicate young root from being torn apart during growth through the soil; the root (5) _____ _____ gives rise to cells that differentiate into the three basic tissue systems. The (6) _____ with root hairs surrounds the (7) _____ , which surrounds the (8) _____ _____ . Immediately outside the vascular column lies the innermost cell layer of the cortex, the (9) _____ , which regulates water and solute movements into the vascular column. Just inside the endodermis is the (10) _____ whose cells may divide to produce (11) _____ _____ .

Labeling

Identify each indicated part of the accompanying illustration (p. 226).

(12) _____ _____   (17) _____

(13) _____ _____   (18) _____

(14) _____ _____   (19) _____

(15) _____ _____ _____   (20) _____

(16) _____ _____

20-V
(pp. 302–304)

WOODY PLANTS
Herbaceous and Woody Plants Compared
Tissue Formation During Secondary Growth
Early and Late Wood
SUMMARY

Summary

Most monocots and some dicots are *herbaceous plants,* lacking wood. There is little or no secondary growth during their life cycle. Many dicots and all gymnosperms are *woody plants,* which means that they exhibit secondary growth for

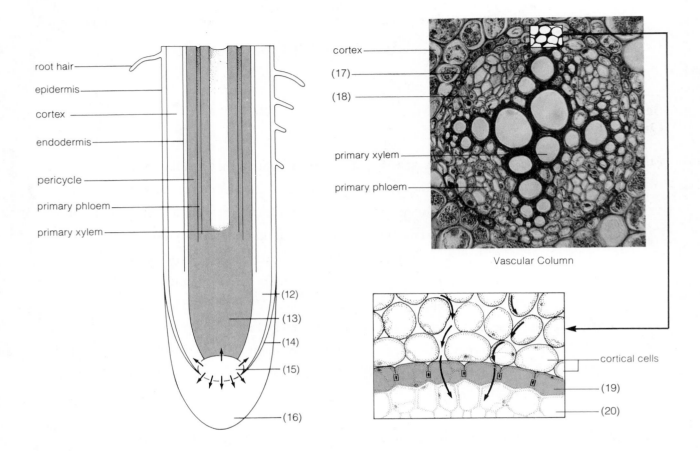

root hair
epidermis
cortex
endodermis
pericycle
primary phloem
primary xylem

(12)
(13)
(14)
(15)
(16)

cortex
(17)
(18)
primary xylem
primary phloem

Vascular Column

cortical cells
(19)
(20)

two or more seasons. Herbaceous and woody plants are characterized as annuals, biennials, and perennials.

Older roots and stems accumulate wood and often great mass through two active meristems, the *vascular cambium* and the *cork cambium*. The meristematic cells of the vascular cambium are only a few cells thick; these cells give rise to secondary xylem tissues on their inner faces and secondary phloem tissues on their outer faces. Season after season the larger amount of xylem increases and crushes the thin-walled phloem cells from the previous season. The xylem mass may rupture the cortex and phloem; as the cortex splits, epidermis may be lost. The cork cambium produces corky periderm to replace the lost epidermis. The term *bark* refers to all tissues outside of the vascular cambium (includes phloem). Thus, living phloem is a very thin film of cells beneath the periderm. Cutting away a strip of phloem all the way around the tree will result in plant death—the roots cannot obtain food.

Early wood represents the first xylem cells with large diameters produced early in the growing season. Further into the growing season, xylem cell diameters become smaller; this is the *late wood.* One band of early wood alternates with one band of late wood in one annual growth layer, or *tree ring.* In tropical regions, the growing seasons are continuous and it is difficult to find growth rings in wood.

Key Terms

herbaceous plants	perennials	bark
woody plants	vascular cambium	early wood
annuals	cork cambium	late wood
biennials	cork	tree ring

Objectives

1. Distinguish between *herbaceous* plants and *woody* plants; give examples of each.
2. Define and distinguish among *annuals, biennials,* and *perennials.*
3. Describe the activity of the *vascular cambium* and the *cork cambium.*
4. Describe the cells of *early wood* and *late wood* and tell the cause of those differences.

Self-Quiz Questions

Fill-in-the-Blanks

(1) _____ plants show little or no secondary growth during their life cycle. Plants that complete their life cycle in two growing seasons are (2) _____ plants. Annual increments of xylem are added to woody plants by the meristematic activity of the (3) _____ cambium. The xylem in a tree ring having the largest diameter conducting cells is known as (4) _____ _____ . Trees in tropical regions have poorly defined tree rings due to a continuous (5) _____ _____ .

Labeling

Identify each indicated part of the accompanying illustration.

(6) _____ _____
(7) _____
(8) _____
(9) _____
(10) _____

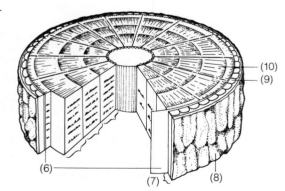

___ (1) _____ develops into the plant's surface layers.

    (a) Ground tissue
    (b) Dermal tissue
    (c) Vascular tissue
    (d) Pericycle

___ (2) Which of the following is *not* considered a ground cell type?

    (a) epidermis
    (b) parenchyma
    (c) collenchyma
    (d) sclerenchyma

___ (3) The _____ produces secondary xylem growth.

    (a) apical meristem
    (b) lateral meristem
    (c) cork cambium
    (d) endodermis

___ (4) The _____ is a leaflike structure that is part of the embryo; monocot embryos have one, dicot embryos have two.

    (a) shoot tip
    (b) root tip
    (c) cotyledon
    (d) apical meristem

___ (5) Leaves are differentiated and buds develop at specific points along the stem called _____ .

    (a) nodes
    (b) internodes
    (c) vascular bundles
    (d) cotyledons

___ (6) Which of the following structures is *not* considered to be meristematic?

    (a) vascular cambium
    (b) lateral meristem
    (c) cork cambium
    (d) endodermis

___ (7) Which of the following statements about monocots is *false*?

    (a) They are usually herbaceous.
    (b) They develop one cotyledon in their seeds.
    (c) Their vascular bundles are scattered throughout the ground tissue of their stems.
    (d) They have a single central vascular cylinder in their stems.

___ (8) Of the following tissues, the ground meristem would *not* give rise to _____ .

    (a) epidermis
    (b) leaf mesophyll
    (c) cortex
    (d) pith

___ (9) Vascular bundles called _____ form a network through a leaf blade.

    (a) xylem
    (b) phloem
    (c) veins
    (d) stomata

___ (10) A primary root and its lateral branchings represent a(n) _____ system.

    (a) lateral root
    (b) adventitious root
    (c) taproot
    (d) branch root

___ (11) Plants whose vegetative growth and seed formation continue year after year are _____ plants.

    (a) annual
    (b) perennial
    (c) biennial
    (d) herbaceous

___ (12) The _____ layer of a root divides to produce lateral roots.

    (a) endodermis
    (b) pericycle
    (c) xylem
    (d) cortex

INTEGRATING AND APPLYING KEY CONCEPTS

List the specific behavioral restrictions that might be imposed if the human body resembled the plant body in having (1) open growth with apical meristematic regions, (2) stomata in the epidermis, (3) cells with chloroplasts, (4) excess carbohydrates stored primarily as starch rather than as fat, and (5) dependence on the soil as a source of water and inorganic compounds.

21

PLANT NUTRITION AND TRANSPORT

NUTRITIONAL REQUIREMENTS
WATER ABSORPTION
WATER TRANSPORT AND
CONSERVATION
    Transpiration
    Control of Water Loss
MINERAL UPTAKE AND ACCUMULATION

TRANSPORT THROUGH THE PHLOEM
    Storage and Transport Forms of Organic
    Compounds
    Translocation
    Pressure Flow Theory
SUMMARY

**General
Objectives**

1. Know which elements are essential to plant health.
2. Explain how water is absorbed, transported, used, and lost by a plant.
3. Describe how the intake of CO_2 is connected with water loss.
4. Explain how essential mineral ions are taken up by a plant.
5. Know how translocation of organic substances occurs, according to the pressure flow theory.

21-I
(pp. 306–309)

NUTRITIONAL REQUIREMENTS

Summary

Plants use only three of their sixteen required essential elements—oxygen, carbon, and hydrogen—as the main building blocks for carbohydrates, lipids, proteins, and amino acids. The thirteen remaining required essential elements are available to plants as *mineral ions*. Six are *macronutrients* and the remainder are *micronutrients*. Plant growth suffers when even one of the essential elements is missing. For example, plants cannot use gaseous nitrogen; nitrogen-fixing bacteria convert nitrogen to forms plants require for their metabolism. Legumes like clover and beans have root nodules in which nitrogen-fixing bacteria live. In this *symbiotic* relationship bacteria have access to the plant's organic molecules while providing usable nitrogen for the plant.

    Plants sometimes have adaptations for survival when an environmental resource is present in very low concentrations. Carnivorous plants such as the Venus flytrap thrive in areas of low nitrogen concentration; they are able to capture and digest animals to obtain nitrogen.

Key Terms

plant physiology	micronutrients	symbiosis
mineral ions	nodules	nitrogen-fixing bacteria
macronutrients	legumes	

Objectives

1. State the number of essential elements plants require.
2. List the three elements plants use as their main metabolic building blocks.
3. State the number of elements available to plants as dissolved salts or mineral ions.
4. Define *macronutrients* and *micronutrients* and explain their role in plant metabolism.
5. Describe the role of nitrogen-fixing bacteria in plant metabolism and explain how some plants have evolved a symbiotic relationship with these bacteria.
6. Give one example of how plants have evolved adaptations to survive in areas where an environmental resource is scarce.

Self-Quiz Questions

Fill-in-the-Blanks

The three essential elements that plants use as their main metabolic building blocks are oxygen, carbon, and (1) _____ . The thirteen essential elements available to plants as dissolved salts are known as (2) _____ _____ . Six dissolved salts become significantly incorporated in plant tissues and are known as (3) _____ . The remainder of the dissolved salts occur in very small amounts in plant tissues and are known as (4) _____ . (5) _____ - _____ bacteria convert nitrogen to forms usable by plants. Some plants have a(n) (6) _____ relationship with the bacteria that live in nodules on their (7) _____ . The Venus flytrap is an example of a plant that has adapted to live in environments low in (8) _____ .

Matching

Refer to Table 21.1 in the main text; choose at least one and, in most cases, no more than two letters per blank.

(9) ___ boron	A. Macronutrient
(10) ___ calcium	B. Micronutrient
(11) ___ chlorine	C. Component of nucleic acids, ATP, and phospholipids
(12) ___ copper	D. Component of chlorophyll molecule
(13) ___ iron	E. Activation of enzymes; role in maintaining water-solute balance
(14) ___ magnesium	F. Component of enzyme used in nitrogen metabolism
(15) ___ manganese	G. Component of proteins, nucleic acids, coenzymes
(16) ___ molybdenum	H. Involved in chlorophyll synthesis and electron transport
(17) ___ nitrogen	I. Needed in cementing cell walls and in forming mitotic spindles
(18) ___ phosphorus	
(19) ___ potassium	
(20) ___ sulfur	
(21) ___ zinc	

**WATER ABSORPTION
WATER TRANSPORT AND CONSERVATION**
    Transpiration
    Control of Water Loss

Summary

The direction and amount of plant root growth occurs in response to water availability. Water and solutes are absorbed into the root through the *root hairs,* extensions of the thin-walled epidermal cells. *Mycorrhiza,* the symbiotic association between roots and fungi, enhances water and mineral absorption for some plants. Water absorbed into the root from the soil moves through the cells of the cortex region until it reaches the inner layer of cortex, the endodermis. Casparian strips on the walls of endodermal cells force water and solutes to pass through the plasma membranes and cytoplasm of those cells. This mechanism of membrane transport affords the plant control over types of solutes absorbed.

    Transpiration is defined as water loss through evaporation from stems, leaves, and other plant parts. Water moves to the tops of the tallest trees through dead conducting cells of the xylem by negative tensions that extend from leaves to roots. Several factors contribute to the negative tensions: (1) water evaporates from the walls of photosynthetic leaf cells but is replaced by water molecules from the cell cytoplasm; water from leaf veins replaces that lost from cells; (2) water from stem xylem replaces that lost from veins; (3) water continually enters root xylem. A pulling action (due to tension) exists between moving water molecules attracted to each other by hydrogen bonds; thus, a xylem fluid column extends throughout the plant. This explanation, offered first by Henry Dixon, is known as the *cohesion theory of water transport.*

    It is imperative that plant water loss be controlled; if transpiration exceeds water uptake by roots, plants may wilt and die. *Stomata* are small openings in the cuticle-covered leaf epidermis where transpiration and carbon dioxide uptake occurs. Two kidney-shaped cells, the *guard cells,* surround a stoma; water and carbon dioxide content of the two guard cells control the opening and closing of a stoma. Light directly affects the leaf cells and triggers active transport of potassium ions into guard cells; this is followed by water movement into guard cells. During the day, as water enters, the guard cells swell and produce the stoma opening through which water is lost and carbon dioxide enters the leaf. When the sun goes down, photosynthesis ceases and carbon dioxide from aerobic respiration collects in cells. Potassium leaves the guard cells, followed by water. This collapses the two guard cells, which come together quite tightly, thus producing a closed stoma; transpiration and water loss are reduced at night.

Key Terms

root hairs	tension	stoma, -ata
mycorrhiza	cohesion theory of water	guard cells
Casparian strip	transport	potassium ions
transpiration	cuticle	

Objectives

1. Explain how water passes from dry soil into the vascular column of the root.
2. Define *transpiration* and list the main points of the cohesion theory of water transport.
3. Compare the amount of water that gets stored or used in metabolism with the amount that is transpired.

4. Explain how land plants regulate water loss as environmental conditions change. List the plant structures that participate in regulating water loss.

Self-Quiz Questions

Fill-in-the-Blanks

Water absorption by roots is enhanced by epidermal cell extensions, the (1) _____ _____ . "Fungus roots" known as (2) _____ also enhance water absorption. This association between root and fungus benefits both partners; it is thus a(n) (3) _____ relationship. (4) _____ is evaporation of water from stems and leaves. The (5) _____ theory of water transport suggests that (6) _____ _____ allow water molecules to cohere tightly enough to keep from breaking apart as they are pulled up through the plant body. Transpiration occurs mainly at (7) _____ surrounded by two kidney-shaped (8) _____ _____ . Light acting on guard cells results in (9) _____ entering these cells. Water entry follows, the guard cells swell, and the stomata (10) _____ . At night, photosynthesis ceases and potassium leaves the guard cells. Water also then leaves the guard cells; this (11) _____ the stomata.

21-III
(pp. 312–314)

MINERAL UPTAKE AND ACCUMULATION
TRANSPORT THROUGH THE PHLOEM
    Storage and Transport Forms of Organic Compounds
    Translocation
    Pressure Flow Theory
SUMMARY

Summary

Growth results from coordination of solute absorption and accumulation. Solute absorption and accumulation occur as energy from ATP (from photosynthesis and aerobic respiration) drives the membrane pumps involved in actively transporting substances into cells. ATP for active transport is formed in nonphotosynthetic cells by mitochondria involved in aerobic respiration.

Sucrose and transportable forms of other organic compounds are distributed through the plant by *translocation* from one plant organ to another via the system of living tubes in the phloem tissue of vascular bundles. Numerous wall perforations in the tubes allow quick transport of water and organic compounds. We have learned much about translocation from small insects called aphids. According to the *pressure flow theory*, translocation is driven by differences in pressure gradients from one region of the phloem pipeline to another. Phloem movements follow a "source-to-sink plan." *Source* regions are sites of photosynthesis; *sink* regions are plant parts where the organic compounds are needed for nutrition or stored. Sucrose and other organic molecules are actively transported into phloem tubes in the leaf; water follows osmotically due to the increased solute concentration. Water and its dissolved components seek areas of lower water pressure. Young leaves, stems, fruits, seeds, and roots are all regions of lower water pressure because cells there convert the food molecules delivered to them into substances they store or need for cellular respiration. Food (sucrose) therefore tends to flow from the leaves (the *source*) to the young leaves, stems, fruits, seeds, and roots (*sink* regions).

Key Terms

solute	phloem tubes	pressure flow theory
membrane pumps	source regions	pressure gradients
translocation	sink regions	

Objectives

1. Explain how plants absorb and accumulate solutes.
2. Discuss the evidence that leads us to believe in the pressure flow theory.
3. Describe the process of translocation and list the key points of the pressure flow theory of translocation. Explain what causes sucrose to be transported from one plant organ to another.

Self-Quiz Questions

Fill-in-the-Blanks

The coordination of solute absorption and (1) _____ affects plant growth. Cells expend ATP energy to actively accumulate (2) _____ , especially dissolved (3) _____ _____ , through operation of membrane pumps. (4) _____ , the dominant food storage product in plants, generally is converted into (5) _____ for transport through the plant body to another plant organ; this transport process is known as (6) _____ . Insects known as (7) _____ have given us much information about this process. The (8) _____ _____ theory suggests a mechanism by which food can be transported from one plant region to another. The main (9) _____ _____ are the sites of photosynthesis; (10) _____ _____ are plant parts that require organic compounds for nutrition or storage. Sucrose is actively transported into (11) _____ _____ , (12) _____ follows osmotically, and (13) _____ _____ builds up in the leaves. Sucrose flows from the leaves to young leaves, fruits, seeds, and (14) _____ because all are regions of lower pressure.

UNDERSTANDING AND INTERPRETING KEY CONCEPTS

___ (1) The _____ theory of water transport states that hydrogen bonding allows water molecules to maintain a continuous fluid column as water is pulled from roots to leaves.

  (a) pressure flow
  (b) evaporation
  (c) cohesion
  (d) abscission

___ (2) The three elements present in carbohydrates, lipids, proteins, *and* nucleic acids are _____ .

  (a) oxygen, carbon, and nitrogen
  (b) oxygen, hydrogen, and nitrogen
  (c) oxygen, carbon, and hydrogen
  (d) carbon, nitrogen, and hydrogen

___ (3) Macronutrients are the six mineral ions that _____ .

  (a) play vital roles in photosynthesis and other metabolic events
  (b) occur in only small traces in plant tissues
  (c) become heavily incorporated in plant tissues
  (d) can function only without the presence of micronutrients
  (e) both a and c

___ (4) Without _____ , plants would rapidly wilt and die during hot, dry spells.

  (a) a cuticle
  (b) a mycorrhiza
  (c) phloem
  (d) cotyledons

___ (5) Water inside all the xylem cells is being pulled upward by _____ .

  (a) turgor pressure
  (b) tension and negative pressure
  (c) osmotic gradients
  (d) pressure-flow

___ (6) Gaseous nitrogen is converted to a plant-usable form by _____ .

  (a) root nodules
  (b) mycorrhiza
  (c) nitrogen-fixing bacteria
  (d) Venus flytraps

___ (7) Symbiotic relationships between fungi and young roots are _____ .

  (a) root nodules
  (b) mycorrhizae
  (c) root hairs
  (d) Venus flytraps

___ (8) _____ prevent(s) water from moving past the abutting walls of the root endodermal cells.

(a) Cytoplasm
(b) Plasma membranes
(c) Casparian strips
(d) Osmosis

___ (9) Most of the water moving into a leaf is lost through _____ .

(a) osmotic gradients being established
(b) transpiration
(c) pressure-flow forces
(d) translocation

___ (10) _____ causes transpiration.

(a) Hydrogen bonding
(b) The drying power of air
(c) Cohesion
(d) Turgor pressure

___ (11) Stomata remain _____ during daylight, when photosynthesis occurs, but remain _____ during the night when carbon dioxide accumulates through aerobic respiration.

(a) open; open
(b) closed; open
(c) closed; closed
(d) open; closed

___ (12) By control of _____ levels inside the guard cells of stomata, the activity of stomata is controlled when leaves are losing more water than roots can absorb.

(a) oxygen
(b) potassium
(c) carbon dioxide
(d) ATP

___ (13) Large pressure gradients arise in sieve-tube systems by means of _____ .

(a) vernalization
(b) abscission
(c) osmosis
(d) transpiration

___ (14) Leaves represent _____ regions; growing leaves, stems, fruits, seeds, and roots represent _____ regions.

(a) source; source
(b) sink; source
(c) source; sink
(d) sink; sink

INTEGRATING AND APPLYING KEY CONCEPTS

How do you think maple syrup is made from maple trees? Which specific systems of the plant are involved, and why are maple trees only tapped at certain times of the year?

22
PLANT REPRODUCTION AND DEVELOPMENT

REPRODUCTIVE MODES
GAMETE FORMATION IN FLOWERS
    Floral Structure
    Microspores to Pollen Grains
    Megaspores to Eggs
POLLINATION AND FERTILIZATION
    Pollination
    Fertilization and Endosperm Formation
    Commentary: Coevolution of Flowering Plants and Their Pollinators
EMBRYONIC DEVELOPMENT
    Seed Formation

Fruit Formation and Seed Dispersal
PATTERNS OF GROWTH AND DEVELOPMENT
    Seed Germination and Early Growth
    Effects of Plant Hormones
    Plant Tropisms
    Biological Clocks and Their Effects
    Commentary: From Embryo to Mature Oak
SUMMARY

General Objectives

1. Describe the typical life cycle pattern in flowering plants.
2. Draw and label the parts of a perfect flower. Explain where gamete formation occurs in the male and female structures.
3. Define and distinguish between *pollination* and *fertilization*.
4. Trace embryonic development from zygote to seedling.
5. Cite two examples of coevolution of flowering plants and their pollinators.
6. Describe the general pattern of plant growth and list the factors that cause plants to germinate.
7. List the various chemical messengers that regulate growth and metabolism in plants. Explain how plants respond to changes in their environment.
8. Know the factors that cause a plant to flower, to age, and to enter dormancy. Describe each process.

22-I
(pp. 316–317)

REPRODUCTIVE MODES

Summary

Flowering plants reproduce sexually with rather complicated reproductive systems involving sperm and egg cells. Developing embryos are protected in female flower structures. Pollinators bring sperm and egg together in a flower. Many plants also are capable of asexual reproduction; new individuals are produced by mitosis that are genetically identical to the single plant that produced them.

Familiar plants such as radishes, cacti, and elm trees are usually diploid, vegetative, flower-producing *sporophyte* generations. Meiosis occurs in the diploid cells of *flowers* to produce haploid *gametophyte* generations. Sperm-producing male gametophytes are released (as pollen grains) from flowers of sporophytes; egg-producing female gametophytes embedded in floral tissues are parasitic on sporophytes. Pollen grains move sperm-producing cells to female flower parts.

Although sexual reproduction is the predominate means of reproduction in most flowering plants, sporophytes may reproduce asexually by horizontal, aboveground stems called *runners*, special buds as in onions and lilies, and underground stems called *rhizomes*. Clones of asexually produced plants may be developed from cuttings (navel oranges) or buds of a parent plant.

Key Terms

pollinators   sporophyte   gametophyte
sexual reproduction   flower   clone
asexual reproduction

Objectives

1. State the role of pollinators in flowering plant life cycles.
2. Distinguish sexual from asexual reproduction; state the advantage of sexual reproduction to a species.
3. Define *sporophyte, flower,* and *gametophyte.*
4. List several means by which sporophytes reproduce asexually.

Self-Quiz Questions

Fill-in-the-Blanks

(1) _____ reproduction requires the formation of gametes, followed by fertilization. (2) _____ reproduction occurs by mitosis and produces individuals that are genetically (3) _____ . Familiar plants such as radishes, cactus plants, and elm trees are usually diploid (4) _____ generations; they produce reproductive shoots called (5) _____ . Diploid cells within flowers undergo (6) _____ to produce haploid (7) _____ . Male gametophytes produce (8) _____ ; female gametophytes produce (9) _____ . Runners, special buds, rhizomes, and cuttings are all methods by which sporophytes may reproduce (10) _____ .

22-II
(pp. 317–318)

GAMETE FORMATION IN FLOWERS
    Floral Structure
    Microspores to Pollen Grains
    Megaspores to Eggs

Summary

A flower is the modified end of a floral shoot attached to a receptacle. Flower parts are carpels, stamens, petals, and sepals. Female parts are closed *carpels* containing a chamber, the *ovary*, where egg development, fertilization, and seed maturation occur. The style and stigma form portions of the carpel. *Stamens* are the male flower parts; each stamen consists of an anther (containing one or more pollen sacs) supported by a slender stalk. Flowers with both male and female parts are *perfect* while those with only male or female parts are *imperfect*. Flowers have petals (collectively, corolla) attached to the receptacle below the male and female flower parts. Corolla colors, perfume, and nectar may attract pollinators. The often green, leaflike sepals enclose the other floral parts and protect the developing flower bud.

Pollen grains form within young anthers growing inside the developing flower bud. In the young anther, mitosis produces four masses of diploid mother cells; each divides by meiosis to produce four haploid *microspores*. All microspores become enclosed within four pollen sacs. Haploid microspores divide by mitosis to produce *pollen grains* that contain two cells. One cell divides to produce two sperm; the other develops into a pollen tube, which will transport sperm to ovary.

Within the carpel of the flower, dome-shaped masses on the ovary inner wall develop into *ovules,* or immature seeds. Integuments develop around each new ovule. Inside each ovule a diploid mother cell undergoes meiosis to produce four haploid *megaspores*. Three megaspores disintegrate; one undergoes three successive mitotic divisions to form a single cell containing eight haploid nuclei. These nuclei migrate, cytoplasmic divisions occur, and the female gametophyte (the embryo sac) develops, which contains seven cells. One cell within the embryo sac (the endosperm mother cell) has two nuclei and will assist in forming stored food (endosperm) that will support the embryo. Another cell within the embryo sac is the egg.

Key Terms

carpel	imperfect flowers	ovule
ovary	microspore	megaspores
stamen	pollen grain	embryo sac
pollen sacs	pollen tube	endosperm
perfect flowers		

Objectives

1. List the parts of a flower and explain how they are arranged on a receptacle.
2. Describe the structure of the carpel and ovary; state their general functions.
3. Describe the complete structure of a stamen.
4. Distinguish *perfect* from *imperfect* flowers.
5. Explain the development of microspores and pollen grains.
6. Identify the parts of the flower in Figure 22.4 of your main text and locate the sites of the (a) megaspore mother cell, (b) megaspore, (c) female gametophyte (embryo sac), (d) egg cell, (e) pollen grain, (f) pollen tube, (g) microspore, (h) pollen release, and (i) pollen tube (mature male gametophyte) with two sperm cells.
7. Tell how events that occur in an ovule produce an embryo sac containing seven cells. Tell what this represents.

Self-Quiz Questions

Matching

Select the one most appropriate answer for each term.

(1) ___ carpel

(2) ___ pollen tube

(3) ___ ovary

(4) ___ perfect flower

(5) ___ ovule

(6) ___ pollen grain

(7) ___ stamen

A. Chamber in the carpel where egg formation and fertilization occur
B. Will become a seed, especially if fertilization occurs
C. Female sex organs
D. Male sex organs
E. Has both male and female parts
F. Transports sperm to the ovary of the flower
G. Immature male gametophyte

22-III
(pp. 318–322)

POLLINATION AND FERTILIZATION
   Pollination
   Fertilization and Endosperm Formation
   Commentary: **Coevolution of Flowering Plants and Their Pollinators**

Summary

Pollination, accomplished by insects, birds, or other agents, is the transfer of pollen grains to floral stigmas. Floral structures have coevolved with their pollinators in that changes in one affected changes in the other. The ovules of flowering plants are protected within carpels of a broad diversity of flower types. More elaborate pollinating schemes are required than for exposed gymnosperm ovules. Flower color, color patterns, and odors are correlated with various kinds of pollinator sensory apparatus. Flower size, shape, structure, and location of reproductive parts are correlated with pollinator body size and shape of body parts (feeding apparatus). Nectar composition may be correlated with pollinator diets, and the timing of flowering may be correlated with pollinator foraging behavior (day, night, or season). Many flowering plants accomplish their life cycles with wind pollination.

Following pollen grain deposit on the proper stigma, the pollen tube begins growth toward the ovule. Within the pollen tube are two sperm cells formed before or during tube growth. Eventually, the pollen tube reaches the ovule, penetrates the embryo sac, and releases two sperm within. *Double fertilization* occurs in flowering plants. One sperm nucleus fuses with the egg to produce a diploid zygote; the other sperm nucleus fuses with the diploid endosperm mother cell to form the triploid *primary endosperm cell*. This cell divides to form nourishing tissues for the young dependent sporophyte that will develop from a seed to a seedling.

Key Terms

pollination	<u>double fertilization</u>	triploid
pollinator	<u>diploid zygote</u>	primary endosperm cell
coevolution		

Objectives

1. Define *pollination;* list typical pollinators.
2. Completely describe double fertilization as it occurs in flowering plants.
3. Describe the formation of the *primary endosperm cell* and give its function.

Self-Quiz Questions

Fill-in-the-Blanks

Pollen grains are transferred from (1) _____ to (2) _____ ; this is called (3) _____ . Flowering plants and their (4) _____ have coevolved. After a pollen grain is deposited on a stigma, the (5) _____ _____ begins to grow toward the ovules. (6) _____ sperm cells form inside the pollen tube. The sperm cells are released after the pollen tube penetrates the (7) _____ _____ inside the ovule. (8) _____ fertilization occurs in flowering plants. One sperm nucleus fuses with the egg nucleus; this forms a diploid (9) _____ . The other sperm nucleus fuses with the two nuclei of the endosperm mother cell, forming a(n) (10) _____ nucleus that produces the primary (11) _____ cell. This cell divides to form tissues that will nourish the young (12) _____ until its leaves form.

Labeling

Identify each indicated part of the accompanying illustration.

(13) _____

(14) _____

(15) _____

(16) _____

(17) _____ _____

(18) _____

(19) _____ _____ _____

(20) _____ _____ _____

(21) _____

(22) _____ _____

(23) _____ _____

(24) _____

(25) _____

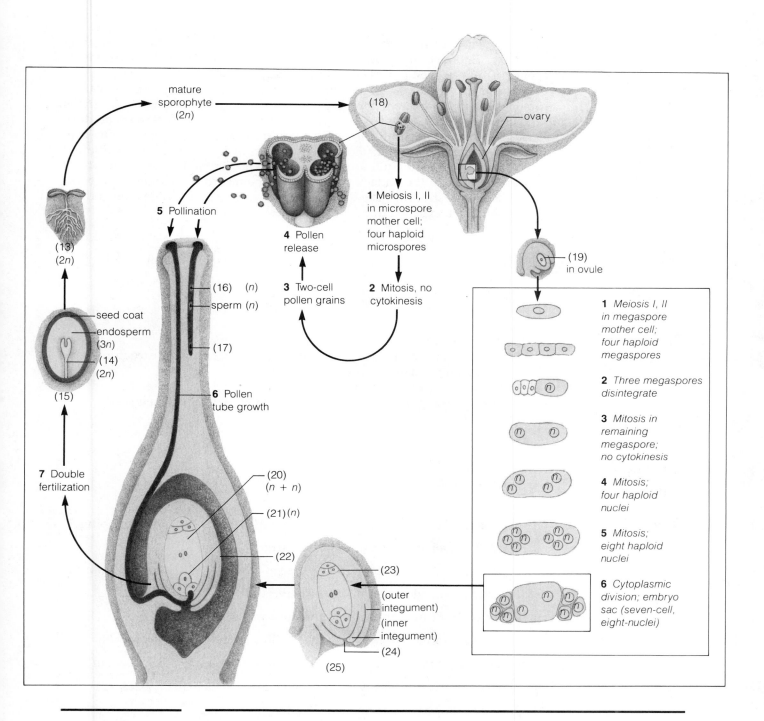

mature
sporophyte ──────
(2n)

(18)

ovary

5 Pollination

4 Pollen
release

1 Meiosis I, II
in microspore
mother cell;
four haploid
microspores

3 Two-cell
pollen grains

2 Mitosis, no
cytokinesis

(19)
in ovule

(13)
(2n)

(16)   (n)
sperm (n)

(17)

seed coat
endosperm
(3n)
(14)
(2n)

(15)

6 Pollen
tube growth

7 Double
fertilization

(20)
(n + n)

(21)(n)

(22)

(23)

(outer
integument)

(inner
integument)

(24)

(25)

1 Meiosis I, II
in megaspore
mother cell;
four haploid
megaspores

2 Three megaspores
disintegrate

3 Mitosis in
remaining
megaspore;
no cytokinesis

4 Mitosis;
four haploid
nuclei

5 Mitosis;
eight haploid
nuclei

6 Cytoplasmic
division; embryo
sac (seven-cell,
eight-nuclei)

22-IV
(pp. 323–325)

EMBRYONIC DEVELOPMENT
   Seed Formation
   Fruit Formation and Seed Dispersal

Summary

Following formation, the zygote within the ovule is attached to the parent
sporophyte plant. The first cell divisions produce a short row of cells for transfer
of nutrients from the parent plant to the embryo. Further cell divisions of the
young embryo produce the tissues of the mature embryo. Seed leaves or *cotyledons* form as part of the embryo. Plants with large cotyledons function in
endosperm absorption and food storage; those with thin cotyledons may pro-

duce enzymes for transfer of stored food from endosperm to the germinating seedling. Endosperm expands as the embryo develops; ovule integuments develop into the seed coat. The mature ovule is a *seed*.

The mature ovary with its ovule(s) is a *fruit*. The nature of fruits varies. Some, like apples and tomatoes, are juicy and fleshy. Others are dry like grains and nuts. Mature walnuts have a dry, intact fruit wall. Multiple flowers produce a fruit like pineapple. Fruits protect seeds and aid dispersal in specific environments. Winged maple fruits move seeds to sites away from the competition of the parent plant. Some fruits are equipped with hooks, spines, hairs, and sticky surfaces by which they attach to birds and animals that function to disperse them. Edible fleshy fruits are eaten, but their seeds survive digestion and are dispersed when expelled from animal bodies.

Key Terms

embryo	integuments	fleshy fruit
cotyledons	seed	dry fruit
endosperm	fruit	dispersal

Objectives

1. Explain how the developing embryo obtains nutrients.
2. Define *cotyledons* and give the location and function of these structures.
3. Contrast the floral origin of *seeds* and *fruits*.
4. Tell what forms the seed coat.
5. Give some examples of how fruits are adapted to disperse their seeds.

Self-Quiz Questions

Fill-in-the-Blanks

A fully matured ovule is a(n) (1) _____ ; the (2) _____ surrounding the seeds(s) develops into most or all of a fruit. The (3) _____ of the ovule harden and thicken to become the seed coats. (4) _____ are embryo structures that function to provide the embryo with nutrition. (5) _____ and nuts are dry fruits; apples and tomatoes are (6) _____ fruits. Pineapples are formed from (7) _____ flowers whose ovaries remain together. Fruits function in (8) _____ protection and (9) _____ in specific environments. Fleshy fruits such as cherries have their seeds dispersed as they are eaten and expelled by (10) _____ .

22-V
(pp. 326–328)

PATTERNS OF GROWTH AND DEVELOPMENT
    Seed Germination and Early Growth
    Effects of Plant Hormones

Summary

Embryonic growth slows greatly before or after seed dispersal. Germination occurs when the embryo absorbs water, resumes growth, and emerges through the seed coat. *Germination* is influenced by several environmental factors. Mature seeds usually require an external water source to initiate embryo metabolism, seed swelling, and seed coat rupture. Seed coat rupture allows oxygen entry so embryo cells may begin aerobic respiration; metabolic rate increases as cells divide more rapidly. When the first root emerges from the seed coat, germination is completed. Cell divisions in seedling tissues leading to a growth pattern occur in different planes and directions under genetic control. The patterns of growth and development characteristic of a monocot plant differ from those of a dicot plant. The first root emerging from the seed coat bends downward. The plant is equipped with hormones that control growth and development as interactions with the environment occur.

The mechanism by which powerful and various *hormones* influence plant growth and development is not well understood. Hormones are molecules secreted by certain cells that "signal" in that they alter activities of other target cells equipped with chemical receptors for certain hormones. Known plant hormones are auxins, gibberellins, cytokinins, abscisic acid, and ethylene. *Auxins* promote stem elongation; one synthetic auxin, 2,4-D, is used as a selective weed herbicide. *Gibberellins* also promote stem elongation. Genetically short-stemmed corn plants grow longer with gibberellin application. *Cytokinins* stimulate cell division, promote leaf expansion, and retard leaf aging. *Abscisic* acid promotes seed and bud dormancy; it also acts on guard cells to close stomata in short, dry periods. *Ethylene* stimulates ripening of fruit and initiates abscission of flowers, fruits, and leaves. There are more hormones than those listed. One unknown shoot tip hormone inhibits lateral bud growth, an effect known as *apical dominance*. An unidentified hormone, florigen, may initiate flowering.

Key Terms

germination	hormones	abscisic acid
growth patterns	auxins	ethylene
monocot	gibberellins	apical dominance
dicot	cytokinins	

Objectives

1. Describe the events occurring in a seed that lead to germination.
2. Tell what event terminates germination.
3. State what plant hormones are and, in general, what they do.
4. Identify the five principal plant hormones (or hormone groups) and state the known or suspected effects of each.
5. Describe the cause and effect of apical dominance.

Self-Quiz Questions

Fill-in-the-Blanks

The embryo absorbs water, resumes growth, and breaks through the seed coat during (1) _____ . Embryonic metabolism increases when oxygen is available and (2) _____ respiration occurs. When the first root is visible through the seed coat, (3) _____ is completed. Monocot and dicot (4) _____ and (5) _____ patterns differ. Plant (6) _____ are information-carrying messenger molecules that cause certain cell types to change their activities in response to environmental conditions. (7) _____ and (8) _____ both promote elongation of cells in stems. (9) _____ _____ promotes the closure of stomata and may bring about the end of (10) _____ dormancy. (11) _____ promotes the ripening of fruit. An unknown hormone from shoot tips inhibits lateral bud growth and causes an effect known as (12) _____ _____ .

Identification

Name the hormone that is primarily responsible for the effects described after each blank.

(13) _____ Promote stem elongation (especially in dwarf plants); might help break dormancy of seeds and buds

(14) _____ Arbitrary designation for as yet unidentified hormone (or hormones) thought to cause flowering

(15) _____ Promote cell elongation in coleoptiles and stems; thought to be involved in phototropism and gravitropism

(16) _____ _____ Promotes stomatal closure; might trigger bud and seed dormancy

(17) _____ Promotes fruit ripening; promotes abscission of leaves, flowers, and fruits

(18) _____ Promote cell division; promote leaf expansion and retard leaf aging

---

22-VI
(pp. 328–329)

PATTERNS OF GROWTH AND DEVELOPMENT (cont.)
Plant Tropisms

Summary

Tropisms are growth responses to environmental factors that are poorly understood. Plant stems will curve toward the side of the plant where the light intensity is greatest; leaves turn to have their flat surfaces receive the maximum amount of light. Directional and growth movements by plants in response to light is termed *phototropism*. Grass coleoptiles exhibit phototropism. Coleoptile tips contain a growth promoting substance named *auxin* by Frits Went. Auxin moves down from the tip to cells receiving less light; those cells grow longer and the tip bends toward the light. *Gravitropism*, a growth response to gravity, is demonstrated by germinating seedlings; the first root always bends down and the stem always bends up. Increased rate of cell divisions on the lower side of a stem turns it upward, even in the absence of light. Root gravitropism may be initiated by a growth-inhibiting hormone produced by the root cap. Roots positioned horizontally will not curve downward if the root cap is removed; replacement of the root cap permits normal root growth. *Thigmotropism* is defined as unequal growth due to physical contact with solid objects. This tropism is illustrated by peas, beans, and other climbing vines. A plant that curls its stem around a tree or post does so due to cessation of cell elongation on the side of the stem making contact; a curling of the stem results.

Key Terms

tropisms   gravitropism   thigmotropism
phototropism

Objectives

1. Define the term *tropism.*
2. Describe the meanings of the following concepts and give an example of each: *phototropism, gravitropism,* and *thigmotropism.*
3. Try to indicate how each of the above plant responses promotes the survival of a plant.

Self-Quiz Questions

Matching

Choose the one most appropriate answer for each.

(1) —— gravitropism

(2) —— phototropism

(3) —— thigmotropism

    A. A response to electric shock caused by potassium redistribution

    B. Negative response in shoot; positive response in root due to growth inhibitor produced by the root cap

    C. A response to physical contact evident in climbing vines

    D. Known to be controlled by auxin and light

22-VII
(pp. 329–334)

PATTERNS OF GROWTH AND DEVELOPMENT (cont.)
    Biological Clocks and Their Effects
    Commentary: **From Embryo to Mature Oak**
SUMMARY

Summary

Biological clocks are internal time-measuring mechanisms exhibited by plants and all other organisms. Some plants must measure time and show "sleep" movements of their leaves in the evening, but open more fully to light during the day. Circadian rhythms refer to activities that occur regularly in about twenty-four-hour cycles. *Photoperiodism* refers to responses plants make to seasonal environmental changes such as changing daylength. A blue-green molecule, *phytochrome,* seems to act as an eye for the plant to detect such changes. Phytochrome reversibility is shown by the following: at sunrise, absorption of red wavelengths converts phytochrome to an active form, Pfr; at sunset, far red wavelengths convert phytochrome to an inactive form, Pr. Pfr may control production of enzymes in certain cells that are necessary for different growth responses. Seed germination, stem elongation and branching, leaf expansion, and formation of flowers, fruits, and seeds are influenced by Pfr. Plants grown in the dark show the effects of Pfr deficiency; stems elongate exceptionally and branch less, and leaves expand less.

Plants apparently measure day and night lengths to adjust to seasonal changes; this is reflected by adjustments in growth, development, and reproductive patterns. In the flowering process, plants are keyed to changes in daylength throughout the year. Daylength cues flowering and reproduction in many species. Long-day plants are adapted to long days and flower in the spring when daylength becomes longer than some critical value. Short-day plants are attuned to the shorter daylengths and flower in late summer or early autumn when daylength becomes shorter than some critical value. Day-neutral plants flower when mature, almost independently of daylength. Spinach, a long-day plant, will not flower unless exposed to 14 hours of light daily for two weeks. Cocklebur, a short-day plant, is extremely sensitive. A single night longer than 9 1/2 hours triggers flowering, but if even a minute or two of light occurs during the dark period, flowering will not occur. Red wavelengths have the greatest effect on short-day plant phytochrome to inhibit flowering. A relationship between the pigment phytochrome and an elusive hormone, florigen, is supposed to exist but has never been demonstrated conclusively.

All the processes that lead to the death of a plant or any of its organs are called *senescence*. One stimulus for senescence could be the drain of nutrients during the growth of reproductive organs, but some sort of death signal that forms during short days is also suspected. *Abscission* is the dropping of leaves, flowers, fruits, or other plant parts; the process may be triggered by ethylene formed by cells near break points. When any plant stops growing under physical conditions that are actually quite suitable for growth, it is said to have entered a period of *dormancy*. Short days and long nights may trigger dormancy. Between fall and spring, a dormancy-breaking process may operate that involves abscisic acid and gibberellins. Exposure to low winter temperatures may be involved.

An organism such as the coast live oak develops and grows in an ecosystem according to interactions between environmental cues and its own genotype. Humans, in their ignorance of the requirements of other organisms, often sentence their intended neighbors to death.

Key Terms

biological clock	Pr, inactive phytochrome	florigen
circadian rhythms	seasonal adjustments	abscission
photoperiodism	long-day plants	senescence
phytochrome	short-day plants	dormancy
Pfr, active phytochrome	day-neutral plants	

Objectives

1. Explain the concept of *biological clocks*.
2. Describe how sunlight influences plant growth, by explaining how phytochrome responds to specific colors of light and how seed germination, stem elongation, leaf expansion, stem branching, and flowering are affected by the activated form of phytochrome.
3. Distinguish among long-day, short-day, and day-neutral plants and give an example of each.
4. Present the reasons some botanists believe there must be a flowering hormone named florigen.
5. Define *abscission, senescence,* and *dormancy* and list possible cues that trigger these processes.
6. Identify factors that help a plant to break out of dormancy.

Fill-in-the-Blanks

Many plants show (1) _____ , a response made when days or nights become longer or shorter than some set length. (2) _____ , a blue-green molecule, seems to be an on-off switch for hormone activities that govern leaf expansion, stem branching, stem length, and (in many plants) seed (3) _____ and flowering. Before the sun rises, phytochrome exists mainly in an (4) [choose one] ( ) active ( ) inactive form that absorbs (5) [choose one] ( ) red ( ) far-red light and is converted to the (6) [choose one] ( ) active ( ) inactive form. In the shade, at sunset, or at night, (7) [choose one] ( ) Pr is converted to Pfr ( ) Pfr is converted to Pr. (8) _____ is a highly reliable cue for flowering and reproduction of many species. (9) _____ -day plants flower in spring when daylength becomes longer than some critical value. (10) _____ -day plants flower in late summer or early autumn when daylength becomes shorter than some critical value. (11) _____ -neutral plants flower when mature without regard to daylength. Dropping of plant parts such as leaves, flowers, or fruits is called (12) _____ . (13) _____ is the total of the processes which lead to the death of plant parts or the entire plant. When plants cease growth under suitable growing conditions, they are said to enter (14) _____ . (15) _____ days and (16) _____ nights are strong cues for dormancy. Two compounds that may play a role in breaking dormancy are (17) _____ and (18) _____ _____ .

CHAPTER TEST

UNDERSTANDING AND INTERPRETING KEY CONCEPTS

___ (1) A stamen is _____ .

    (a) composed of a stigma
    (b) the mature male gametophyte
    (c) the site where microspores are produced
    (d) part of the vegetative phase of an angiosperm

___ (2) A gametophyte is _____ .

    (a) a gamete-producing plant
    (b) haploid
    (c) both a and b
    (d) the plant produced by the fusion of gametes

___ (3) The process during which the diploid set of chromosomes becomes haploid is _____ .

    (a) metastasis
    (b) fertilization
    (c) cleavage
    (d) meiosis

___ (4) An immature fruit is a(n) _____ and an immature seed is a(n) _____ .

    (a) ovary; megaspore
    (b) ovary; ovule
    (c) megaspore; ovule
    (d) ovule; ovary

___ (5) The joint evolution of flowers and their pollinators is known as _____ .

    (a) adaptation
    (b) coevolution
    (c) joint evolution
    (d) covert evolution

___ (6) In flowering plants, one sperm nucleus fuses with that of an egg, and a zygote forms that develops into an embryo. Another sperm fuses with _____ .

    (a) a primary endosperm cell to produce three cells, each with one nucleus
    (b) a primary endosperm cell to produce one cell with one triploid nucleus
    (c) both nuclei of the endosperm mother cell, forming a primary endosperm cell with a single triploid nucleus
    (d) one of the smaller megaspores to produce what will eventually become the seed coat

___ (7) Plant stems sometimes grow in a curved form to wrap themselves around a solid object such as a post; this is known as _____ .

    (a) geotropism
    (b) phototropism
    (c) thigmotropism
    (d) thigmomorphogenesis

___ (8) _____ may be involved in the plant response called *thigmotropism*.

    (a) Cytokinins
    (b) Gibberellins
    (c) Abscisic acid
    (d) Ethylene

___ (9) _____ is the principal substance that causes phototropism in stems or leaves.

    (a) Ethylene
    (b) Growth-inhibiting hormone
    (c) Auxin
    (d) Cytokinin

___ (10) 2,4-D, a potent dicot weed killer, is a synthetic _____ .

    (a) auxin
    (b) gibberellin
    (c) cytokinin
    (d) phytochrome

___ (11) All the processes that lead to the death of a plant or any of its organs are called _____ .

    (a) dormancy
    (b) vernalization
    (c) abscission
    (d) senescence

___ (12) Which of the following is *not* promoted by the active form of phytochrome?

    (a) seed germination
    (b) stem elongation
    (c) leaf expansion
    (d) none of the above

___ (13) Gibberellins, abscisic acid, and exposure to low winter temperatures are probably involved in _____ .

    (a) pollination
    (b) dormancy
    (c) senescence
    (d) breaking dormancy

INTEGRATING AND APPLYING KEY CONCEPTS

Unlike the growth pattern of animals, the vegetative body of vascular plants continues growing throughout its life by means of meristematic activity. Explain why most representatives of the plant kingdom probably would not have survived and grown as well as they do if they had followed the growth pattern of animals.

An oak tree has grown up in the middle of a forest. A lumber company has just cut down all of the surrounding trees except for a narrow strip of woods that includes the oak. How will the oak be likely to respond as it adjusts to its changed environment? To what new stresses will it be exposed? Which hormones will most probably be involved in the adjustment?

Crossword
Number Six

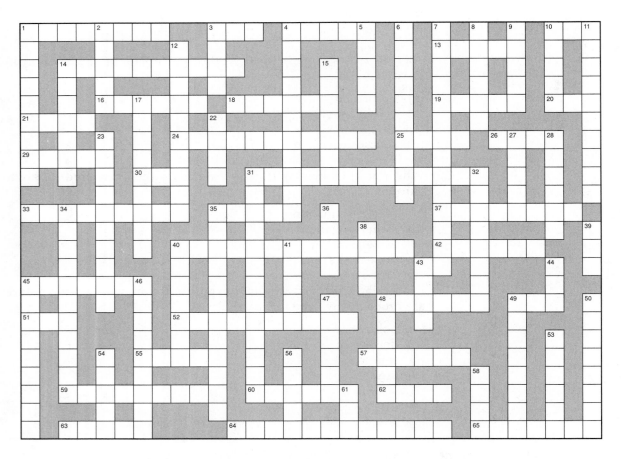

Across

1. sticking together
3. root _____
4. cells bordering each stoma
10. the result of addition
13. large water-dwelling mammal (abbreviated form)
14. a haploid cell that can give rise to a male gametophyte (pollen grain) in heterosporous plants
16. person who has left his native country (French)
18. split
19. female egg-containing structure
20. dense, dark, oily substance that can be extracted from coal
21. underground part of a plant responsible for anchorage and absorption of water and dissolved minerals
24. the transfer of immature male gametophytes from anther to stigma
25. a fully mature ovule that contains a plant embryo within a protective covering
26. an ancient Egyptian goddess of fertility; also, sister and wife of Osiris
29. any positively charged electrode, as of a storage battery
30. a narrow beam of radiation
31. the response or motion of an organism to direct contact with an object or surface

33. a haploid cell that can give rise to a female gametophyte in heterosporous plants
35. a small hard fruit or seed of a cereal grass
37. a state of being relatively inactive in which some processes are slowed down or suspended
40. in vascular plants, the transport of soluble food molecules from one plant organ to another by way of phloem tissue
42. a group of interacting elements that function together
43. centimeter (abbr.)
45. noncellular surface coating composed of waxes and cutin associated with shoot epidermal cells
48. the food-conducting tissue of vascular plants
49. *very important person*
51. _____ root
52. not woody; green and leaflike
55. casparian _____
57. the reproductive structure of an angiosperm prior to fertilization
59. the outermost layer of cells or protective covering of a plant or part of a plant
60. queasiness
62. point on a stem from which a leaf, bud, or other organ grows
63. conductive tissue that moves water and dissolved minerals from root to stem and leaves
64. growth movement of a plant in response to light
65. a signalling molecule released from one cell, transported to (typically) nonadjacent target cells

Down

1. strip that prevents water from entering the vascular column of a root except by way of endodermal cytoplasm
2. _____ *bleu!* (French exclamation)
3. type of cambium that produces suberized cells that die at maturity
4. sprouting of a seed
5. an abbreviated term for a subgroup of angiosperms
6. leaf (or fruit or flower) drop after hormonal action
7. relative exposure of an organism to light as a proportion of an entire day as the exposure affects growth and other plant functions
8. pertaining to a root or shoot tip
9. relating to the tough, fibrous cellular substance that makes up the xylem of trees and shrubs
10. system consisting of stems and leaves that usually grow aboveground
11. a combination of plant structures and fungal hyphae that provides benefits for both participants
12. the diploid phase of a plant life cycle that produces haploid cells by meiosis
14. a subgroup of angiosperms with one cotyledon and parallel veins in leaves
15. one of two types of meristematic tissue responsible for secondary growth; for example, vascular _____
17. the distance between two adjacent regions of outward growth from a stem
22. the pressure _____ theory explains how translocation works
23. a process that directs a plant (or part of a plant) toward rapid aging and death
27. an opening defined by two guard cells that permit gas exchange
28. transport of sugar water in plants occurs from _____ to sink
31. evaporative water loss from stem and leaves
32. implants a mass of self-perpetuating cells that can divide by mitosis and are not yet specialized
34. the haploid phase of a plant life cycle that produces eggs and/or sperms

35. directional growth of a coleoptile, root, or shoot in response to forces originating beneath Earth's crust
36. generally, the principal photosynthetic organ of a plant
38. sucrose dissolved in water travels from source to _____ in a vascular plant
39. small protuberance on a stem or branch, often enclosed in protective scales, and containing an undeveloped shoot, leaf, or flower
40. instructor
41. immature seed
43. the basic structural unit of any living organism
44. fluid that carries food (mainly sucrose) and other substances to plant tissues
45. "seed leaf"; often contains stored nutrients used during early plant growth
46. nutritive tissue used by a plant embryo
47. a vascular _____ in a monocot stem contains xylem and phloem, among other structures
48. a _____ grain is an immature male gametophyte
49. pertaining to conductive tissue
50. the _____ flow theory suggests how translocation occurs in a plant
53. an organism in early development before it resembles the adult form of a species
54. pertaining to main stem or root or central part of a plant body
56. the ripened ovary or ovaries of a seed-bearing plant together with accessory parts
58. centrally located parenchyma tissue
61. breath of fresh _____

23

ANIMAL TISSUES, ORGAN SYSTEMS, AND HOMEOSTASIS

OVERVIEW OF ANIMAL TISSUES
   Epithelial Tissue
   Connective Tissue
   Muscle Tissue
   Nervous Tissue

MAJOR ORGAN SYSTEMS
HOMEOSTASIS AND SYSTEMS CONTROL
   The Internal Environment
   Mechanisms of Homeostasis
SUMMARY

General Objectives

1. Understand the various levels of animal organization (cells, tissues, organs, and organ systems) and be familiar with the anatomical terms provided.
2. Know the characteristics of the various types of animal tissues. Know the types of cells that compose each tissue type and cite some examples of organs that contain significant amounts of each tissue type.
3. Explain how the human body maintains a rather constant internal environment despite changing external conditions.

23-I
(pp. 336–337)

OVERVIEW OF ANIMAL TISSUES
Epithelial Tissue

Summary

Animals' bodies (1) keep tolerable internal working conditions even though outside conditions change, (2) find and take in nutrients and other raw materials, distribute them through the body and dispose of wastes, (3) protect themselves from harmful agents, and (4) reproduce, feed, and protect newly developing individuals. Animals' bodies also monitor and move through their environment.

The trillions of cells in a complex animal are organized into *tissues* or groups of cells and intercellular substances that function together, carrying out a common purpose; different tissues divide the work of being alive into different tasks so that the animal can survive. Tissues are organized into *organs;* two or more organs work together as an *organ system* that performs a specific task.

Epithelium is a tissue composed of densely packed cells of one or more layers; it forms a continuous barrier between body parts it covers and the surrounding medium. One surface of epithelium is exposed; the other adheres to a basement membrane. Skin is a type of epithelium that covers exterior body surfaces and lines cavities, ducts, or tubes. Substances diffuse or are actively transported across epithelia. Some epithelia contain glands; endocrine glands secrete substances such as hormones into the bloodstream. Exocrine glands secrete substances onto an exposed epithelial surface (saliva, digestive enzymes, milk) or into a tubular system. Cells in epithelial tissues may resemble flat floor tiles, cubes, or columns. Stratified epithelium protects underlying tissues.

Key Terms

tissue

organ

organ system

somatic cell

germ cell

epithelium

endocrine gland

exocrine gland

Objectives

1. List four major tasks every animal body is adapted to accomplish.
2. Distinguish *tissues* from *organs* and give an example of each structure.
3. Explain how, if each cell can perform all of its basic activities, organ systems contribute to cell survival.
4. Define what is meant by *epithelial tissue* and list its functions.
5. Describe how the general structure of epithelial tissues permits it to carry out its principal functions.
6. Explain the meaning of the term *gland*, distinguish between *endocrine* and *exocrine* glands, and cite examples of products secreted by each.

Self-Quiz Questions

Fill-in-the-Blanks

Animals maintain tolerable (1) _____ operating conditions even though external changes occur; locate and take in (2) _____ , distribute them in the (3) _____ , dispose of (4) _____ ; protect themselves against injury or attack from viruses and (5) _____ ; reproduce and help (6) _____ and protect new developing individuals. In complex animals, trillions of cells are organized into (7) _____ . Different tissues become organized in specific ways that form (8) _____ . Two or more organs interact and perform a common task when organized as a(n) (9) _____ _____ . All of the diverse body parts found in different animals may be assembled from a few (10) _____ types, though there are variations in the way they are combined and arranged. (11) _____ cells constitute the physical structure of the animal body; they become differentiated into the components of four main types of tissue: (12) _____ , (13) _____ , (14) _____ , and (15) _____ . (16) _____ cells are the only cells in the body that give rise to sperms and eggs. (17) _____ cells are organized as one or more layers that cover and protect external body surfaces. Epithelial glands are of two types: (18) _____ , which secrete substances like saliva into tubes or ducts and a surface, and (19) _____ that secrete substances such as hormones into the bloodstream.

OVERVIEW OF ANIMAL TISSUES (cont.)
   Connective Tissue

Summary

The four categories of *connective tissue* are connective tissue proper, cartilage, bone, and blood. The first two are composed of a jellylike *ground substance* of cells and strong proteins, collagen or elastin fibers, or both. *Connective tissue proper* includes dense, loose, or fatty tissues. *Dense* connective tissue contains tension-resisting collagen fibers; it connects different body parts such as attachment of muscle to bone. *Loose* connective tissue has both collagen and elastin fibers; it supports epithelia and other organs and encloses blood vessels and nerves. *Adipose* tissue is composed of fatty cells that serve as reserve energy stores and padding for organs.

Cartilage is an elastic connective tissue that cushions and shapes body parts. Cartilage resists compression; it is found at the ends of bones, in the nose and external ear, and between individual bones of the backbone.

Bones are variously shaped and harder than cartilage; they support and protect body parts. Bones function with muscles to form lever systems that move the body parts. Bone marrow (central cavities in bone) produces blood cells. Bone is hardened by deposits of mineral salts, but it also contains small spaces holding living bone cells.

Blood is a connective tissue that supplies body cells with oxygen and transports their wastes away; it contains clotting agents and defense mechanisms against disease. Hormones and enzymes are moved by blood to the various body parts.

Key Terms

connective tissue	collagen	cartilage
connective tissue proper	loose connective tissue	bone
ground substance	elastin	bone marrow
dense connective tissue	adipose (fatty) tissue	blood
tendon		

Objectives

1. Describe the basic features of connective tissue and explain how they enable connective tissue to carry out its various tasks.
2. List each type of connective tissue and characterize each in terms of its function and body location.
3. Explain how bone works with other body parts to accomplish movement.
4. State the functions of marrow and adipose tissue.
5. List three functions of blood.

Self-Quiz Questions

Fill-in-the-Blanks

Cells of the connective tissues are scattered throughout an extensive jellylike background, the (1) _____ _____ . (2) _____ connective tissue provides strong connections between various body parts. (3) _____ connective tissue has both (4) _____ and (5) _____ fibers and supports epithelia and organs. Energy reserve and organ padding is the function of (6) _____ tissue. The shape and flexibility of the ear and nose are maintained by a connective tissue called (7) _____ . Support, protection, and movement are functions of (8) _____ . (9) _____ transports cellular oxygen and wastes, provides protection against disease, and forms clots.

23-III
(pp. 339–341)

OVERVIEW OF ANIMAL TISSUES (cont.)
    Muscle Tissue
    Nervous Tissue

Summary

Nervous tissue is composed of nerve cells, or neurons, that extend throughout the body and function as a communication network. Different types of neurons exist; some (sensory neurons) detect changes inside and outside the body, while others (motor neurons) coordinate muscles and glands as they respond to these changes.
    Muscle tissue is composed of cells capable of contracting (shortening) with stimulation and then returning to a relaxed state (lengthening). Muscles move the body and its parts. *Smooth* muscle tissue (involuntary muscle) is composed of spindle-shaped cells bound by connective tissue. It is found in blood vessel walls, the stomach, and other internal organs. Long cylindrical cells wrapped in bundles make up *skeletal* muscle tissue. A muscle is several bundles wrapped in a connective tissue sheath. The contractile tissue of the heart is *cardiac* muscle. Adjacent cardiac muscle cells are fused at their plasma membranes and connected by cell junctions; this allows all cardiac muscle cells to react as one unit.

Key Terms

nervous tissue	contract	skeletal muscle
neurons	smooth muscle	muscle
muscle tissue	involuntary muscle	cardiac muscle

Objectives

1. Define *neuron* and describe the basic features of a nervous system.
2. Describe the relationships among muscle cells, muscle bundles, and muscles.
3. Distinguish among smooth, skeletal, and cardiac muscle tissue in terms of location, structure, and function.

**Self-Quiz
Questions**

True-False

If false, explain why.

___ (1) Muscle bundles are identical to skeletal muscle cells.

_____ (2) Body muscles are constructed of cardiac muscle.

_____ (3) Cardiac muscle cells are fused, end-to-end, at regions called *muscle bundles.*

_____ (4) Smooth muscle tissue is composed of spindle-shaped cells enclosed in bundles.

_____ (5) Smooth muscle tissue contracts voluntarily.

_____ (6) Neurons relay signals to other neurons or to muscles or glands.

23-IV
(pp. 341–344)

MAJOR ORGAN SYSTEMS
HOMEOSTASIS AND SYSTEMS CONTROL
    The Internal Environment
    Mechanisms of Homeostasis
SUMMARY

Summary

The four basic types of tissues are organized into the organ systems of vertebrates. Coordination of these systems at the organism level works for the survival of all the cells in the organism. The major organ systems of the human body are integumentary, muscular, skeletal, nervous, endocrine, circulatory, lymphatic, respiratory, digestive, urinary, and reproductive.

The internal environment of the body is one in which each of the trillions of cells must be bathed with fluid that brings nutrient supplies and removes metabolic wastes. Fluid outside of the cells is known as *extracellular fluid.* Interstitial fluid is that found between cells. Plasma is the fluid portion of blood enclosed within the circulatory system. It is the interstitial fluid that exchanges substances between blood and the cells it bathes. The composition and volume of extracellular fluid are important to the survival of individual cells of the body. Individual cells, organs, and organ systems work to maintain a stable internal environment to ensure survival of individual cells. *Homeostasis* refers to the stable chemical and physical operating conditions of the internal environment.

Organisms maintain internal conditions through systems of homeostatic control, which try to preserve physical and chemical aspects of the body's interior within specific ranges of tolerance. In *negative feedback mechanisms,* deviations from stable conditions activate reactions that work to counteract or reverse the change. *Positive feedback mechanisms,* in which some disturbance to the homeostatic state activates a chain of events that throws things even further off balance, are also called into play in such situations as the formation of a blood clot or the birth process. Homeostatic feedback mechanisms have three basic components: receptors, integrators, and effectors. *Receptors* detect specific types of stimuli (energy changes in the environment). *Integrators* are control points where responses to the stimulus are selected and signals are sent to the organism's *effectors* (the muscles and glands), which respond to the stimuli.

Key Terms

extracellular fluid	negative feedback	receptor
interstitial fluid	mechanism	integrator
plasma	positive feedback	effector
homeostasis	mechanism	

Objectives

1. Tell why a stable internal environment is important to animal life.
2. Define *homeostasis* and cite an example.
3. Distinguish between *negative* and *positive feedback mechanisms*. Give an example of each in human physiology.
4. Describe the relationships among receptors, integrators, and effectors in a negative feedback system.

**Self-Quiz
Questions**

Fill-in-the-Blanks

Each (1) _____ of an animal's body carries on basic (2) _____ activities to ensure its own survival. At the same time, cells of a particular (3) _____ accomplish one or more activities that contribute to the survival of the entire (4) _____ . Collective contributions of cells, organs, and organ systems maintain the (5) _____ internal environment of extracellular fluid necessary for the survival of each (6) _____ in the organism. The term (7) _____ refers to stable operating conditions in the internal environment of the organism. A(n) (8) _____ feedback mechanism occurs when sensing a change in the internal environment provokes a response to return conditions to the original condition. Sexual arousal is an example of a(n) (9) _____ feedback mechanism in which the original condition is intensified. Homeostatic control mechanisms operate with three components: (10) _____ , which detect environmental changes, an (11) _____ , where various pieces of information are analyzed and a response is selected, and (12) _____ such as muscles or glands that accomplish the response.

True-False

If false, explain why.

___ (13) The process of childbirth is an example of a negative feedback mechanism.

___ (14) The human body's receptors are, for the most part, muscles and glands.

___ (15) An integrator is constructed in such a way that it detects specific energy changes in the environment and relays messages about them to a receptor.

— (1) Chemical and structural bridges link groups or layers of like cells, uniting them in structure and function as a cohesive _____ .

(a) organ
(b) organ system
(c) tissue
(d) cuticle

— (2) The cells making up the four kinds of tissues in an animal body are collectively called _____ cells; the only cells that give rise to sperms and eggs in an animal's body are called _____ cells.

(a) somatic; somatic
(b) germ; somatic
(c) germ; germ
(d) somatic; germ

— (3) Which of the following is the correct order of organization?

(a) cells → organs → tissues → organ systems
(b) cells → tissues → organs → organ systems
(c) tissues → cells → organ systems → organs
(d) tissues → organ systems → cells → organs

— (4) Which of the following is *not* included in connective tissues?

(a) bone
(b) blood
(c) cartilage
(d) collagen
(e) skeletal muscle

— (5) Which of the following is *not* included in connective tissue proper?

(a) adipose tissue
(b) epithelium
(c) dense tissue
(d) loose tissue

— (6) Blood is considered to be a(n) _____ tissue.

(a) epithelial
(b) muscular
(c) connective
(d) none of the above _____

— (7) From smallest structure to largest, which group is arranged correctly?

(a) muscle cells, muscle bundle, muscle
(b) muscle cells, muscle, muscle bundle
(c) muscle bundle, muscle cells, muscle
(d) none of the above

___ (8) Involuntary muscle whose cells are spindle-shaped and found in blood vessel walls is _____ .

(a) cardiac
(b) skeletal
(c) striated
(d) smooth

___ (9) Endocrine and exocrine glands are associated with _____ .

(a) connective tissues
(b) muscle tissues
(c) epithelial tissues
(d) nervous tissue

___ (10) The secretion of tears, milk, sweat, and oil are functions of _____ tissue.

(a) epithelial
(b) loose connective
(c) lymphoid
(d) nervous

___ (11) Connective tissue that cushions and helps maintain shape of body parts such as the ears and the nose is _____ .

(a) fatty tissue
(b) connective tissue proper
(c) cartilage
(d) bone

___ (12) A stable internal fluid environment is due to the collective results of the concept known as _____ .

(a) monitoring
(b) chemical control
(c) homeostasis
(d) physical control

___ (13) An animal that feels heated from the sun moves to an environment that tends to cool its body. This is an example of _____ .

(a) intensifying an original condition
(b) positive feedback mechanism
(c) positive phototropic response
(d) negative feedback mechanism

INTEGRATING AND APPLYING KEY CONCEPTS

Explain the responses of your body when you suddenly encounter very cold or very warm environmental temperatures.

24

PROTECTION, SUPPORT, AND MOVEMENT

INTEGUMENTARY SYSTEM
  Functions of Skin
  Structure of Skin
SKELETAL SYSTEM
  Functions of Bones
  Characteristics of Bone
  Skeletal Structure
  Joints
  Commentary: On Runner's Knee

MUSCULAR SYSTEM
  Comparison of Muscle Tissues
  Skeletal Muscle Contraction
  Control of Contraction
  Commentary: Athletes and Anabolic
  Steroids
SKELETAL-MUSCULAR INTERACTIONS
SUMMARY

General Objectives

1. Explain the function and structure of the integumentary system.
2. Describe the skeletal system and its major components; identify human bones by name and location as shown in Figure 24.8 in the main text.
3. Describe the structural causes of *runner's knee.*
4. Tell how skeletal and muscular systems work together.
5. Explain in detail the structure of muscles, from the molecular level to the organ systems level. Then explain how biochemical events occur in muscle contractions and how antagonistic muscle action refines movements.
6. Contrast the benefits and risks involved in the use of anabolic steroids.

24-I
(pp. 346–348)

INTEGUMENTARY SYSTEM
  Functions of Skin
  Structure of Skin

Summary

The *integumentary system* is composed of the skin and structures derived from it (for example, hair, nails, and oil, milk, and sweat glands). The skin: (1) protects the body from abrasion and microbial attack, (2) controls internal temperature, (3) serves as an emergency reservoir of blood, (4) produces vitamin D used in calcium metabolism, and (5) houses nerve endings that provide the brain with information from inside and outside the body.

The skin has two discrete layers, a covering *epidermis,* and an underlying cushioning layer of dense connective tissue, the *dermis.* Beneath these layers is the hypodermis, an anchoring, flexible region that stores some fat, provides insulation, and protects some internal organs. Skin structure varies greatly in vertebrates. The thickness of the skin varies on different areas of the human body. The skin of fishes may be covered with hard scales and may secrete mucus. The epidermis of the skin may produce scales, feathers, hair, beaks, hooves, horns, claws, nails, or quills in various vertebrate animals.

The skin surface loses millions of cells daily by abrasion. The top layer of skin is replaced every thirty-five to forty-five days. After being newly formed, cells in the mid-epidermal regions die and undergo *keratinization*. Such cells are filled with a water-insoluble protein, keratin. Rapid cell division that produces new skin cells repairs abrasions, cuts, or burns rather quickly. Keratinized cells at the skin surface protect against drying, bacterial attack, and poisonous substances.

Melanin, hemoglobin, and carotene are three pigments that provide the skin with its color. Melanin, formed in deep epidermis and brown-black, protects the skin against ultraviolet radiation. Pale skin, with little melanin, receives a pinkish cast from hemoglobin. Carotene is a yellow-orange pigment abundant in the skin of most Asians; it is formed in the upper epidermis.

Excessive exposure to ultraviolet radiation is damaging to skin. The skin changes its structure and appears leathery; the immune system is suppressed, and proto-oncogenes may be activated to initiate epidermal cancers.

The dense connective tissue of the dermis cushions the body. Dermal tears cause the abdominal "stretch marks" resulting from pregnancy. Hairs are structured of overlapping, dead, keratinized cells; they grow from dermis-embedded follicle cavities. The average scalp has about 100,000 hairs. The quality of hair growth is influenced by nutrition and hormones. Skin becomes thinner and more susceptible to injury as we age; the skin becomes drier due to less glandular secretion, and elasticity is lost as collagen and elastin fibers break in the dermis. Environmental effects of wind and tobacco smoke may also accelerate skin aging.

Key Terms

skin	keratinization	carotene
integumentary system	melanin	collagen
epidermis	hemoglobin	elastin fibers
dermis		

Objectives

1. Explain what is included in the *integumentary system*.
2. List the functions of the integumentary system.
3. Give the names of the three skin regions, their structure, and function.
4. Explain *keratinization* and its value.
5. List the three pigments that contribute to skin color and tell how each can influence different skin colors.
6. Describe the structure of hair and tell what factors can influence hair quality.
7. List factors that may cause the "aging" of skin.

Self-Quiz Questions

Matching

Match the terms about the skin appropriately.

(1) __ integumentary system

(2) __ epidermis

(3) __ dermis

(4) __ hypodermis

(5) __ keratinization

(6) __ melanin, hemoglobin, and carotene

(7) __ hair

(8) __ dermal tears

(9) __ skin aging

A. Dying cells filled with a water-insoluble protein
B. Cushion just under the epidermis
C. Consists of dead, overlapping, keratinized cells
D. Skin becomes thinner, proteins break down
E. Skin pigments
F. Skin and its derivatives
G. Anchor skin, a layer beneath the dermis
H. Skin covering
I. Stretch marks

24-II
(pp. 348–352)

SKELETAL SYSTEM
   Functions of Bones
   Characteristics of Bone
   Skeletal Structure
   Joints
   Commentary: **On Runner's Knee**

Summary

The skeletal system provides for: (1) movement, (2) protection, (3) support, (4) mineral storage, and (5) blood cell formation. Human bones vary greatly and are classified as long, short (or cubelike), flat, and irregular.

Bone is a connective tissue composed of living cells and collagen fibers hardened by calcium salts; these are distributed in a ground substance. *Spongy bone* is made up of tiny, needlelike parts but is firm. *Red marrow*, a major site of blood cell formation, fills the spongy tissue. *Yellow marrow* produces red blood cells when blood loss is severe and is also found in the interior cavities of most mature bones. Dense, compact bone surrounds the spongy tissue, resists mechanical shock, and forms the shaft of long bones. *Haversian canals*, surrounded by concentric layers of mineralized matrix, form channels for blood vessels and nerves.

Long bones form in embryonic cartilage models through osteoblast deposition of calcium. Cartilage cavities merge to form the marrow cavity; osteoblasts become trapped in their own secretions and become osteocytes.

Mineral turnover is characteristic of living bone; minerals are constantly deposited and withdrawn. Such turnover is characteristic of young forming bones that change their shape during growth. Bone mineral turnover helps maintain body calcium levels but turnover ability can deteriorate with age; this may leave some areas short in calcium. *Osteoporosis* (bone mass decrease and collapse) is part of the syndrome of aging.

Both bone and cartilage are found in living vertebrate skeletons. Humans have a total of 206 bones in two different parts of the skeleton. The *axial skeleton* includes the skull, vertebral column, ribs, and sternum. The *appendicular skeleton* includes the bones of arms, hands, legs, feet, pelvic girdle, and pectoral girdle. The backbone, extending from the skull to the pelvic girdle, is curved, flexible, and supported by muscles and ligaments. The backbone transmits the torso weight to the lower limbs. The spinal cord is found in a cavity formed by the column of individual bony vertebrae. *Intervertebral disks* of cartilage serve as shock absorbers between vertebrae. Herniated disks are those that slip out of place and cause painful, undue pressure on the spinal cord or its branching nerves.

The rather weak pectoral girdle consists of two large, flat shoulder blades, and two long, slender collarbones that connect to a breastbone.

Joints are areas of contact between bones. Freely movable *synovial joints* are the most common. A flexible capsule of dense connective tissue holds the joint bones close; cells lining the capsule secrete a lubricating fluid into the joint. Freely movable joints such as the human knee are easily damaged. *Osteoarthritis* is excessive wear of the end-bone cartilage of freely movable joints. Genetic *rheumatoid arthritis* is a degenerative disorder where the capsule lining of the joint thickens and becomes inflamed, the cartilage degenerates, and bone is deposited in the joint. Joints between vertebrae and those between ribs and breastbone are *cartilaginous joints,* where only slight movements are permitted due to cartilage filling the space between bones. Fibrous tissue unites the bones without cavities. Loosely connected, flat, fetal skull bones are connected by *fibrous joints;* these "soft spots" harden during adulthood.

"Runner's knee" refers generally to various disruptions of the bone, cartilage, muscle, tendons, and ligaments at the knee joint. The knee joint permits considerable movement. Wedges of cartilage between the femur and tibia add stability and act like shock absorbers for the weight placed on the joint. When the knee joint is hit hard or twisted too much, its cartilage can be torn; the cartilage is surgically removed to prevent arthritis. The seven ligaments that strap the femur and tibia together are also vulnerable to injury. A ligament is composed of many connective tissue fibers. Severed ligaments must be surgically repaired within ten days of injury. The fluid that lubricates the knee joint contains phagocytic cells that remove debris from day-to-day wear and tear. These same cells can work on torn ligaments and turn them to mush.

Key Terms

bone	bone turnover	synovial joint
red marrow	osteoporosis	osteoarthritis
yellow marrow	axial skeleton	rheumatoid arthritis
Haversian canal	appendicular skeleton	cartilaginous joint
osteoblasts	intervertebral disk	fibrous joint
osteocyte		

Objectives

1. List the major functions of bones.
2. Describe the structure of bone.
3. Generally describe how bones develop.
4. Explain what is meant by *bone turnover* and explain its possible effects in the human aging process.
5. Distinguish between the *axial skeleton* and the *appendicular skeleton.*
6. Describe the function of *intervertebral disks.*
7. List characteristics and give an example of the *synovial joint,* the *cartilaginous joint,* and the *fibrous joint;* explain why the structure of the synovial joint is easily disrupted (see *Commentary,* p. 352).
8. Distinguish between *osteoarthritis* and *rheumatoid arthritis.*

Self-Quiz
Questions

Fill-in-the-Blanks

The parts of the skeleton function in (1) _____ , (2) _____ ,

(3) _____ , (4) _____ storage, and (5) _____ _____

formation. Human bones have various sizes and (6) _____ . Bone is a(n)

(7) _____ tissue; it has living cells and (8) _____ fibers distributed

throughout a(n) (9) _____ substance. In many bones, (10) _____

marrow fills the spongy tissue and is a major site of (11) _____ _____

formation. (12) _____ marrow also is found in the interior cavities of

mature bones; it produces (13) _____ blood cells in cases of severe

blood loss. Long bones form in cartilage models in the (14) _____ .

(15) _____ secrete material inside the shaft of the cartilage model.

Osteoblasts trapped in their own secretions are called (16) _____ .

Minerals are constantly deposited and withdrawn from living bone; this is

called bone tissue (17) _____ . In aging persons, particularly women,

deterioration of bone tissue turnover can result in a condition known as

(18) _____ . Humans have (19) _____ (number) bones and the

skeleton is divided into the (20) _____ skeleton and the (21) _____

skeleton. Pieces of cartilage found between the bony parts of the vertebrae in

a backbone are called (22) _____ _____ . The most common type of

joint is the (23) _____ joint. (24) _____ joints occur between

vertebrae and between breastbone and ribs. (25) _____ joints loosely

connect the flat skull bones of a fetus.

Labeling

Identify each indicated part of the accompanying illustration (p. 268).

(26) _____

(27) _____

(28) _____

(29) _____

(30) _____

(31) _____

(32) _____

(33) _____

(34) _____

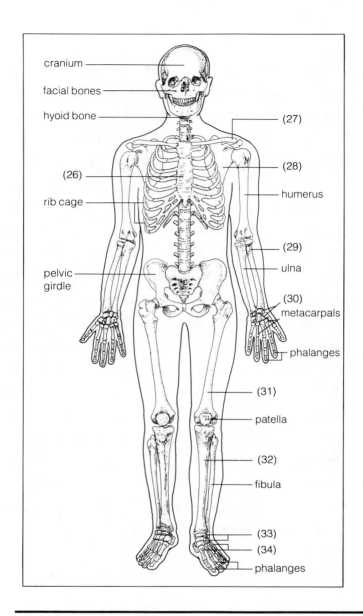

cranium

facial bones

hyoid bone

(26)

rib cage

pelvic
girdle

(27)

(28)

humerus

(29)

ulna

(30)

metacarpals

phalanges

(31)

patella

(32)

fibula

(33)

(34)

phalanges

24-III
(pp. 353–359)

MUSCULAR SYSTEM
    Comparison of Muscle Tissues
    Skeletal Muscle Contraction
    Control of Contraction
    Commentary: **Athletes and Anabolic Steroids**
SKELETAL-MUSCULAR INTERACTIONS
SUMMARY

Summary

The three types of muscle tissues are skeletal, cardiac, and smooth muscle. All three types of muscle tissues are excitable, contractile, and elastic. *Excitable* cells can reverse electric gradients in their plasma membranes if stimulated adequately. *Contractile* cells shorten in response to adequate stimulation. *Elastic* cells can be extended to regain the original shape they had prior to contraction. In vertebrates, smooth muscle occurs primarily in the walls of internal organs; cardiac muscle makes up the heart, and skeletal muscle interacts with the skeleton to accomplish body movement.

A skeletal muscle of the human body is composed of thousands of muscle cells; each contains many threadlike *myofibrils.* The protein filaments *actin* and *myosin* are packed in each myofibril and thus are part of each *sarcomere* or unit of contraction. Skeletal muscles have a striped appearance due to boundaries between sarcomeres, the Z lines. A muscle shortens because sarcomeres within muscle cells shorten.

The *sliding-filament model* explains how sarcomeres alternately contract and relax. Actin filaments slide over myosin filaments (actin moves toward the center of the sarcomere during contraction and away during relaxation). Sliding movement depends on forming cross-bridges between "head" ends of myosin molecules and the binding sites on actin. ATP energy and a series of power strokes by myosin heads in each sarcomere cause a complete muscle to contract. Without ATP, cross-bridges never detach and muscles become rigid (rigor). *Rigor mortis* following death is explained by such locked cross-bridges.

A multitude of mitochondria in muscle cells carry on aerobic respiration to produce enough ATP to support extended contraction periods; oxygen supply must be adequate. Glycogen, a starch, is the major fuel supply for cells consuming ATP rapidly. When oxygen levels decline during short periods of strenuous muscle activity, glycolysis is used to synthesize ATP.

The nervous system signals skeletal muscles to contract. Action potentials travel along the plasma membrane of the muscle cell, entering the cell by a system of tubes that connect with the *sarcoplasmic reticulum,* a system of membranes that surrounds the cell's myofibrils. Sarcoplasmic reticulum stores calcium ions, but releases them in response to action potentials and they then bind to actin sites. Cross-bridges then form. Following contraction, calcium ions are taken up and stored in the sarcoplasmic reticulum; the muscle then relaxes. The nervous system controls muscle contraction by controlling the action potentials that reach the sarcoplasmic reticulum and thus control muscle calcium ion levels.

By the 1960s, many athletes began using anabolic steroids to produce enhanced masculinizing effects; the practice is still going on today. The advantages cited by athletes include increased muscle mass and strength, increased oxygen-carrying capability, and an increase in aggressive behavior. Present evidence indicates that use of the drugs does not enhance athletic performance. Nonusers handily outperform users in athletic endeavors requiring fine muscle coordination and endurance. Physicians say the risks outweigh the benefits. Use of steroids causes bloated faces, shriveled testes, infertility, liver damage and liver cancer, and changes in blood cholesterol levels that predispose long-term users to coronary heart disease. In addition, it now appears that the psychiatric hazards of anabolic steroids may be equally threatening; about one-third of anabolic steroid users have serious mental problems.

The human body has more than 600 muscles arranged as pairs or in groups. Some work together but some oppose the action of others. Contracting muscles transmit force to the bones they are attached to. The skeleton and attached muscles form lever systems in which rigid rods (bones) move about at fixed points (joints). Rotation of a limb around a joint can occur because of arrangements of muscle pairs or groups. Biceps and triceps of the arm are an example of muscles arranged in antagonistic pairs.

Key Terms

skeletal muscle	contract	myosin
cardiac muscle	elastic	sarcomere
smooth muscle	myofibril	sliding-filament model
excitability	actin	rigor mortis

glycogen   action potential   antagonistic muscle pair
sarcoplasmic reticulum   calcium ions

Objectives

1. List the three types of muscle tissue, where they occur in the human body, and the three properties they have in common.
2. Draw a picture of the fine structure of a muscle cell; identify terms such as *myofibril, sarcomere, Z-lines, actin,* and *myosin.*
3. List, in sequence, the biochemical and fine structural events that occur during the contraction of a skeletal muscle fiber. Then explain how the fiber relaxes.
4. Describe what fuels muscle contraction during nonstrenuous and strenuous activity.
5. Explain how the sarcoplasmic reticulum, calcium ions, and cross-bridges are involved in muscle contraction.
6. In general, understand how pairs of muscles attached to bones act like lever systems and what this accomplishes for the body.

Self-Quiz Questions

Fill-in-the-Blanks

The three types of muscle tissue are (1) _____ , (2) _____ , and (3) _____ . Muscle cells have three properties in common: they are excitable, they contract, and they are (4) _____ . Each muscle cell contains many threadlike (5) _____ . In these there are two types of protein filaments, (6) _____ and (7) _____ . These protein filaments are components of (8) _____ , the basic units of contraction. Sarcomeres alternately contract and relax according to the (9) _____ -_____ model. The sliding movement depends on the formation of (10) _____ -_____ between actin and myosin. An input of energy from (11) _____ breakdown causes the myosin heads to tilt in a short power stroke. The (12) _____ filaments attached to the myosin heads also move toward the center of the sarcomere. Many muscle cells have abundant (13) _____ , which produce ATP through aerobic respiration. During strenuous activity, when oxygen levels are low in muscle tissues, the rate of (14) _____ increases and ATP is synthesized. A membranous system of tubes, the (15) _____ _____ , stores calcium ions and releases them in response to action potentials; calcium ions bind to sites on the (16) _____ filaments, and this allows cross-bridges to form. Together, the skeleton and the muscles attached to it are like a system of (17) _____ in which rigid rods (bones) move about at fixed points (joints).

Labeling

Identify each indicated part of the accompanying illustration.

(18) _____

(19) _____   _____

(20) _____

(21) _____

(22) _____   _____

(23) _____   _____

(24) _____

(25) _____   _____

(26) _____   _____

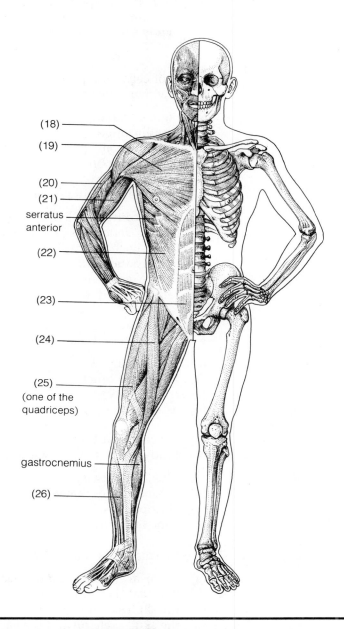

(18) ―
(19) ―
(20) ―
(21) ―
serratus anterior ―
(22) ―
(23) ―
(24) ―
(25) ― (one of the quadriceps)
gastrocnemius ―
(26) ―

CHAPTER TEST

UNDERSTANDING AND INTERPRETING KEY CONCEPTS

____ (1) Which of the following is *not* part of the integumentary system?

    (a) skin
    (b) hair
    (c) nails
    (d) oil and sweat glands
    (e) all of the above *are* part of the integumentary system

____ (2) From outside to inside, which is the correct order of skin regions?

  (a) epidermis $\rightarrow$ hypodermis $\rightarrow$ dermis
  (b) dermis $\rightarrow$ hypodermis $\rightarrow$ epidermis
  (c) epidermis $\rightarrow$ dermis $\rightarrow$ hypodermis
  (d) hypodermis $\rightarrow$ dermis $\rightarrow$ epidermis

____ (3) Death of cells in mid-epidermal regions accompanied by subsequent filling with a water-insoluble protein is called _____ .

  (a) pigmentization
  (b) keratinization
  (c) melanization
  (d) epidermization

____ (4) Which of the following is *not* true?

  (a) A hair consists of living keratinized cells.
  (b) Protein deficiency causes hair to thin.
  (c) Excessive hairiness in women may result from abnormal testosterone production.
  (d) Severe fever, emotional stress, and excessive vitamin A intake can cause hair thinning.

____ (5) In many bones, _____ fills the spongy tissue, which serves as a major site of blood cell formation.

  (a) yellow marrow
  (b) Haversian canals
  (c) red marrow
  (d) osteocytes

____ (6) The concept that bone tissue is like a bank from which minerals are constantly deposited and withdrawn is known as _____ .

  (a) osteocyte activity
  (b) red marrow exchange
  (c) osteoporosis
  (d) bone tissue turnover

____ (7) The two major divisions of your skeleton are _____ .

  (a) the axial skeleton and the backbone skeleton
  (b) the axial skeleton and the appendicular skeleton
  (c) the skull skeleton and the appendicular skeleton
  (d) the skull skeleton and the backbone skeleton

____ (8) The pectoral girdle is rather weak and its most frequently broken bone is the _____ .

  (a) collarbone
  (b) breastbone
  (c) shoulder blade
  (d) long bone of the upper arm

For questions 9 to 11, choose from these answers:

(a) cartilaginous
(b) fibrous
(c) synovial

___ (9) Vertebral disks with small amounts of movement are examples of _____ joints.

___ (10) Nonmoving joints between skull bones are examples of _____ joints.

___ (11) Freely movable bones that are separated by a fluid-filled cavity compose a _____ joint.

___ (12) The basic unit of muscle contraction is _____ .

(a) the sarcomere
(b) the myofibril
(c) actin
(d) myosin
(e) none of the above

INTEGRATING AND APPLYING KEY CONCEPTS

If humans had an exoskeleton rather than an endoskeleton, would they move differently from the way they do now? Name any advantages or disadvantages that having an exoskeleton instead of an endoskeleton would present in human locomotion and growth.

25

DIGESTION AND HUMAN NUTRITION

DIGESTIVE SYSTEM
    Human Digestive System: An Overview
    Into the Mouth, Down the Tube
    The Stomach
    The Small Intestine
    The Large Intestine
HUMAN NUTRITIONAL REQUIREMENTS
    Energy Needs and Body Weight

Commentary: Extreme Eating
Disorders—Anorexia Nervosa and
Bulimia
Carbohydrates
Lipids
Proteins
Vitamins and Minerals

General Objectives

1. Give a complete definition of *nutrition*.
2. Give the general functions of a digestive system.
3. List the parts of the human digestive system in order.
4. Discuss ways the digestive system controls its operation in response to volume and composition of food passing through.
5. Generally cite the digestive functions accomplished by the various parts of the human digestive system.
6. Explain what factors maintain acceptable weight and overall health.

25-I
(pp. 361–362)

DIGESTIVE SYSTEM
    Human Digestive System: An Overview

Summary

The *digestive system* breaks down foodstuffs to molecules small enough to move into the internal environment of the animal body. The digestive system functions with the circulatory system, the respiratory system, and the urinary system to meet the body's metabolic needs. Most animals have a digestive system composed of an internal tube or cavity that serves the following functions: movement, secretion, digestion, and absorption.

    The adult human digestive system is a tube composed of specialized regions that stretches out to 6.5 to 9 meters (21 to 30 feet) in length. The regions are the mouth (oral cavity), pharynx, esophagus, stomach, small intestine, large intestine (colon), rectum, and anus (the term *gut* includes parts from the stomach down). Secretions containing enzymes and other digestive substances are produced by the salivary glands, stomach, liver, gallbladder, small intestine, and pancreas. The digestive tube is composed of two smooth muscle layers enclosed in an outer sheath of connective tissue and an inner lining. *Peristalsis* mixes and moves food through the digestive tube. Rings of muscle, *sphincters*, in the tube wall mark the beginning and end of the stomach and other special regions of the digestive tube. Sphincters also control forward food movement and prevent backflow. Digestive system action is controlled by the nervous and endocrine systems, as well as by food volume.

Key Terms

nutrition	pharynx	salivary glands
digestive system	esophagus	liver
movement	stomach	gallbladder
secretion	small intestine	pancreas
digestion	large intestine	peristalsis
absorption	rectum	sphincter
mouth	anus	

Objectives

1. List the general functions of a digestive system.
2. List all parts of the digestive tube beginning with the mouth; cite the names of structures that secrete enzymes and other substances into the digestive tube.
3. Describe peristalsis and tell what it accomplishes.
4. Define *sphincters*, their locations, and functions.

Self-Quiz Questions

Fill-in-the-Blanks

The (1) _____ system reduces foodstuffs to molecules small enough to enter the animal's internal environment. The (2) _____ system transports absorbed nutrients throughout the body. The (3) _____ system helps cells use the nutrients by supplying them with oxygen and carrying away carbon dioxide waste. The (4) _____ system helps maintain volume and composition of extracellular fluid. The digestive system has four major functions: (5) _____ is the breakdown, mixing, passage, and elimination of ingested food; (6) _____ refers to the release of enzymes and other substances required for digestive function; (7) _____ is the chemical breakdown of nutrients; (8) _____ is the passage of digested nutrients across the digestive tube wall into blood or lymph. Contractions of the muscle layers in the digestive tube that mix and move food are called (9) _____ . (10) _____ are controlling rings of muscle in the digestive tube that mark the beginning and ending of special regions.

Sequence

Arrange in order from the head to the opposite end of the digestive tube.

(11) ___   A. Small intestine
(12) ___   B. Pharynx
(13) ___   C. Rectum

(14) ___   D. Mouth

(15) ___   E. Anus

(16) ___   F. Stomach

(17) ___   G. Large intestine

(18) ___   H. Esophagus

25-II
(pp. 363–364)

DIGESTIVE SYSTEM
Into the Mouth, Down the Tube
The Stomach

Summary

Polysaccharide digestion begins in the *mouth* as teeth accomplish chewing; only humans and other mammals chew food. The thirty-two adult human teeth are extremely sturdy with a calcium-hardened enamel coat, a bonelike dentine layer, and an inner pulp with nerves and blood vessels. Incisor teeth bite off food chunks, cuspids tear, and molars grind. Saliva (containing amylase, bicarbonate, and mucins) from the *salivary glands* mixes with food in the mouth. Carbonate buffers the mouth pH between 6.5 and 7.5; mucins bind food bits into a lubricated ball. Tongue muscles contract and move the food ball into the muscular *pharynx*, which connects with the *esophagus*. The stimulated pharynx triggers swallowing contractions that continue through the esophagus to move food to the stomach. The trachea also opens into the pharynx; the epiglottis closes over the trachea and prevents choking as a person swallows food.

The stomach, with muscular, stretchable walls, stores and mixes food, secretes substances that dissolve and degrade food, and controls the movement of food into the small intestine. The stomach daily produces 2 liters of secretions that include hydrochloric acid (dissolves food and kills microorganisms), pepsinogens, and mucus. Food is mixed with fluid in the stomach and converted to a slurry called *chyme*. Brain responses to food stimuli initiate stomach secretion, which continues with food in the stomach; food stretching the stomach stimulates increased secretions of substances. Protein digestion begins in the acid environment of the stomach where pepsinogens are converted to pepsins that break down proteins. Protein fragments stimulate secretion of gastrin that in turn stimulates additional HCl secretion to accommodate additional protein intake. The stomach lining is protected by mucus and buffering (bicarbonate) molecules from being digested. When these normal protective controls fail and the stomach lining breaks down, H^+ diffuses into the stomach lining and triggers the release of histamine from tissue cells. Histamine affects local blood vessels and HCl secretion increases; tissue damage and bleeding may result in a *peptic ulcer*.

Peristaltic waves along the stomach wall mix chyme in contractile movements toward a sphincter between the stomach and small intestine; strong contractions close the sphincter, but small amounts continue to move into the small intestine. The volume and composition of chyme control the rate of stomach emptying. Food is not moved along faster than the stomach can process it.

Key Terms

mouth
salivary glands
pharynx

esophagus
stomach

epiglottis
peptic ulcer

Objectives

1. Explain the role of the mouth, teeth, and salivary glands in the human digestive tract.
2. Tell how a ball of food enters the pharynx and moves to the esophagus.
3. Describe the mechanism that prevents food from entering the trachea.
4. Explain the chemical and mechanical role of the stomach in the digestion of food.
5. Describe the mechanisms that prevent the stomach from digesting itself.
6. Explain how the rate at which food is passed from the stomach to the small intestine is regulated.

Self-Quiz Questions

Fill-in-the-Blanks

Digestion begins in the (1) _____ . Adult humans normally chew food with (2) _____ (number) teeth. Saliva contains an enzyme called salivary (3) _____ that digests starch. The pH of the mouth remains between 6.5 and 7.5 due to (4) _____ in saliva. (5) _____ in saliva bind bits of food into a softened, lubricated ball. Muscular contractions of the tongue force the ball of food into the (6) _____ whose contractions move it into a tube, the (7) _____ , that delivers food to the stomach. Choking on food is prevented because a flaplike valve, the (8) _____ , closes off the trachea while food moves into the esophagus. HCl, pepsinogens, and mucus make up the (9) _____ fluid in the stomach. Food stimuli received by the brain and the presence of food in the stomach promote the occurrence of stomach (10) _____ . (11) _____ and (12) _____ molecules prevent the stomach from digesting itself. A(n) (13) _____ ulcer results from tissue damage and bleeding of the stomach lining. At the junction of the stomach and the small intestine, a(n) (14) _____ plays a role in controlling the emptying of the stomach.

25-III
(pp. 364–366)

DIGESTIVE SYSTEM (cont.)
The Small Intestine
The Large Intestine

Summary

The *small intestine* is the area of the digestive tube where digestion is completed and most nutrients are absorbed. Secretions from the *pancreas* and *liver* enter the small intestine through a common duct. Approximately nine liters of fluid enter the small intestine daily from the stomach, liver, and pancreas.

Enzymes from the pancreas act on carbohydrates, fats, proteins, and nucleic acids; bicarbonate from the pancreas buffers stomach HCl. Insulin and glucagon are pancreatic hormones that play a role in nutrition but not in digestion.

Bile is secreted by the liver; bile salts assist in the breakdown of fats and their absorption and enhance digestion by allowing fat-degrading enzymes greater access to more triglycerides. Bile is stored and concentrated in the *gallbladder* between meals.

Most of the protein, lipid, and carbohydrate molecules are broken down into smaller molecules by the time they have moved about midway through the small intestine. These smaller molecules include monosaccharides, amino acids, fatty acids, and monoglycerides that can move across the intestinal lining. Absorptive surface area is increased by *villi* that cover the densely folded lining of the small intestine; surface epithelial cells have crowns of *microvilli* that further increase absorptive surface. At each villus, glucose and most amino acids cross epithelial cell plasma membranes by active transport; this is followed by diffusion into extracellular fluid and then to blood vessels in villi. Free fatty acids and monoglycerides diffuse through membranes of epithelial cells and then recombine to form glycerides; droplets leave the cells, enter lymph vessels, and then enter the general blood circulation. Water and mineral ions are absorbed by the villi.

The *large intestine* (colon) receives material not absorbed in the small intestine; here the undigested and unabsorbed material is concentrated by active transport of sodium ions and stored as *feces*. The colon, about 1.2 meters long, leaves the small intestine as a cup-shaped pouch; the *appendix*, with no known functions, projects from the colon. The colon ascends on the right side of the body's abdomen, then to the left side and descends to connect to the *rectum*. The nervous system controls the rectal wall and contractions of a muscle sphincter at the *anus* to expel feces from the body. Additionally, the wall of the large intestine absorbs water and salts. Bacteria in the colon not only synthesize some of the B complex vitamins but they also ferment some of the residual carbohydrates. Fecal material is generally brown because red and green bile pigments have been mixed in with the chyme. The average American diet does not include enough bulk (fiber volume). *Appendicitis* and colon cancer appear to be the consequences of inadequate fiber in the human diet.

Key Terms

small intestine	gallbladder	appendix
pancreas	villus, villi	rectum
liver	microvillus, microvilli	anus
bile	large intestine (colon)	appendicitis
bile salts		

Objectives

1. Describe the digestive and absorptive processes occurring in the small intestine; include the roles of the pancreas and liver.
2. Explain the structure of the large intestine and relate its role in the digestive process by describing processes taking place there.

Self-Quiz Questions

True-False

If false, explain why.

___ (1) Digestion is completed in the large intestine.

_____ (2) Bile is a secretion produced by the liver.

_____ (3) The epithelial cells lining the inside of the small intestine have surfaces covered with threadlike villi.

_____ (4) Bile salts assist in protein breakdown and absorption.

_____ (5) The colon concentrates and stores feces, a mixture of undigested, unabsorbed material, water, and bacteria.

Labeling

Identify each of the numbered structures of the accompanying illustration (p. 280).

(6) _____ _____   (12) _____ _____

(7) _____ _____   (13) _____

(8) _____   (14) _____

(9) _____   (15) _____

(10) _____   (16) _____

(11) _____ _____

25-IV
(pp. 366–371)

HUMAN NUTRITIONAL REQUIREMENTS
Energy Needs and Body Weight
Commentary: Extreme Eating Disorders—Anorexia Nervosa and Bulimia
Carbohydrates
Lipids
Proteins
Vitamins and Minerals

Summary

Body growth and maintenance depend on a supply of energy and materials from certain foods in certain amounts. Nutritionists measure energy in calories; the kilocalorie is 1,000 calories of energy.

Physical, environmental, and genetic factors influence how much metabolic activity and exercise an individual requires in order to balance incoming calories and maintain satisfactory body weight and function; studies of twins support this statement. In most adults, energy input balances output to maintain constant body weight. Imbalances between caloric intake and energy output may cause *obesity*, an excess of fat in the body's adipose tissues. Insurance company charts identify the *ideal* weight by age and height groups; people twenty-five percent heavier than the ideal are viewed as obese. Some researchers studying causes of death suggest that the ideal is actually 10 to 15 pounds heavier than the ideal. Obsessive dieting may lead to an eating disorder of women in their teens and early twenties: anorexia nervosa. This involves self-induced starvation and, frequently, overexercise. Bulimia is another eating disorder affecting mostly college-age women; food intake goes out of control with hour-long eating binges. Uncontrolled, both disorders may lead to death.

Complex carbohydrates are digested into glucose molecules, which are further broken down during cellular respiration. Energy released along the

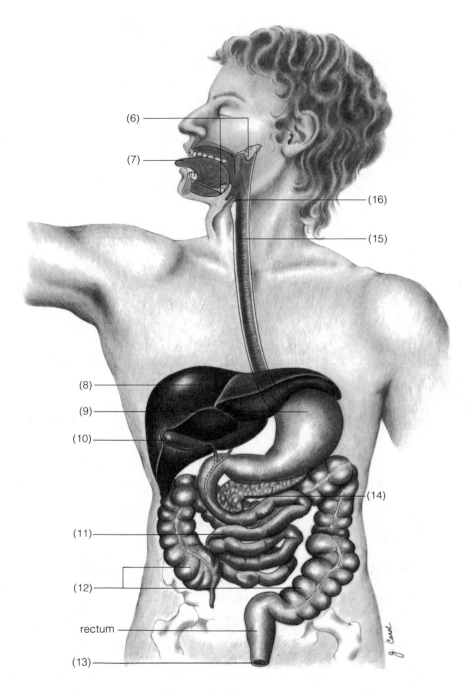

(6)

(7)

(16)

(15)

(8)

(9)

(10)

(14)

(11)

(12)

rectum

(13)

respiratory pathways is quickly captured and stored in the chemical bonds of ATP and used by cells to do necessary work later. Fleshy fruits, cereal grains, legumes, and other fibrous carbohydrates supply glucose and a diet with desirable amounts of fiber. Eating only table sugar (sucrose) supplies calories but not fiber. The average American eats 128 pounds of sucrose each year.

Lipids are important to the body. Phospholipids and cholesterol are components of animal cell membranes. Fats serve as an energy reserve, help absorb fat-soluble vitamins, and cushion various body organs. Lipids make up forty percent of the average American diet; it should be less than thirty percent. The body can manufacture most of its own fats from protein and carbohydrates. An exception is linoleic acid, one of the *essential fatty acids* that must be supplied by the diet. Saturated fats like butter and other animal fats raise blood cholesterol, which contributes to circulatory system disorders.

Protein intake should be about twelve percent of the total human diet. Digested proteins make amino acids available for the body's own protein building programs. Diet must supply the eight *essential amino acids*. Most animal proteins are "complete" in that they contain large amounts of all essential amino acids; plant proteins are "incomplete," which means several different plant sources must be eaten to supply the essential amino acids. Net protein utilization (NPU) is used to compare protein from different sources for essential amino acid content. The proper structure and function of the body depends on proteins. Without intake of sufficient amounts of essential amino acids, abnormal body growth and development are likely; the body cannot construct enzymes and other necessary proteins. Protein deficiency can affect brain growth and development as well as physical performance.

Plant cells manufacture most vitamins; animal cells have lost this ability. Humans need at least thirteen different vitamins that are necessary in specific chemical reactions. *Inorganic minerals* are also necessary for metabolism. Calcium, magnesium, potassium, and iron are examples.

A well-balanced diet of carbohydrates, lipids, and proteins supplies essential vitamins and minerals. Excessive intake of vitamins and minerals may be wasted or harmful to health. Excess vitamin C is eliminated in urine; excess vitamins A and D can cause serious disorders; these are fat-soluble vitamins that accumulate in tissues and interfere with normal metabolism. Excessive intake of sodium in table salt can lead to or aggravate high blood pressure.

Key Terms

calorie	essential fatty acid	calcium
kilocalorie	saturated fats	magnesium
obesity	proteins	potassium
ideal weight	essential amino acids	iron
anorexia nervosa	net protein utilization (NPU)	vitamin C
bulimia	vitamins	vitamin A
carbohydrates	fat-soluble vitamin	vitamin D
lipids	mineral	sodium

Objectives

1. Tell what is meant by *calorie* and *kilocalorie*.
2. Describe what is required for maintaining acceptable body weight and normal function.
3. Give the cause of obesity as discussed in the text.
4. Explain what is meant by *balanced diet* and discuss its importance in maintaining good health.
5. Explain what is meant by *essential fatty acids* and *essential amino acids*; explain their importance in terms of health.
6. Tell the value of a measure nutritionists use called *net protein utilization* (NPU).
7. Distinguish *vitamins* from *minerals*.
8. Explain the role of various vitamins and minerals in healthy metabolic function; give the effects of shortage or massive intake of vitamins and minerals.

Self-Quiz Questions

Fill-in-the-Blanks

The unit equivalent to 1,000 calories of energy is the (1) _____ . To maintain an acceptable weight and keep normal body function, (2) _____ intake must be balanced with (3) _____ output. (4) _____ is an excess of fat in the body's adipose tissues caused by imbalances between (5) _____ intake and (6) _____ output. Complex (7) _____ are the body's main energy sources. Brain, muscle, and other tissues depend on (8) _____ as a primary energy source. Fleshy fruits, cereal grains, legumes, and other (9) _____ carbohydrates should make up fifty-eight to sixty percent of daily caloric intake. Lipids are important to the body as (10) _____ reserves; fat deposits serve as (11) _____ for many body organs. A potentially fatal eating disorder that involves obsessive dieting is called (12) _____ _____ . (13) _____ is an eating disorder involving uncontrollable food intake. A fatty acid such as linoleic acid that cannot be produced by the body is known as a(n) (14) _____ fatty acid. Butter and other animal fats are (15) _____ fats and their intake may raise the level of blood (16) _____ ; this contributes to (17) _____ system disorders. Eight of the twenty common amino acids must be obtained from food; they are the (18) _____ amino acids. Most animal proteins are "complete" in that they contain the essential amino acids, whereas plant proteins are (19) "_____ ." Animal cells have lost the ability to produce organic substances necessary for proper metabolism, the (20) _____ ; animals must obtain them from (21) _____ . Human cells need at least thirteen different (22) _____ , each with specific metabolic roles. Proper metabolism also requires inorganic substances such as calcium and magnesium that are called (23) _____ . Excessive intake of sodium (as from table salt) can lead to high (24) _____ _____ .

25-V
(pp. 371–372)

NUTRITION AND METABOLISM SUMMARY

Summary

Once absorbed into the body, nutrient molecules are shuffled and reshuffled into various metabolic pathways. Carbohydrates, lipids, and proteins are torn apart and their parts are reconstructed into new molecules over and over again. There is a rapid turnover of these molecules in the body. Eating gives the body its supply of organic molecules. Extra carbohydrate molecules are converted to fat, which is stored in adipose tissue; some is converted to glycogen in liver and muscle. Glucose is the main energy source for cells well-supplied with nutrients; brain cells have a priority on available glucose. Between meals blood glucose levels are maintained by glycogen stored in the liver that is converted to glucose and by digested proteins whose amino acids are converted to glucose. These nutritional adjustments are controlled by the nervous and endocrine systems. Many cells use fats (glycerol plus fatty acids) as the main energy source between meals. Glycerol is converted to glucose in the liver and fatty acids may be used to produce ATP.

The liver is important to storage and interconversion of absorbed carbohydrates, lipids, and proteins. The liver also removes toxic substances from the blood and maintains concentrations of organic substances in the blood; it also inactivates hormones before excretion in urine. Ammonia is a cellular waste product from amino acid breakdown; it is carried to the liver in blood where conversion to less toxic urea occurs. Urea leaves the body through the kidneys and urine.

Key Terms

molecular turnover	glycogen	fatty acids
fats	liver	ammonia
adipose tissue	glycerol	urea

Objectives

1. Understand the idea that most carbohydrates, lipids, and proteins undergo a rapid turnover in the body.
2. Explain what happens to excess carbohydrate molecules that are eaten.
3. Discuss the chemical means the body uses to maintain blood glucose levels between meals.
4. List four functions of the liver.
5. List two examples of how chemical functioning of the kidney assists chemical functioning of the liver.

Self-Quiz Questions

Fill-in-the-Blanks

After being absorbed into the body, most carbohydrates, lipids, and proteins are broken down continually, with their component parts picked up and used again in new (1) _____ . After a meal, excess carbohydrates and other dietary molecules are mostly converted to (2) _____ , which is stored in adipose tissue, although some is converted to (3) _____ in the liver and muscle tissue. Between meals there is a shift in the type of food molecules used to support cell activities. A major reason for this is the need to provide brain cells with (4) _____ . At this time, glycogen stored in the liver can be converted to (5) _____ and body proteins are broken down to (6) _____ _____ , which are sent to the liver for conversion to (7) _____ that is released into the blood. Most cells use (8) _____ as the main energy source between meals. Fats stored in adipose tissue are broken down into (9) _____ and (10) _____ acids. The liver helps maintain concentrations of the blood's (11) _____ _____ and removes many (12) _____ substances from it. The liver inactivates most (13) _____ molecules and sends them to the kidneys for excretion. Toxic ammonia is formed when cells break down (14) _____ _____ ; blood carries ammonia to the liver where it is converted to less toxic (15) _____ , which leaves the body with urine.

CHAPTER TEST **UNDERSTANDING AND INTERPRETING KEY CONCEPTS**

___ (1) The process that moves nutrients into the blood or lymph is _____ .

    (a) ingestion
    (b) absorption
    (c) assimilation
    (d) digestion
    (e) none of the above

___ (2) The enzymatic digestion of proteins begins in the _____ .

    (a) mouth
    (b) stomach
    (c) liver
    (d) pancreas
    (e) small intestine

___ (3) The enzymatic digestion of starches begins in the _____ .

    (a) mouth
    (b) stomach
    (c) liver
    (d) pancreas
    (e) small intestine

___ (4) The greatest amount of absorption of digested nutrients occurs in the _____ .

    (a) stomach
    (b) pancreas
    (c) liver
    (d) large intestine
    (e) small intestine

—— (5) Which of the following represents the correct order of the areas of the human digestive system?

    (a) mouth → pharynx → stomach → esophagus → small intestine → large intestine

    (b) mouth → pharynx → esophagus → stomach → small intestine → large intestine

    (c) mouth → pharynx → esophagus → small intestine → stomach → large intestine

    (d) mouth → esophagus → pharynx → stomach → small intestine → large intestine

—— (6) Which of the following is not found in bile?

    (a) lecithin
    (b) salts
    (c) digestive enzymes
    (d) cholesterol
    (e) pigments

—— (7) The colon ascends on the right side of the body's abdominal cavity, crosses to the other side, then descends and connects with a short tube, the _____ .

    (a) appendix
    (b) anus
    (c) rectum
    (d) colon

—— (8) The average American diet does not include enough _____ , which is perhaps related to carcinogenic effects.

    (a) protein
    (b) fiber
    (c) sugar
    (d) amino acids
    (e) starch

—— (9) Obesity is defined as being _____ percent above the ideal weight as developed by insurance companies.

    (a) five
    (b) ten
    (c) fifteen
    (d) twenty
    (e) twenty-five

—— (10) The cells of the body cannot produce proper enzymes for healthy metabolism without nutritional intake of _____ .

    (a) simple sugars
    (b) essential fatty acids
    (c) essential amino acids
    (d) starches
    (e) glycerol

___ (11) Which of the following is *not* an important mineral for human metabolism?

   (a) magnesium
   (b) iron
   (c) calcium
   (d) potassium
   (e) vitamin C

INTEGRATING AND APPLYING KEY CONCEPTS

Suppose that you could not eat solid food for two weeks and that you had only water to drink. Try to list in correct sequential order the measures that your body would take to try to preserve your life. Think about photographs you have seen of starving prisoners in concentration camps.

26

CIRCULATION

CIRCULATORY SYSTEM: AN OVERVIEW
CHARACTERISTICS OF BLOOD
    Functions of Blood
    Blood Volume and Composition
HUMAN CIRCULATORY SYSTEM
    Blood Circulation Routes
    The Human Heart
    Blood Pressure in the Vascular System

Commentary: On Cardiovascular
Disorders
    Hemostasis
    Blood Typing
LYMPHATIC SYSTEM
    Lymph Vascular System
    Lymphoid Organs
SUMMARY

General Objectives

1. Describe the composition and functions of blood.
2. Describe the structure of the human cardiovascular system and tell how it functions.
3. Explain the factors that cause blood to exist under different pressures.
4. Describe the composition and function of the lymphatic system.

26-I
(pp. 374–378)

CIRCULATORY SYSTEM: AN OVERVIEW
CHARACTERISTICS OF BLOOD
    Functions of Blood
    Blood Volume and Composition

Summary

A *circulatory system* and the blood it transports through the animal body carries vital material to cells, carries products and wastes from them, and helps maintain an internal environment that is favorable for cell activities. The circulatory system of invertebrates and vertebrates has three components: blood, heart, and blood vessels. The open circulatory systems of many invertebrates (including arthropods and mollusks) pump blood from the heart into one or more blood vessels that open directly into tissue spaces. The blood bathes nearby cells and then seeps sluggishly back to the heart. The closed circulatory systems of most animals contain blood enclosed within the walls of one or more hearts and blood vessels. The heart continuously pumps a constant amount of blood with adjustments in flow rate. Rapid blood flow in large-diameter vessels slows in *capillary beds,* which enable materials to diffuse slowly outward to the body cells; here also fluid filters out of capillaries to surrounding tissues. Animals with closed circulatory systems often have a supplementary network of tubes, the *lymph vascular system,* which recovers and purifies excess tissue fluid and proteins and returns them to the circulatory system.

Blood carries oxygen and nutrients to cells and carries secretions and metabolic wastes away. Phagocytic cells scavenge and fight infection. Blood helps stabilize internal pH levels. Blood in birds and mammals transfers heat between body regions; this helps equalize temperature in different body regions.

The blood volume of an adult human male of average weight is about 5 liters. Human blood consists of red blood cells, white blood cells, platelets, and plasma (fifty to sixty percent of total volume).

Blood *plasma* is mostly water with hundreds of different dispersed *plasma proteins* of various functions. Ions, various nutrients, and dissolved gases (oxygen, carbon dioxide, and nitrogen) are also found in plasma. *Red blood cells* (erythrocytes) contain the iron-containing protein hemoglobin, which transports oxygen by binding to it; hemoglobin also transports some carbon dioxide. Red blood cells continually form in bone marrow, lack nuclei at maturity, and function for about 120 days. White blood cells (leukocytes) scavenge dead or worn-out cells; they also respond to tissue damage and invasion by bacteria, viruses, and other foreign agents. White blood cells form from stem cells in bone marrow. The five types of white blood cells are lymphocytes, neutrophils, monocytes, eosinophils, and basophils. "B" and "T" lymphocytes are central to immune responses. Fragments of ruptured stem cells are called *platelets;* they have a role in forming blood clots.

Key Terms

circulatory system	lymph vascular system	leukocytes
blood	plasma	stem cells
heart	plasma proteins	B cells
blood vessels	red blood cell (erythrocyte)	T cells
closed circulation system	erythrocytes	platelets
open circulation system	hemoglobin	megakaryocytes
capillary beds	white blood cell (leukocyte)	

Objectives

1. List the main components of most circulatory systems.
2. Distinguish between *open* and *closed* circulation systems.
3. State what the lymph vascular system does.
4. Generally describe the functions of blood.
5. Name the major plasma components and their functions.
6. Describe the origin and general function of red blood cells, white blood cells, and platelets.

Self-Quiz Questions

Fill-in-the-Blanks

In most invertebrates and all vertebrates, the circulatory system is composed of blood, heart, and (1) _____ _____ . Most animals have a(n) (2) _____ circulatory system with heart and blood vessels continuously connected. Blood flow rate is slowed in (3) _____ _____ where diffusion of substances occurs. Organisms with (4) _____ circulation systems generally also have a supplementary (5) _____ _____ system that recovers protein and excess tissue fluid and returns them to the circulatory system. Blood helps stabilize internal (6) _____ and equalize internal temperature throughout an animal's body. The total volume of blood in an adult human male of average weight is about (7) _____ (number) liters. (8) _____ is the portion of blood that is composed of water, proteins of various functions, and nutrients. In (9) _____ _____ cells, oxygen binds with the (10) _____ molecule in a form that can be transported in the blood. Red blood cells form in (11) _____ _____ and function for about (12) _____ (number) days. White blood cells (13) _____ dead or worn-out cells, respond to tissue (14) _____ and invasion by (15) _____ , viruses, and other foreign agents. All white blood cells arise from immature cells in (16) _____ _____ . Fragments of ruptured stem cells form (17) _____ ; they have roles in (18) _____ formation.

26-II
(pp. 378–383)

HUMAN CIRCULATORY SYSTEM
Blood Circulation Routes
The Human Heart

Summary

The division of the human heart into right and left halves is the basis for two body cardiovascular circuits. The *pulmonary circulation circuit* pumps blood from the right half of the heart to the lungs where carbon dioxide is given up and oxygen is absorbed; the blood then returns to the left half of the heart. The oxygen-laden blood is then pumped through the remainder of the body through the *systemic circulation circuit* where oxygen is given up to cells and carbon dioxide is picked up; blood then flows to the right half of the heart. Blood in both circuits begins to flow from the heart in arteries and then continues with arterioles, capillaries, venules, and veins. Exchange of gases, nutrients, and waste products occurs in capillary beds.

The durable pump called the human heart beats about 2.5 billion times during a seventy-year life span. Cardiac muscle makes up most of the heart; connective tissue and endothelium line the chambers. An upper *atrium* and a lower *ventricle* make up each half of the heart. A membranous flap, the *AV valve,* is located between atrium and ventricle; the *semilunar valve,* another membranous flap, is found between the ventricle and an artery leading away from the heart. These valves keep blood moving in one direction without backflow. Blood supply to heart tissue is provided by the coronary circulation in which two small arteries branch from the *aorta* to the heart.

With each heartbeat in the *cardiac cycle* the four chambers contract (systole) and relax (diastole): relaxed atria fill, AV valves open, ventricles fill. AV valves shut when filled ventricles contract, ventricular pressure rises, semilunar valves open, and blood flows out of the heart. The driving force for blood circulation is *ventricular contraction.*

Cardiac muscle cells connect with each other at their narrowest boundaries where *communication junctions* occur. Each heartbeat signals a rapid spread of contraction across junctions; cells contract as a single unit. The cardiac muscle cells are self-excitatory; the nervous system only adjusts contraction rate and strength. Excitation begins in the SA node of the right atrium; seventy or eighty waves are generated per minute. Each wave spreads over both atria; they contract and then reach the AV node; the wave slows before the signal spreads over the ventricles.

Key Terms

pulmonary circuit	veins	cardiac cycle
systemic circuit	atrium	ventricular contraction
arteries	ventricle	communication junctions
arterioles	AV valve	sinoatrial (SA) node
capillaries	semilunar valve	atrioventricular (AV) node
venules	aorta	

Objectives

1. Trace the path of blood in the pulmonary and systemic circulations of the human body.
2. Draw and identify the structures in the human heart.
3. Describe the sequence of heart muscle contraction and relaxation known as a *cardiac cycle* and name the driving force for blood circulation.
4. Explain why the cardiac muscle cells of the heart react as a single unit during a heartbeat.
5. Explain what causes a heart to beat; include functions of the SA node and AV node.

Self-Quiz Questions

Fill-in-the-Blanks

In the (1) _____ circulation, blood from the (2) _____ half of the heart is pumped to the lungs, where it picks up oxygen and gives up (3) _____ _____ ; then it flows to the (4) _____ half of the heart. In the (5) _____ circulation, the (6) _____ -enriched blood is pumped through the remainder of the body, then it flows to the (7) _____ half of the heart. Both circuits begin with vessels called (8) _____ , then continue with (9) _____ , capillaries, (10) _____ , and finally veins. The two upper chambers of the heart are called (11) _____ ; the two lower chambers are called (12) _____ . Membrane flaps between atria and ventricles are the (13) _____ _____ ; flaps located between ventricles and arteries leading from the heart are the (14) _____ _____ . Blood circulation to the heart itself is known as the (15) _____ circulation. The phases of contraction and relaxation occurring with each heartbeat is the (16) _____ cycle. Atria fill with blood; atrial pressure rises and forces the (17) _____ _____ to open and fill the ventricles. The (18) _____ _____ snap shut when filled ventricles contract; (19) _____

_____ leading to arteries open and blood flows out of the heart.

(20) _____ contraction is the driving force for blood circulation.

(21) _____ junctions occur where cardiac muscle cells are fused together. Contraction signals from each heartbeat spread rapidly; the cardiac muscle cells contract (22) _____ as though they were one unit. Cardiac muscle cells are self-excitatory; excitation begins in the (23) _____ node. Each excitation wave spreads over the atria and they contract; the wave then reaches the (24) _____ node where it spreads more slowly before signaling the (25) _____ to contract.

Labeling

Identify each indicated part of the accompanying illustration; use Figure 26.6 of the main text as a reference.

(26) _____

(27) _____ _____ _____

(28) _____ _____ _____

(29) _____ _____

(30) _____ _____ _____

(31) _____ _____ _____

(32) _____ _____ _____

(33) _____ _____ _____

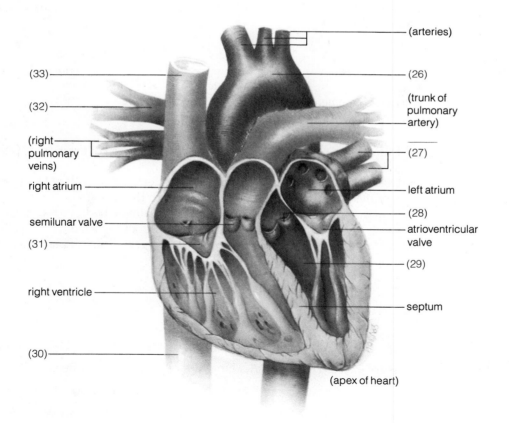

HUMAN CIRCULATORY SYSTEM (cont.)
Blood Pressure in the Vascular System

Summary

The pressure exerted on fluids resulting from heart contractions is known as *blood pressure;* initial pressures are high but drop as blood leaves and returns to the heart.

Arteries have thick muscular walls and large diameters; they "smooth out" the blood pressure changes of the cardiac cycle and send blood on its way by recoil after receiving pressures from ventricular contractions. Blood pressure is measured with a sphygmomanometer. Arteries branch into smaller diameter *arterioles;* their walls have rings of smooth muscle cells that respond to signals from nervous and endocrine systems as well as to local chemical stimuli. Arteriole control of blood to diffusion zones is accomplished by changing their diameters. Capillary beds serve as zones of diffusion for exchange of materials between blood and interstitial fluid. *Capillary walls* are thin; they consist of only one layer of flat epithelial cells. Capillaries are so narrow that red blood cells move in single file. Blood flows slowly in capillaries as it is channeled into smaller pathways, but the total volume in a capillary bed pathway exceeds the volume in the arterioles that enter the capillary bed, so there is little drop in blood pressure here. Capillaries merge into *venules,* which permit some diffusion; venules merge to form large-diameter, thin-walled veins containing one-way valves that prevent backward blood flow. *Veins* also serve as reservoirs of blood, containing fifty to sixty percent of the total blood volume. Smooth muscle cells in the walls of veins can contract to force more blood toward the heart when more rapid circulation is required. Venous blood pressure generally rises with body movements.

Key Terms

blood pressure	pulse pressure	venule
artery	arteriole	vein
sphygmomanometer	capillary	

Objectives

1. Define *blood pressure;* describe how the structures of arteries, capillaries, and veins differ.
2. Describe blood pressure in arteries, arterioles, capillary beds, veins, and venules; cite the factors that contribute to that pressure.
3. Understand the events happening in capillary beds and explain how you think those events change the composition of the blood.
4. Explain how veins and venules can act as reservoirs of blood volume.

Self-Quiz Questions

Fill-in-the-Blanks

Pressure exerted on fluids developed by heart contractions is (1) _____ _____ . Pressures on blood leaving the heart are normally high but then (2) _____ along circuits proceeding away from and back to the (3) _____ . A(n) (4) _____ is constructed of thick, muscular walls and carries blood away from the heart. Heart contractions expand arterial walls that then (5) _____ and force blood onward. Arteries branch into (6) _____ with smaller diameters; the walls have rings of smooth muscle that serve as control points where adjustments in blood (7) _____ can be made for service to diffusion zones. A(n) (8) _____ is a blood vessel with such a small diameter that red blood cells must flow through it in single file; its walls consists of no more than a single layer of (9) _____ cells. In each (10) _____ _____ , small molecules move between the bloodstream and the (11) _____ fluid. Capillaries merge into (12) _____ ; their walls are slightly thicker than those of capillaries. Venules merge into large-diameter (13) _____ . (14) _____ that prevent backflow are found inside veins. When body activities increase, smooth muscle cells in venous walls (15) _____ ; this increases venous (16) _____ and drives more blood to the (17) _____ . Movement and rapid breathing increase venous (18) _____ . Both (19) _____ and (20) _____ serve as temporary reservoirs for blood volume.

True-False

If false, explain why.

—— (21) Capillary beds present more total resistance to blood flow than the combined diameters of arterioles leading into them.

—— (22) Because the total volume of blood remains constant in the human body, blood pressure must also remain constant throughout the circuit.

26-IV
(pp. 381–387)

HUMAN CIRCULATORY SYSTEM (cont.)
Commentary: **On Cardiovascular Disorders**
Hemostasis
Blood Typing

Summary

Cardiovascular disorders claim about a million American lives each year. Several risk factors have been identified as contributing to cardiovascular disorders (several are avoidable). *Hypertension* is defined as sustained high blood pressure; treatment is by medication, changes in diet, and regular exercise. *Atherosclerosis* is a condition in which arteries thicken and lose elasticity; the condition is worsened by deposits of cholesterol, calcium salts, and fibrous tissue, or *athero-sclerotic plaque*, in arterial walls. Plaque formation is related to cholesterol intake; high levels of LDL (low-density lipoproteins) cholesterol carriers are associated with atherosclerosis and heart trouble, while high levels of HDL (high-density lipoproteins) are associated with healthier arteries and low cholesterol. Platelets torn by plaques initiate clotting; a *thrombus* is a stationary clot and an *embolus* is a traveling clot. The two disorders, hypertension and atherosclerosis, are the leading causes of most *heart attacks* (death of heart muscle due to blood supply interruption) and can also cause *strokes* (brain damage due to interruption of blood supply to it). *Arrhythmia* refers to irregular or abnormal heart rhythms; detection is by electrocardiogram (EKG).

The arresting of bleeding is called *hemostasis;* the process requires spasms in blood vessels, platelet plug formation, blood coagulation, and other mechanisms.

Your cell surfaces display membrane-bound proteins that identify your body; they serve as "self" markers. Proteins in your body called *antibodies* recognize markers on foreign cells and act against them. In ABO blood typing, type A blood has A markers on red blood cells and type B has B markers, type AB has both markers, and type O has neither A nor B markers. Type A blood carries antibodies against B markers but not against A markers; type B blood carries antibodies against A markers but not against B markers; type AB has neither A nor B antibodies; type O has antibodies against both A and B markers that act against one or both of those markers. When blood from different types of donors and recipients is mixed, *agglutination* or clumping occurs as antibodies react against foreign markers and clump cells. Clumping can cause severe damage to the circulatory system or even death.

The Rh marker can also cause agglutination responses. Blood cells of Rh⁺ persons have the marker, those of Rh⁻ persons do not. People do not usually have the antibodies that act against Rh markers unless they have received a transfusion of Rh⁺ blood; they will produce antibodies against the marker. Rh⁻ females made pregnant by Rh⁺ males may have an Rh⁺ fetus. Should such fetal red blood cells enter the mother's bloodstream during pregnancy or birth, it stimulates her body to produce antibodies against the fetal Rh⁺ markers. This birth occurs without difficulty but when a second pregnancy of the same woman occurs, Rh antibodies may enter the fetal bloodstream, causing swelling and rupture of red blood cells. Hemoglobin is released into the bloodstream. This disorder is called *erythroblastosis fetalis.* If too many cells are destroyed the fetus may die before birth. If born alive, all blood may have to be replaced. More recently, Rh⁻ mothers may be treated with a drug immediately after the first pregnancy, which inactivates any Rh antibodies in her bloodstream.

Concept Aid

Agglutination (clumping) involves a reaction between antibody and marker; this differs from coagulation (clotting), which involves a series of proteins and Ca⁺⁺ cooperating to seal a wound and prevent hemorrhage.

Key Terms

hypertension	LDL (low-density	antibodies
heart attack	lipoproteins)	ABO blood typing
stroke	HDL (high-density	A markers
atherosclerosis	lipoproteins)	B markers
cardiovascular disorders	cholesterol	agglutination
atherosclerotic plaque	arrhythmia	Rh blood typing
thrombus	electrocardiogram (EKG)	Rh⁺, Rh⁻
embolus	hemostasis	Rh antibodies
angiography	coagulation	erythroblastosis fetalis
coronary bypass surgery		

Objectives

1. Distinguish a *stroke* from *atherosclerosis* and a *stroke* from a *heart attack.*
2. Describe the cause and treatment of hypertension.
3. Describe the chemical makeup of atherosclerotic plaque and its relationship to LDLs and HDLs.

4. Define *arrhythmia* and tell how it is detected.
5. Describe *hemostasis* and list factors that bring it about.
6. Describe how blood is typed for the ABO blood group and for the Rh factor.
7. Explain the cause and characteristics of a child with *erythroblastosis fetalis*.

Self-Quiz Questions

Fill-in-the Blanks

One cause of (1) _____ is the rupture of one or more blood vessels in the brain. Sustained high blood pressure is known as (2) _____ . Death of heart muscle due to interruption in blood supply can bring about a(n) (3) _____ _____ . (4) _____ _____ is a term for a formation that can include cholesterol, calcium salts, and fibrous tissue. It is not healthful to have a high concentration of (5) _____ -density lipoproteins in the bloodstream. A(n) (6) _____ is a stationary clot; a(n) (7) _____ is a traveling clot. Irregular or abnormal heart rhythms are referred to as (8) _____ . Bleeding is stopped by several mechanisms that are collectively referred to as (9) _____ ; the mechanisms include spasms in blood vessels, (10) _____ _____ formation, and blood (11) _____ . Your cells carry surface membrane proteins that serve as "self" (12) _____ . Your body also has proteins called (13) _____ that recognize (14) _____ on foreign cells. If you are blood type B, you have antibodies against (15) _____ but not against (16) _____ markers. If you are blood type (17) _____ , you have antibodies against A and B markers. When antibodies are mixed with "foreign" cells, a response may occur called (18) _____ . (19) _____ persons have the Rh marker. Without intervention, Rh⁻ mothers bearing more than one child from Rh⁺ fathers put these children at risk for (20) _____ _____ .

26-V
(pp. 387–388)

LYMPHATIC SYSTEM
Lymph Vascular System
Lymphoid Organs
SUMMARY

Summary

The *lymphatic system,* consisting of transport vessels and lymphoid organs, supplements the circulatory system by returning escaped tissue fluid to the bloodstream; it is also vital to the body's defense mechanisms. Tissue fluid in the lymph vascular system is called *lymph.*

The *lymph vascular system* accumulates, transports, and returns escaped fluid and a small amount of proteins to the bloodstream from which they came. It also transports fats absorbed from the digestive tract to the bloodstream and carries foreign particles and cellular debris to disposal centers (lymph nodes). The lymph vascular system has capillaries, which are "blind-end" tubes that merge with larger lymph vessels whose walls have smooth muscles. Movements of breathing and body muscles move fluid through lymph vessels, which converge into collecting ducts and drain into the lower neck.

The *lymphoid organs* include lymph nodes, spleen, thymus, tonsils, adenoids, and tissue patches in the small intestine and appendix; here some infection-fighting cells, including lymphocytes, are formed. Lymphocytes derived from bone marrow stem cells move in the blood to lymphoid organs. Divisions of derivative cells in blood and lymphoid organs produce most new lymphocytes.

All lymph trickles through at least one *lymph node* enroute to the bloodstream; here macrophages help clear lymph of bacteria, cellular debris, and other substances. The *spleen* is the largest lymphoid organ; it filters blood and retains lymphocytes. The red pulp of the spleen holds red blood cells and macrophages; the red blood cells of human embryos are produced in the red pulp. The thymus secretes hormones concerned with lymphocyte activity, reproduction, and maturation.

Key Terms

lymphatic system	lymphoid organs	plasma cells
lymph	lymph nodes	macrophages
lymph vascular system	tonsils	spleen
lymph capillaries	adenoids	thymus
lymph vessels	lymphocytes	

Objectives

1. Describe the general functions of the lymphatic system.
2. List the components of the lymph vascular system and four of its functions.
3. Describe the structures involved in collecting and returning fluid that has escaped from the circulatory system and explain how they work.
4. Explain the general functions of lymphoid organs.
5. Give some specific functions of lymph nodes, the spleen, and the thymus.

Self-Quiz Questions

Labeling

Identify each indicated part of the accompanying illustration.

(1) _____

(2) _____ _____ _____

(3) _____

(4) _____ _____

(5) _____

(6) _____ _____

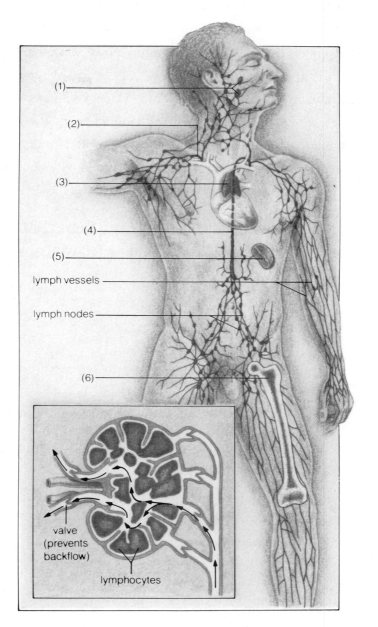

(1)

(2)

(3)

(4)

(5)

lymph vessels

lymph nodes

(6)

valve
(prevents
backflow)

lymphocytes

Fill-in-the-Blanks The lymphatic system supplements the circulatory system by returning fluids that bathe (7) _____ to the bloodstream. The lymphatic system is also vital to the body's (8) _____ . Fluid in the lymph vascular system is called (9) _____ . The (10) _____ _____ system includes lymph capillaries, lymph vessels, and ducts that drain the processed fluid back into the circulatory system. Lymph (11) _____ are "blind-end" tubes that merge with larger lymph (12) _____ , which have (13) _____ muscles in their walls; lymph (14) _____ converge into collecting ducts, which drain into veins in the lower (15) _____ . The (16) _____ organs include the lymph nodes, spleen, thymus, tonsils, adenoids, and tissue patches in the small intestine and appendix; these areas are production centers for infection-fighting cells, including some types of (17) _____ . Most new (18) _____ are produced by divisions in the blood and lymphoid organs, not in bone (19) _____ . All lymph trickles through at least one lymph (20) _____ before being delivered to the bloodstream. The (21) _____ is the largest lymphoid organ; it serves as a filtering station for (22) _____ and a holding station for (23) _____ . The (24) _____ secretes hormones concerned with lymphocyte activity.

CHAPTER TEST

UNDERSTANDING AND INTERPRETING KEY CONCEPTS

___ (1) Most of the oxygen in human blood is transported by _____ .

    (a) plasma
    (b) serum
    (c) platelets
    (d) hemoglobin
    (e) leukocytes

___ (2) Which of the following is *not* a function of blood?

    (a) carries oxygen and nutrients to cells
    (b) returns fluid leaked from blood vessels and reclaimed proteins to the circulatory system
    (c) carries hormones and metabolic wastes
    (d) equalizes body temperature through heat transfer
    (e) contains phagocytic cells that scavenge and fight infection

___ (3) Open circulatory systems generally lack _____ .

    (a) a heart
    (b) arterioles
    (c) capillaries
    (d) veins
    (e) arteries

___ (4) Red blood cells originate in the _____ .

    (a) liver
    (b) spleen
    (c) yellow bone marrow
    (d) thymus gland
    (e) red bone marrow

___ (5) Hemoglobin contains _____ .

    (a) copper
    (b) magnesium
    (c) sodium
    (d) calcium
    (e) iron

___ (6) In the human heart, excitation begins in the _____ .

    (a) sinoatrial (SA) node
    (b) semilunar valve
    (c) communication junction
    (d) AV valve
    (e) atrioventricular (AV) node

___ (7) During systole, _____ .

    (a) oxygen-rich blood is pumped to the lungs
    (b) the heart muscle tissues contract
    (c) the atrioventricular valves suddenly open
    (d) oxygen-poor blood from all parts of the human body, excepting the lungs, flows toward the right atrium
    (e) none of the above

___ (8) _____ are reservoirs of blood pressure in which resistance to flow is low.

    (a) Arteries
    (b) Arterioles
    (c) Capillaries
    (d) Veins

___ (9) Beginning with the heart, which is the correct sequence?

    (a) heart→ veins → venules→ capillaries→ arteries→ arterioles
    (b) heart→ arterioles→ arteries → capillaries→ venules→ veins
    (c) heart→ arteries → arterioles→ capillaries→ venules→ veins
    (d) heart → arteries→ arterioles → capillaries→ veins →venules
    (e) heart → arteries → veins→ capillaries → arterioles →venules

___ (10) Which one of the following statements is *incorrect*?

    (a) In the systemic circuit, blood moves from the heart to the lungs and back to the heart.
    (b) The AV valve is found between the atrium and the ventricle.
    (c) Coronary circulation serves heart muscle.
    (d) Ventricular contraction is the driving force for blood circulation.
    (e) The aorta is the major artery carrying oxygen-enriched blood away from the heart.

___ (11) If you are blood type A, _____ .

    (a) you do not carry antibodies against A markers
    (b) you have antibodies against B markers
    (c) you have no antibodies against A or B markers
    (d) you have antibodies against A but not B markers
    (e) both a and b are correct

___ (12) The lymphatic system is the principal avenue in the human body for transporting _____ .

    (a) blood
    (b) wastes
    (c) carbon dioxide
    (d) amino acids
    (e) fats and fluid that bathes body tissues

INTEGRATING AND APPLYING KEY CONCEPTS

You observe that some people appear as though fluid had accumulated in their lower legs and feet. Their lower extremities resemble those of elephants. You inquire about what is wrong and are told that the condition is caused by the bite of a mosquito that is active at night. Construct a testable hypothesis that would explain (1) why the fluid was not being returned to the torso, as normal, and (2) what the mosquito did to its victims.

27

IMMUNITY

NONSPECIFIC DEFENSE RESPONSES
    Barriers to Invasion
    Phagocytes
    Complement System
    Inflammation
SPECIFIC DEFENSE RESPONSES: THE
IMMUNE SYSTEM
    Overview of the Defenders
    Recognition of Self and Nonself
    Primary Immune Response
    Commentary: Cancer and the Immune
    System
    Antibody Diversity and the Clonal
    Selection Theory

    Secondary Immune Response
IMMUNIZATION
ABNORMAL OR DEFICIENT IMMUNE
RESPONSES
    Allergies
    Autoimmune Disorders
    Deficient Immune Responses
    Commentary: AIDS—The Immune
    System Compromised
    Case Study: The Silent, Unseen Struggles
SUMMARY

General Objectives

1. Describe typical external barriers that organisms present to invading organisms.
2. Distinguish between nonspecific and specific defense responses.
3. Understand the process involved in the nonspecific inflammatory response.
4. Understand how vertebrates (especially humans) recognize and discriminate between self and nonself tissues.
5. Distinguish between antibody-mediated and cell-mediated patterns of warfare.
6. Describe some examples of immune failures and identify as specifically as you can which weapons in the immunity arsenal failed in each case.

27-I
(pp. 391–394)

NONSPECIFIC DEFENSE RESPONSES
    Barriers to Invasion
    Phagocytes
    Complement System
    Inflammation
SPECIFIC DEFENSE RESPONSES: THE IMMUNE SYSTEM
    Overview of the Defenders
    Recognition of Self and Nonself

Summary

The body is able to defend itself against many pathogens, which are disease-causing viruses, bacteria, fungi, and protozoans. Nonspecific defense responses by the vertebrate body against invasion include: (1) intact skin, (2) ciliated mucous membranes in the respiratory tract, (3) exocrine gland secretions in the skin, mouth, and elsewhere (lysozyme destroys bacterial cell walls), (4) acidic gastric fluid that destroys pathogens, and (5) the normal microbial flora of the gut and vagina that keep the growth of pathogens in check.

Mobilization of phagocytic white blood cells occurs when physical barriers to invasion are breached; they engulf and destroy foreign agents. For example, when the skin is cut, phagocytes destroy foreign cells. Phagocytes are found within blood vessels and enter damaged or invaded tissues by passing through capillary walls; others are found in the lymph nodes, spleen, liver, kidneys, lung, joints, and brain.

The *complement system* is a set of twenty plasma proteins that enhance nonspecific and specific defense responses. Complement becomes activated when it encounters foreign cells; it attracts phagocytes and can promote lysis of foreign cells.

The *inflammatory response* is a series of events involving cells and substances (including complement) that destroy invaders and restore tissues and operating conditions to normal. These events occur during both nonspecific and specific defense responses. While complement proteins are being activated, basophils and mast cells secrete histamine; histamine dilates blood vessels and fluid seeps into the area. Clotting mechanisms keep blood vessels intact and wall off the infected or damaged area; warmth, redness, and swelling result.

Specific defense responses include *immune system* mechanisms; certain white blood cells are mobilized against a particular invader, not invaders in general. Phagocytes summoned to an inflammation site are indiscriminate engulfers and may not be enough to check the spread of an invader. Macrophages and T and B lymphocytes of the vertebrate immune system may be needed. Interactions among these cells are the basis of the immune system, a system that shows specificity (attacks specific invaders) and memory (attacks the same invaders upon return).

One of every 100 cells in the body is a white blood cell. White blood cells that carry out the immune response are: (1) *macrophages* (alert helper T cells to specific foreign agents); (2) *helper T cells* (stimulate rapid division of B cells and killer T cells); (3) *B cells* (produce antibodies for specific targets and tag them for destruction); (4) *killer T cells* and *NK cells* (directly destroy infected body cells); (5) *suppressor T cells* (lymphocytes that slow down or prevent immune responses); and (6) *memory cells* (B cells and T cells not used in battle but that circulate and respond to new attacks by the same invader). T cells carry out cell-mediated responses; B cells carry out antibody-mediated responses (helper T cells and suppressor T cells are also involved in antibody-mediated responses).

MHC (major histocompatibility complex) *markers* are cell surface receptors that are unique to each individual; such markers allow the immune system defense mechanisms to distinguish *self* (the body's own cells) from *nonself*. Rarely is this distinction blurred, but when it is, lymphocytes may trigger *autoimmune responses* by attacking cells bearing self *antigens*. Antigens are usually large protein or polysaccharide molecules with distinct shapes that trigger immune responses.

Key Terms

Jenner	exocrine glands	inflammatory response
vaccination	lysozyme	histamine
pathogens	phagocytes	immune system
nonspecific defense response	complement system	specificity
	lysis	memory

macrophages	suppressor T cells	nonself markers
helper T cells	memory cells	autoimmune response
B cells	cell-mediated response	MHC markers
killer T cells	self markers	antigen
natural killer (NK) cells		

Objectives

1. List and discuss five nonspecific defense responses that serve to exclude invaders from the body.
2. Explain the role of phagocytes when invasion barriers are breached.
3. Explain the function of the complement system in tissue invasion; describe how complement may be related to an inflammatory response.
4. List the events occurring in an inflammatory response.
5. List the three general types of white blood cells whose interactions form the basis of the immune system.
6. Describe the physical basis that establishes self cells and nonself cells.

Self-Quiz Questions

Fill-in-the-Blanks

Disease-causing agents such as viruses, bacteria, fungi, and protozoans are generally called (1) _____ . Intact skin, ciliated mucus membranes, exocrine gland secretions, gastric fluid, and normal bacterial inhabitants are barriers to invasion that are (2) _____ _____ responses. When physical barriers to invasion are breached, phagocytic (3) _____ _____ cells are mobilized. Tissue invasion by certain bacterial or fungal cells activate reactions of (4) _____ interacting plasma proteins, the (5) _____ system. These defensive reactions attract (6) _____ , that (7) _____ the invader, and help kill the pathogen by promoting (8) _____ . Warmth, redness, fluid seeping to cause local swelling, phagocytosis, and tissue repair are all events of the (9) _____ response. The interactions of three white blood cells, macrophages, (10) _____ lymphocytes, and (11) _____ lymphocytes, form the basis of the (12) _____ system. (13) _____ alert helper T cells to the presence of foreign agents; (14) _____ _____ cells stimulate rapid division of B cells and killer T cells; (15) _____ cells are lymphocytes responsible for producing antibodies; (16) _____ _____ cells and NK cells destroy already infected body cells; (17) _____ _____ cells slow down or prevent immune responses; (18) _____ cells are B cells or T cells that respond rapidly to a subsequent attack by a previous invader. Lymphocytes respond to (19) _____ surface cell markers to distinguish self from nonself particles. Lymphocytes will mount an attack when they encounter a foreign (20) _____ .

27-II
(pp. 394–399)

SPECIFIC DEFENSE RESPONSES: THE IMMUNE SYSTEM (cont.)
    Primary Immune Response
    Commentary: **Cancer and the Immune System**
    Antibody Diversity and the Clonal Selection Theory
    Secondary Immune Response
IMMUNIZATION

Summary

The first time macrophages, lymphocytes, and their products encounter an antigen, a *primary immune response* occurs. One type of response to an antigen is the *antibody-mediated immune response*. An *antibody* is a Y-shaped receptor protein molecule with binding sites for specific antigens; antibodies are produced by B cells and their progeny, *plasma cells*. One type of surface antibody (the tail of the Y is embedded) is produced by each B cell maturing in bone marrow; they are released as "virgin" B cells. When a B cell binds with a foreign antigen, it becomes sensitive to macrophage and helper T cell communication signals. Antigen fragments remain on the macrophage surfaces (as antigen-MHC marker complexes) following the engulfment of bacteria during the inflammatory response. Helper T cell surface receptors lock onto the antigen-MHC complexes; with this, macrophages secrete interleukin-1 and helper T cells secrete *lymphokines;* these substances signal antigen-sensitized B cells to begin dividing to form clones of identical B cells, all producing identical antibodies. Some of the cell clones become plasma cells that produce the same antibody (2000 antibody molecules per second may be produced). Circulating antibodies tag the invader for disposal. Several classes of antibodies (immunoglobulins, or Ig) enlist the aid of different immune cells or chemical weapons. As examples, IgM and IgG antibodies bound to antigen enlist the aid of macrophages and complement proteins; IgE antibodies activate histamine-secreting cells. Bacteria, extracellular viral phases, some fungal parasites, and protozoans are the main targets of antibody-mediated responses. Antibodies are only effective if the antigen is circulating in tissues or on cell surfaces.

The key fighters of the *cell-mediated immune response* are the killer T cells and natural killer (NK) cells; they directly destroy already infected, mutant, and cancerous cells. Killer T cells begin development in bone marrow and then move to the thymus where they produce their own antigen-binding receptor molecules. Helper T cells raise the alarm during an invasion; killer T cells with proper receptors puncture and destroy marked (antigen-MHC complexes) infected body cells with perforin holes before internal pathogens can reproduce. Killer cells may be the reason for organ transplant rejection.

Cancer cells (those that uncontrollably divide) may arise in your body at any time. Cancer cells can be destroyed by killer T cells and natural killer cells. At present, surgery, chemotherapy, and irradiation are the only weapons against cancer. *Immune therapy* promises to be useful in defense against cancer. Interferons have destroyed some rare forms of cancer. Labeled monoclonal antibodies may help locate cancers in the body and carry drugs to cancer cells.

Antibody-mediated and cell-mediated responses are regulated by the amount of exposed antigens (binding sites) on pathogens. Fewer antibodies are produced in response to fewer exposed antigens. Suppressor T cell secretions slow the counterattack.

A major question has been how lymphocytes produce enough receptors required to detect an enormous variety of pathogens, each with unique antigens. DNA recombinations occur as each B cell (they have identical genes) matures in bone marrow; recombination produces millions of possible combinations of the polypeptide chains found on the arms of antibodies. In this way, you may have defenses against entirely new antigens encountered.

According to the *clonal selection theory*, activated lymphocytes multiply rapidly; all descendant cells will retain antigen specificity as a clone of cells; each is immunologically identical and specific to the selecting antigen. The clonal selection theory also explains "immunological memory." A subsequent invasion by the same pathogen invokes a more rapid, greater, and longer response, the *secondary immune response*. B and T cells from the primary immune response circulate for years as *memory lymphocytes* that, when they encounter the same type of antigen, divide at once; this releases large clones of active lymphocytes rather quickly.

Immunization involves the deliberate production of memory lymphocytes by a *vaccine* by activating a primary immune response. Killed or weakened bacteria or viruses (Sabin vaccine from a weakened polio virus) may be used as the vaccine preparation that stimulates the primary immune response by the first injection; the second injection (booster shot) elicits the secondary response with production of more antibodies and memory cells for long-lasting protection. Recently, antigen-encoding genes from pathogens have been successfully incorporated into viruses (genetic engineering) that then are used to immunize laboratory animals against serious diseases. For example, the *Vaccinia* virus has been used to immunize animals against hepatitis B, influenza, and rabies. *Passive immunity* is conferred by direct injection of antibodies; protection is short-lived.

Key Terms

primary immune
 response
antibody-mediated
 immune response
antibody
plasma cell
"virgin" B cells
antigen-MHC complexes
interleukin-1

lymphokines
immunoglobulins (Ig)
IgM antibody
IgG antibody
IgE antibody
cell-mediated immune
 response
immune therapy
interferons

monoclonal antibodies
clonal selection theory
clone
secondary immune
 response
memory lymphocytes
immunization
vaccine
passive immunity

Objectives

1. Describe the *primary immune response.*
2. Distinguish between the antibody-mediated immune response and the cell-mediated immune response.
3. Distinguish the roles of T cells from the roles of B cells.
4. Explain how the immune response is controlled.
5. Define *cancer;* describe how *immune therapy, interferons,* and *monoclonal antibodies* are being used in the fight against cancer.
6. Tell how the body can produce millions of different receptors (antibodies) for the great variety of pathogens encountered.
7. Explain what the *clonal selection theory* is and tell what it helps to explain.
8. Contrast the *primary immune response* and the *secondary immune response.*
9. Describe two ways that people can be immunized against specific diseases.

Fill-in-the-Blanks

A(n) (1) _____ _____ response is provoked after a first-time encounter with an antigen. A(n) (2) _____ is a Y-shaped receptor molecule with a binding site for a specific (3) _____ . Only B cells and some of their progeny, (4) _____ cells, produce antibodies. B cells with membrane-bound antibodies that have not yet contacted antigens are (5) _____ B cells. Following engulfment of bacterial cells, macrophages display antigen-MHC complexes at their (6) _____ . Surface receptors on helper T cells lock onto the antigen-MHC complexes; macrophages secrete (7) _____ and helper T cells secrete (8) _____ ; these substances stimulate B cells to begin (9) _____ to produce a clone of identical B cells that produce the same kind of (10) _____ . Circulating antibodies tag invaders for disposal by other means; IgM and IgG antibodies enlist the aid of (11) _____ and (12) _____ proteins. Antibodies are only effective against antigens circulating in (13) _____ or at cell (14) _____ . The key fighters of cell-mediated immune responses are (15) _____ _____ cells and (16) _____ _____ cells. They indirectly destroy cells that are already (17) _____ . Antibody-mediated and cell-mediated responses are regulated by the amount of (18) _____ _____ on pathogens. The (19) _____ _____ theory explains how an individual has immunological memory, which is the basis of a secondary immune response. (20) _____ lymphocytes continue to circulate for years. Deliberately introducing an antigen into the body to provoke an immune response is called (21) _____ ; a(n) (22) _____ is a preparation designed to stimulate the appearance of memory lymphocytes. When people are already exposed to diseases, antibodies may be injected directly to confer (23) _____ immunity.

27-III
(pp. 399–404)

ABNORMAL OR DEFICIENT IMMUNE RESPONSES
    Allergies
    Autoimmune Disorders
    Deficient Immune Responses
    Commentary: **AIDS—The Immune System Compromised**
    Case Study: The Silent, Unseen Struggles
SUMMARY

Summary

Allergies are secondary immune responses to normally harmless substances. Exposure triggers production of IgE antibodies, which cause the release of histamines and prostaglandins. A local inflammatory response results; death can even occur. In *asthma* and hay fever, the resulting symptoms include congestion, sneezing, a drippy nose, and labored breathing.

In an *autoimmune response,* lymphocytes turn against the body's own cells; rheumatoid arthritis is an example.

Deficient immune responses occur when cell-mediated immunity is weakened to render an individual highly vulnerable to opportunistic infections and cancers that are normally not life-threatening. Such a case is AIDS (acquired immune deficiency syndrome), caused by the HIV virus. HIV is a retrovirus with RNA as its genetic material. In the United States, highest transmission is among intravenous drug abusers who share needles and male homosexuals. There is no vaccine against the known forms of the virus (HIV-1 and HIV-2). HIV is transmitted when bodily fluids of an infected person enter another person's tissues.

Key Terms

allergy	hay fever	acquired immune deficiency
IgE antibodies	IgG antibodies	syndrome, AIDS
histamine	autoimmune response	human immune deficiency
prostaglandins	rheumatoid arthritis	virus, HIV
asthma	rheumatoid factor	retrovirus

Objectives

1. Distinguish *allergy* from *autoimmune response.*
2. Describe how AIDS specifically interferes with the human immune system.

Self-Quiz Questions

Fill-in-the-Blanks

A(n) (1) _____ is a secondary immune response to a normally harmless substance that may actually cause injury to tissues. A(n) (2) _____ _____ is a disorder in which the body mobilizes its forces against certain of its own tissues. AIDS follows infection by the (3) _____ _____ _____ virus. In the United States, transmission has occurred most often among intravenous drug abusers who share needles and among (4) _____ _____ . There is no (5) _____ against the known forms of the HIV virus. HIV is transmitted when (6) _____ _____ of an infected person enter another person's tissues.

Matching

Match each letter with its best mate.

(7) ___ allergy

(8) ___ antibody

(9) ___ antigen

(10) ___ macrophage

(11) ___ clone

(12) ___ complement

(13) ___ histamine

(14) ___ MHC marker

(15) ___ plasma cell

(16) ___ killer T cell

A. Begins its development in bone marrow, but matures in the thymus gland

B. Cell that has directly or indirectly descended from the same parent cell

C. A potent chemical that causes blood vessels to dilate and lets protein pass through the vessel walls

D. Y-shaped immunoglobulin

E. A nonself marker

F. The progeny of turned-on B cells

G. A group of about twenty proteins that participate in the inflammatory response

H. An altered secondary immune response to a substance that is normally harmless to other people

I. The basis for self-recognition at the cell surface

J. Principal perpetrator of phagocytosis

CHAPTER TEST

UNDERSTANDING AND INTERPRETING KEY CONCEPTS

___ (1) All the body's phagocytes are derived from stem cells in the _____ .

    (a) spleen
    (b) liver
    (c) thymus
    (d) bone marrow
    (e) thyroid

___ (2) The plasma proteins that are activated when they contact a bacterial cell are collectively known as the _____ system.

    (a) shield
    (b) complement
    (c) IgG
    (d) MHC
    (e) HIV

___ (3) _____ are divided into two groups: T cells and B cells.

    (a) Macrophages
    (b) Lymphocytes
    (c) Platelets
    (d) Complement cells
    (e) Cancer cells

___ (4) _____ produce and secrete antibodies that set up bacterial invaders for subsequent destruction by macrophages.

    (a) B cells
    (b) Phagocytes
    (c) T cells
    (d) Bacteriophages
    (e) Thymus cells

—— (5) Antibodies are shaped like the letter _____ .

    (a) Y
    (b) W
    (c) Z
    (d) H
    (e) E

—— (6) The markers for every cell in the human body are referred to by the letters _____ .

    (a) HIV
    (b) MBC
    (c) RNA
    (d) DNA
    (e) MHC

—— (7) Plasma cells _____ .

    (a) die within a week after they are produced
    (b) develop from B cells
    (c) manufacture and secrete antibodies
    (d) do not divide and form clones
    (e) all of the above

—— (8) Clones of B or T cells are _____ .

    (a) being produced continually
    (b) known as memory cells
    (c) only produced when their surface proteins recognize other specific proteins
    (d) interchangeable
    (e) produced and mature in the bone marrow

—— (9) Whenever the body is reexposed to a specific sensitizing agent as in the case of allergies, IgE antibodies cause _____ .

    (a) prostaglandins and histamine to be produced
    (b) clonal cells to be produced
    (c) histamine to be released
    (d) the immune response to be suppressed
    (e) none of the above

—— (10) In rheumatoid arthritis, lymphocytes are unleashed against the body's own cells; this is an example of a(n) _____ .

    (a) allergic response
    (b) autoimmune response
    (c) human immune deficiency
    (d) excessive number of antigens being released into the bloodstream
    (e) inflammatory response

INTEGRATING AND APPLYING KEY CONCEPTS

Suppose that you have had an allergy for several years and you enter a clinic for this condition. What kinds of procedures do you think would occur to determine the cause of your allergy? Once the cause of your allergy has been determined, what steps could be taken to relieve you of this condition?

28

RESPIRATION

RESPIRATORY SYSTEMS
  Specialized Respiratory Surfaces
  Human Respiratory System
AIR PRESSURE CHANGES IN THE LUNGS
GAS EXCHANGE AND TRANSPORT
  Gas Exchange in Alveoli
  Gas Transport Between Lungs and
  Tissues

Commentary: When the Lungs Break Down
CONTROLS OVER RESPIRATION
  Matching Air Flow to Blood Flow
  Hypoxia
SUMMARY

General Objectives

1. Understand the behavior of gases and the types of respiratory surfaces that participate in gas exchange.
2. Understand how the human respiratory system is related to the circulatory system, to cellular respiration, and to the nervous system.
3. List some of the things that go awry with the respiratory system and describe the characteristics of the breakdown.

28-I
(pp. 406–409)

RESPIRATORY SYSTEMS
  Specialized Respiratory Surfaces

Summary

The overall exchange of gases among cells, the blood, and the environment is known as *respiration*. In small animals with low metabolic rates (flatworms, earthworms, and others), the body surface is used for *integumentary exchange* wherein gases diffuse through a thin, moist epidermis. Animals larger than flatworms cannot depend on the integument alone to provide enough surface area for gas exchange. A few amphibians and some insects have projecting external gills. A typical *gill* has a moist, thin, vascularized layer of epidermis that functions in gas exchange. Adult fishes have internal gills arranged such that water flow is opposite from blood flow. *Tracheas* are air-conducting tubes found in insects, spiders, and some of their relatives; spiracles cover external openings to conserve moisture. *Lungs* contain internal respiratory surfaces shaped as a cavity or sac. Approximately 450 million years ago, primitive lungs began a path of evolution into swim bladders or complex respiratory organs. In all animals with lungs, airways carry gases by bulk flow to and from one side of the respiratory surface; blood vessels of the pulmonary circulation (by bulk flow of blood) carry gases to and from the other side of internal environments (by diffusion between blood and interstitial fluid, then between interstitial fluid and cells).

Key Terms

respiration	trachea	bulk flow of air
integumentary exchange	spiracle	blood vessels
gill	lung	pulmonary circulation
external gills	airways	bulk flow of blood
internal gills		

Objectives

1. Describe what kinds of animals exchange gases by integumentary exchange. Why is this method limited?
2. Explain the flow mechanisms of water and blood that allow a gill to efficiently exchange gases with the environment.
3. Describe how incoming oxygen is distributed to the tissues of insects, and contrast this with the process that occurs in mammal lungs.

Self-Quiz Questions

Fill-in-the-Blanks

A(n) (1) _____ has a thin, moist vascularized epidermis that functions in (2) _____ _____ ; those of some animals are external while others, like fish, are (3) _____ . Water moves (4) _____ fish gills and blood circulates through them in (5) _____ directions. (6) _____ are tubes that branch finely through the insect body and provide a rather self-contained system of gas conduction and exchange; (7) _____ cover the body openings to these tubes. In some fish lineages, the primitive lungs developed into (8) _____ _____ . In all animals with lungs, (9) _____ carry gas to and from one side of the respiratory surface of the lungs, and (10) _____ _____ carry gas to and from the other side.

28-II
(pp. 410–412)

RESPIRATORY SYSTEMS (cont.)
  Human Respiratory System
AIR PRESSURE CHANGES IN THE LUNGS

Summary

Through nasal cavities, air enters or leaves the respiratory system; hair and cilia filter dust and particles. Blood vessels warm the air and mucus moistens the air. Air moves through the pharynx → epiglottis → larynx with vocal cords (open space between the vocal cords is the glottis) → trachea → bronchi → bronchioles → lungs.

　　Human lungs are a pair of organs in the ribcage above the diaphragm. Each lung lies in a pleural sac consisting of a thin membrane. Inside the lungs, airway branches become shorter, narrower, and more numerous. The first branchings that lack cartilage-supported walls are the bronchioles. The terminal respiratory bronchioles have outpouchings called *alveoli* (singular, alveolus). Several clustered alveoli form alveolar sacs, the major sites of gas exchange. A dense network of capillaries surrounds 150 million alveoli in each lung and provides a surface area equivalent to a racquetball court for exchange of gases with the bloodstream.

To breathe, air is inhaled and exhaled. To inhale, the diaphragm contracts, flattens, and moves downward; the volume of the chest cavity increases. Muscles move the ribcage upward and outward; this contributes to an increase in chest cavity volume. Pressure in the narrow space between each lung and the pleural sac becomes lower compared to atmospheric pressure. The pressure difference allows fresh air to flow down the airways. To exhale, the muscles relax and elastic lung tissue recoils, as the chest cavity volume decreases. Alveolar pressure becomes greater than atmospheric pressure. Air follows the gradient and leaves the lungs.

Key Terms

nasal cavities	bronchus, -chi	alveolar sac
pharynx	pleural sac	diaphragm
larynx	bronchioles	chest cavity
epiglottis	respiratory bronchioles	inhale
vocal cord	alveolus, -oli	exhale
glottis		

Objectives

1. Explain how air entering the human respiratory system is filtered, warmed, and moistened.
2. List all the principal parts of the air-conducting portion of the human respiratory system in order beginning with the nasal cavities.
3. Describe the air exchange portion of the human respiratory system.
4. Explain how inhaling and exhaling air is accomplished by the human body.

Self-Quiz Questions

Fill-in-the-Blanks

In succession, air passes through the nasal cavities, pharynx, and (1) _____ , past the epiglottis into the (2) _____ (the space between the vocal cords), into the trachea, and then to the (3) _____ , (4) _____ , and alveoli of the (5) _____ . Exchange of gases occurs across the epithelium of the (6) _____ . During inhalation, the (7) _____ moves downward and flattens, and the (8) _____ _____ moves outward and upward; when these things happen, the chest cavity volume (9) [choose one] ( ) increases ( ) decreases, and the internal pressure (10) [choose one] ( ) rises ( ) drops ( ) stays the same. The (11) _____ _____ surrounds each lung.

Labeling   **Identify each indicated part of the accompanying illustration.**

(12) _____ _____   (17) _____
(13) _____   (18) _____
(14) _____   (19) _____
(15) _____   (20) _____ _____
(16) _____ _____   (21) _____ _____

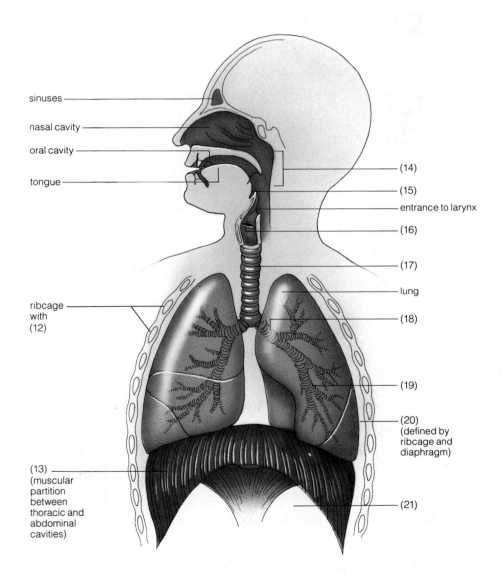

sinuses

nasal cavity

oral cavity

tongue

ribcage with (12)

(13)
(muscular partition between thoracic and abdominal cavities)

(14)

(15)

entrance to larynx

(16)

(17)

lung

(18)

(19)

(20)
(defined by ribcage and diaphragm)

(21)

GAS EXCHANGE AND TRANSPORT
    Gas Exchange in Alveoli
    Gas Transport Between Lungs and Tissues
    Commentary: **When the Lungs Break Down**

Summary

Each alveolus is composed of a single layer of epithelial cells, surrounded by a thin basement membrane. Gas in alveoli diffuses across the alveolar wall, a thin film of interstitial fluid, and the capillary wall. While diffusion can move enough oxygen into the blood, not enough oxygen can dissolve in the blood to meet the needs of the body; hemoglobin increases the oxygen-carrying capacity by seventy times.

Inhaled air that reaches the alveoli contains much more oxygen than carbon dioxide; the opposite is true of blood in the lung capillaries. Oxygen diffuses into the blood plasma, then into red blood cells, where it rapidly binds with hemoglobin. Hemoglobin tends to give up oxygen when the partial pressure of oxygen is low, temperatures are high, and the pH declines—all conditions characteristic of tissues with greater metabolic activity.

The partial pressure of carbon dioxide in metabolically active tissues is greater than it is in blood flowing through the capillaries threading through them. Some remains dissolved in plasma or binds with hemoglobin but most of the carbon dioxide is transported in the form of bicarbonate; carbon dioxide combines with water to form carbonic acid (H_2CO_3), which separates into bicarbonate and hydrogen ions. In red blood cells, an enzyme speeds this reaction by 250 times and maintains a gradient to keep carbon dioxide diffusing from interstitial fluid into the bloodstream. Bicarbonate tends to diffuse out of the red blood cells into the plasma. In the alveoli of the lungs, reactions proceed in reverse because of low partial pressure of CO_2; it is then exhaled from the body.

The defenses of the lungs sometimes break down. A condition called *bronchitis* occurs when smoking and other irritants increase the secretion of mucus and diminish cilia function and numbers. Coughing attempts to clear mucus; scar tissue forms that obstructs the respiratory tract. *Emphysema* results when extensive scar tissue builds up and gas exchange is compromised; it is related to environmental conditions, diet, infections, and genetics. The effects of cigarette smoke prevent cilia from beating and stimulate production of mucus; related coughing can contribute to bronchitis and emphysema. Cigarette smoke kills phagocytic cells in respiratory epithelium; it also contains several carcinogenic compounds. Smoking causes eighty percent of all deaths due to lung cancer.

Key Terms

partial pressure	carbon dioxide transport	bronchitis
gradients	bicarbonate	emphysema
gas transport	carbonic acid	
oxygen transport		

Objectives

1. Explain why oxygen diffuses from alveolar air spaces, through interstitial fluid, and across capillary epithelium. Then explain why carbon dioxide diffuses in the reverse direction.
2. Explain why oxygen diffuses from the bloodstream into the tissues far from the lungs. Then explain why carbon dioxide diffuses into the bloodstream from the same tissues.

3. Describe what happens to carbon dioxide when it dissolves in water under conditions normally present in the human body.
4. State the role played by the enzyme involved in carbon dioxide transport.
5. Distinguish *bronchitis* from *emphysema.* Then explain how lung cancer differs from emphysema.
6. Discuss the effects of smoking on the health of the human body.

Self-Quiz Questions

Fill-in-the-Blanks

Each (1) _____ is only a single layer of epithelial cells, surrounded by a thin basement membrane. Driven by its partial pressure gradient, oxygen diffuses from alveolar air spaces, through (2) _____ fluid, and into the lung (3) _____ . Carbon dioxide, driven by its partial pressure gradient, diffuses in the (4) _____ direction. When oxygen is moved between lungs and tissues, the (5) _____ of red blood cells increases the amount of oxygen transport by (6) _____ (number) times. Hemoglobin molecules tend to give up oxygen to tissues with greater (7) _____ activity. The partial pressure of carbon dioxide in metabolically active tissues is (8) [choose one] ( ) greater ( ) lesser, than it is in blood flowing through the capillaries threading through them. Carbon dioxide diffuses into the capillaries, and then it is transported to the (9) _____ . Some carbon dioxide remains dissolved in plasma or bound to hemoglobin, but most is transported in the form of (10) _____ .

Bicarbonate forms when (11) _____ _____ combines with water in plasma to form (12) _____ _____ , which separates into (13) _____ and hydrogen ions. Much of the carbon dioxide diffuses into (14) _____ _____ cells where an enzyme speeds this reaction by (15) _____ (number) times. The action of this enzyme maintains a gradient that keeps (16) _____ _____ diffusing from interstitial fluid into the (17) _____ . The reactions are reversed in the (18) _____ , where the partial pressure of carbon dioxide is (19) [choose one] ( ) lower ( ) higher than it is in the capillaries. (20) _____ is a disorder that can be brought on by smoking and other forms of air pollution that increase secretions of mucus and interfere with ciliary action in the lungs. When extensive scar tissue builds up in the lung and obstructs the respiratory tract, the condition is called (21) _____ . (22) _____ _____ prevents respiratory cilia from beating, stimulates production of mucus, contributes to

bronchitis and emphysema, and kills phagocytic cells in respiratory
epithelium. Smoking causes (23) _____ percent of lung cancer deaths.

28-IV
(pp. 413–416)

CONTROLS OVER RESPIRATION
    Matching Air Flow to Blood Flow
    Hypoxia
SUMMARY

Summary

Gas exchange in the alveoli is most efficient when air flow equals the rate of
blood flow. Both rates can be adjusted locally (in the alveoli) and through the
entire body. The nervous system controls oxygen and carbon dioxide levels in
arterial blood for the entire body. Control mechanisms in the nervous system
involve adjusting diaphragm and chest wall muscles; this adjusts rate and depth
of breathing. Sensory receptors signal the brain when carbon dioxide levels in
the blood rise. The brain also receives information from receptors in the walls
of certain arteries that detect partial pressures of dissolved oxygen and carbon
dioxide in arterial blood. The brain responds by increasing the rate and depth
of respiration; this increases oxygen delivery and carbon dioxide removal to and
from tissues. Local controls in the lungs adjust imbalances between air flow and
blood flow. The diameter of arterioles increases or decreases in response to
varying needs.

    A deficiency of cellular oxygen is referred to as *hypoxia*. At high altitudes the
partial pressure of oxygen is lower than at sea level, so that hyperventilation
results. Hypoxia can also occur when carbon monoxide poisoning causes the
oxygen content of arterial blood to fall. Carbon monoxide combines with hemoglobin 200 times faster than oxygen does.

Key Terms

air flow	hypoxia	carbon monoxide
blood flow	hyperventilation	poisoning

Objectives

1. Describe how the body adjusts air flow rate to match blood flow rate.
2. Define *hypoxia* and describe conditions that may cause it.

**Self-Quiz
Questions**

Fill-in-the-Blanks

Gas exchange is most efficient when the rate of (1) _____ flow is

matched with the rate of (2) _____ flow. Both rates can be adjusted

locally in the (3) _____ of the lungs and through the body as a whole.

The (4) _____ system controls oxygen and carbon dioxide levels in

(5) _____ blood for the entire body. Sensory receptors in the brain and

walls of arteries detect changes in the partial pressure of (6) _____ and (7) _____ _____ in arterial blood. The brain responds by increasing the rate of (8) _____ , so more (9) _____ can be delivered to affected tissues and more (10) _____ _____ is removed from tissues. Local control of air flow and blood flow imbalances occurs in the (11) _____ by increase or decrease of arteriole diameters. (12) _____ occurs when tissues do not receive enough oxygen. At high altitudes the partial pressure of oxygen is lower than at sea level, so that (13) _____ results. Hypoxia may also occur as a result of poisoning by (14) _____ _____ .

CHAPTER TEST

UNDERSTANDING AND INTERPRETING KEY CONCEPTS

___ (1) Most forms of life depend on _____ to deliver oxygen to and eliminate carbon dioxide from body cells.

 (a) active transport
 (b) bulk flow
 (c) diffusion
 (d) osmosis
 (e) muscular contractions

___ (2) Animals like flatworms and earthworms depend on _____ for diffusion of carbon dioxide and oxygen.

 (a) tracheas
 (b) gills
 (c) integumentary exchange
 (d) lungs

___ (3) With respect to one type of respiratory system, water moves over fish gills and blood circulates through them in _____ directions.

 (a) various
 (b) the same
 (c) opposite
 (d) circular

___ (4) When air is inhaled, air from the larynx moves into the _____ next.

 (a) pharynx
 (b) bronchi
 (c) lungs
 (d) trachea
 (e) epiglottis

___ (5) Immediately before air reaches the alveoli, it passes through the _____ .

   (a) bronchioles
   (b) glottis
   (c) larynx
   (d) pharynx
   (e) trachea

___ (6) During inhalation, the _____ .

   (a) pressure in the chest cavity is less than the pressure within the lungs
   (b) pressure in the chest cavity is greater than within the lungs
   (c) diaphragm moves upward and becomes more curved
   (d) chest cavity volume decreases
   (e) all of the above

___ (7) Hemoglobin _____ .

   (a) releases oxygen more readily to active tissues
   (b) tends to release oxygen in places where the temperature is lower
   (c) tends to hold on to oxygen when the pH of the blood drops
   (d) tends to give up oxygen in regions where partial pressure of oxygen exceeds that in the lungs
   (e) all of the above

___ (8) Oxygen moves from alveoli to the bloodstream _____ .

   (a) whenever the concentration of oxygen is greater in alveoli than in the blood
   (b) by means of active transport
   (c) by using the assistance of hemoglobin bound to carbon dioxide
   (d) principally due to the activity of an enzyme in the red blood cells
   (e) all of the above

___ (9) Hemoglobin releases O_2 when _____ .

   (a) carbon dioxide concentrations are high
   (b) body temperature is lowered
   (c) pH values are high
   (d) hemoglobin approaches tissues with low metabolic activity
   (e) all of the above

___ (10) In humans, most of the carbon dioxide is transported as _____ .

   (a) carbonic acid
   (b) carbon dioxide bound to hemoglobin
   (c) bicarbonate
   (d) carbon dioxide dissolved in plasma

INTEGRATING AND APPLYING KEY CONCEPTS

Consider the amphibians—animals that generally have aquatic larval forms (tadpoles) and terrestrial adults. Outline the respiratory changes that you think might occur as an aquatic tadpole metamorphoses into a land-going juvenile.

29

SOLUTE-WATER BALANCE

CONTROL OF EXTRACELLULAR FLUID
    Water Gains and Losses
    Solute Gains and Losses
    Urinary System of Mammals
    Nephron Structure
URINE FORMATION
    Filtration of Blood

Reabsorption of Water and Solutes
Secretion
Acid-Base Balance
Kidney Failure
Commentary: On Fish, Frogs, and
    Kangaroo Rats
SUMMARY

General Objectives	1. Explain how the vertebrate kidneys function in maintaining the internal environment by balancing intake and output of water and solutes.
	2. Describe urine formation in kidney nephrons, the necessary reabsorption of water and solutes, and secretion.
	3. Explain what occurs in kidney failure.
	4. Compare water and solute balance in animals occupying desert, saltwater, and freshwater environments.

29-I
(pp. 419–421)

CONTROL OF EXTRACELLULAR FLUID
    Water Gains and Losses
    Solute Gains and Losses
    Urinary System of Mammals
    Nephron Structure

Summary

Normally, on a daily basis mammals take in an amount of water that balances the amount of water lost. Gains in water result from absorption in the GI tract and as a product of metabolism. Thirst behavior is controlled by the nervous system and affects water intake. Losses in water occur from urinary excretion, respiratory surface evaporation, evaporation through the skin, sweating, and elimination by the GI tract. Urinary *excretion* from the kidneys (excess water and harmful solutes) is the most important factor in controlling water loss. Evaporation from the respiratory surface and skin evaporation are called "insensible water losses." Sweating is governed by control centers in the nervous system. Water in the GI tract is nearly all reabsorbed by the large intestine.

Solutes are added to the internal environment by absorption, reabsorption, and metabolism. Waste products include CO_2 (the most abundant metabolic waste), ammonia (formed when amino groups are stripped from amino acids), urea, and uric acid (from nucleic acid degradation).

Urine forms in paired *kidneys*; it is composed of water, mineral ions, and organic wastes. Only a small portion of the water and solutes entering the kidneys in the blood leaves as urine. Kidneys regulate the volume and solute concentrations of extracellular fluids. Each kidney is composed of a cortex and medulla. Urine forms in *nephrons*, enters the renal pelvis, travels through the ureter to the urinary bladder, and leaves the body through the urethra. The *urinary system* is composed of two kidneys, two ureters, a urinary bladder, and a urethra. Kidney stones are deposits of uric acid that collect in the renal pelvis and ureters or lodge in the urethra.

Each kidney has more than a million tubular nephrons. The nephron wall is a layer of epithelial cells; some wall regions are highly permeable to water and solutes, others bar solute passage. *Filtration* occurs at the glomerular end of each of these two million nephrons. Each *glomerulus* is a network of capillaries where blood flow suddenly encounters great resistance and where much of the liquid portion (the plasma) leaks through the capillary wall and diffuses across the encasing *Bowman's capsule* wall and into the interior of each capsule. Filtrate then proceeds through the *proximal tubule,* then through the *loop of Henle* and a *distal tubule,* and then through a collecting duct, eventually reaching the renal pelvis. Glomeruli inside Bowman's capsule do not send blood back to the general circulation immediately. Capillaries converge to form an arteriole that emerges from the capsule and branches into another set of capillaries that surround the remainder of the nephron, which recaptures water and essential solutes; they eventually merge to form veins that carry blood from the kidney.

Key Terms

thirst behavior	urine	urine flow
excretion	cortex	kidney stones
nutrients	medulla	filtration
mineral ions	nephrons	glomerulus
waste products	renal pelvis	Bowman's capsule
ammonia	ureter	proximal tubule
urea	urinary bladder	loop of Henle
kidneys	urethra	distal tubule
uric acid	urinary system	

Objectives

1. List two processes that account for water gain in mammals; list five processes of water loss.
2. List successively the parts of the human urinary system that constitute the path of urine formation and excretion.
3. List the parts of a nephron. Begin with the glomerulus and briefly state the function of each part.
4. List three soluble by-products of animal metabolism that are potentially toxic.

Self-Quiz Questions

True-False

If false, explain why.

___ (1) Ordinarily, mammals take in more water than they lose on a daily basis.

___ (2) Thirst behavior, under nervous system control, has no effect on water intake.

___ (3) The most important process in controlling water loss is urinary excretion.

___ (4) The subunits in the human kidney that process urine are called *nephridia*.

___ (5) Of the three principal waste products excreted by vertebrates, ammonia is the most highly toxic.

Fill-in-the-Blanks

Two processes account for daily water gains: absorption of water from liquids and solid foods in the (6) _____ _____ and (7) _____ . The elimination of excess water and excess (or harmful) solutes from the internal environment through the kidneys is called (8) _____ . Water loss by evaporation from the respiratory surface and evaporation through the skin are known as (9) _____ water losses. Other than oxygen, (10) _____ are added to the internal environment by absorption from the GI tract, reabsorption, and metabolism. Other than carbon dioxide, the major metabolic wastes that must be eliminated are (11) _____ , formed by amino groups stripped from amino acids, (12) _____ produced in the liver by reactions linking ammonia to carbon dioxide and, (13) _____ _____ , formed when nucleic acids are degraded. In the mammalian urinary system, a pair of organs called (14) _____ filter water, mineral ions, organic wastes, and other substances from the blood. Only a very small portion of the water and solutes going into the kidneys leaves as a fluid called (15) _____ . Through their action, kidneys regulate the (16) _____ and (17) _____ concentrations of extracellular fluid. Each kidney has two zones, a(n) (18) _____ and a(n) (19) _____ . Each kidney lobe contains many filtration units called (20) _____ ; (21) _____ and solutes enter the nephrons from the blood. Urine flows from each kidney into a(n) (22) _____ , then into a(n) (23) _____ _____ ; it leaves through a tube, the (24) _____ . In nephron function, filtration begins at the (25) _____ where the nephron wall balloons around blood capillaries. (26) The _____ _____ is a cup formed by the nephron wall to receive filtered water and solutes. The filtrate flows through the (27) _____ tubule, then through the loop of (28) _____ and a(n) (29) _____ tubule before entering a collecting duct. Water and essential nutrients are recaptured by (30) _____ that surround the remainder of the nephron.

Labeling

Identify each indicated part of the accompanying illustration.

(31) _____

(32) _____ _____

(33) _____ _____

(34) _____ _____

(35) _____

(36) _____ _____

(37) _____ _____ _____

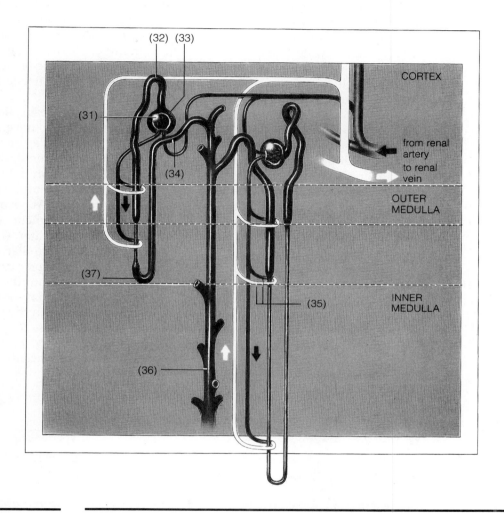

29-II
(pp. 421–424)

URINE FORMATION
    Filtration of Blood
    Reabsorption of Water and Solutes
    Secretion

Summary

Urine forms through three processes, called *filtration, reabsorption*, and *secretion*. Kidneys rapidly filter about 45 gallons of blood every day. Filtration is a process in which essentially protein-free plasma is driven by a relatively high-pressure difference from the blood in the knot of capillaries that makes up each glomerulus across two layers of cells into the cavity of the Bowman's capsule. Filtration is rapid because arterioles delivering blood to glomeruli have a wider diameter (less resistance) than most arterioles; in addition, glomerular capillaries are highly permeable ("leaky") to water and small solutes when compared to other capillaries. With the plasma go substances both useful and not useful to the human body.

As the plasma passes along the length of the nephron tubule, the useful substances (water and most solutes) are reabsorbed and return to general circulation via the surrounding capillaries. *Reabsorption* increases or decreases with water intake.

Large volumes of water are reabsorbed. Water reabsorption is controlled by the hypothalamus through secretion of *antidiuretic hormone* (ADH); *ADH* helps to conserve water when needed. ADH makes the walls of distal tubules and collecting ducts more permeable to water, and thus the urine becomes more concentrated as water leaves to enter the surrounding capillaries. ADH secretion is inhibited when the body must dispose of excess water; urine is more dilute. A thirst mechanism is controlled by the hypothalamus; when needed it stimulates ADH secretion and water-seeking behavior.

The major solute in extracellular fluid is sodium; its reabsorption is controlled by the adrenal cortex. When too much sodium is lost, extracellular fluid volume is reduced, and sensory receptors in blood vessel walls and the heart detect the drop. Then some kidney cells secrete an enzyme, renin; this causes the adrenal cortex to secrete *aldosterone*. Aldosterone stimulates sodium reabsorption in the distal tubules and collecting ducts. Sodium retention is accompanied by water retention; this raises blood pressure. Abnormally high blood pressure, or hypertension, can adversely affect kidney function as well as the vascular system and brain. Substances not useful (excess water, solutes) are secreted from the same capillaries into the nephron. The capillaries secrete hydrogen and potassium ions, foreign substances (for example, drugs), uric acid, products of hemoglobin breakdown, and other wastes into the nephron. *Filtration, reabsorption* of water and solutes, and *secretion* of excess toxic substances by the nephrons (which are surrounded by capillaries in the kidney) all influence the ultimate composition and volume of urine; thus, they influence how much water and solute the body conserves. The longer the loop of Henle, the greater the animal's capacity to conserve water and to concentrate solutes for excretion in urine.

Key Terms

filtration	antidiuretic hormone	aldosterone
reabsorption	(ADH)	hypertension
hypothalamus	renin	secretion

Objectives

1. Describe, in some structural and functional detail, the three processes responsible for urine formation.
2. Explain what mechanisms allow the body to rid itself of excess water and what mechanisms stimulate the body to obtain more water.
3. Summarize the mechanisms by which the body controls reabsorption of sodium and possible problems that may arise.
4. Briefly relate the events of secretion occurring at the nephron.

Fill-in-the-Blanks

Filtration, reabsorption, and secretion are responsible for (1) _____ formation. Filtration occurs when (2) _____ _____ forces filtrate out of the glomerular capillaries into the (3) _____ _____ of a nephron, then into the proximal nephron tubule. Kidneys rapidly filter about (4) _____ gallons of blood daily. Reabsorption occurs as water and surrounding capillaries reclaim water and usable (5) _____ from permeable regions of the (6) _____ . Reabsorption increases or decreases with the level of (7) _____ intake. A part of the brain called the (8) _____ controls water reabsorption through secretion of (9) _____ , which makes the walls of the distal tubules and collecting ducts more (10) _____ to water, thus causing (11) _____ to become more concentrated. When the body must dispose of excess water, ADH secretion is (12) _____ . The major solute in extracellular fluid is (13) _____ . When too much sodium is lost, certain kidney cells secrete (14) _____ , which causes the adrenal cortex to secrete (15) _____ ; this stimulates sodium (16) _____ . Sodium retention is accompanied by (17) _____ retention, which may cause abnormally high blood pressure, or (18) _____ . Secretion occurs at the capillaries surrounding the (19) _____ tubules; the capillaries secrete hydrogen and potassium (20) _____ , foreign substances, uric acid, products of hemoglobin breakdown, and other wastes into the (21) _____ .

29-III
(pp. 423–425)

URINE FORMATION (cont.)
   Acid-Base Balance
   Kidney Failure
   Commentary: **On Fish, Frogs, and Kangaroo Rats**
SUMMARY

Summary

The kidneys help to keep the extracellular environment from becoming too acidic or too basic. Overall acid-base balance is maintained by controls exerted through buffer systems, respiration, and urinary excretion. The pH of human extracellular fluid is normally between 7.35 and 7.45. Kidneys help to buffer the blood by excreting excess hydrogen ions (H^+) and restoring HCO_3^-. Bicarbonate combines with H^+ to form carbonic acid, which dissociates into water and carbon dioxide that may be exhaled. Only the urinary system can eliminate excess amounts of H^+.

An estimated 13 million people in the United States suffer from kidney disorders; diabetes or autoimmune responses damage the glomeruli and interfere with urine formation. Ions and toxic by-products of protein breakdown can accumulate in the bloodstream; sickness and death may occur. Kidney dialysis machines are used to restore proper solute balances; both hemodialysis and peritoneal dialysis are used. With proper treatment and diet, individuals may live fairly normal lives.

Vertebrates are adapted to maintaining water and solute levels in entirely different environments. Bony fish in seawater live in a hypertonic environment and tend to lose water by osmosis; they continually drink and pump out excess solutes through their gills. Freshwater fish tend to gain water from their hypotonic environment by osmosis; their kidneys pump out excess water as dilute urine. Water enters the frog body by osmosis through the skin; water leaves in large volumes of dilute urine from well-developed kidneys. Kangaroo rats in the desert gain most of their water through metabolic oxidation of seeds; their urine and feces contain very little water.

Key Terms

acid-base balance	kidney failure	hemodialysis
extracellular pH	kidney dialysis machine	peritoneal dialysis

Objectives

1. Explain how the kidney helps to buffer the extracellular fluids of the body and maintains the internal pH between 7.35 and 7.45.
2. List two kidney disorders and explain what can be done if kidneys become too diseased to work properly.
3. Describe how fish living in seawater and those living in freshwater control the solute and water balance of their internal environments.

Self-Quiz Questions

Fill-in-the-Blanks

Controls exerted through buffer systems, respiration, and kidney excretion maintain an overall bodily (1) _____ - _____ balance. The (2) _____ of human extracellular fluid is normally between 7.35 and 7.45. The (3) _____ help to buffer the blood by excreting excess (4) _____ ions and reabsorbing (5) _____ . Only the urinary system can eliminate excess amounts of (6) _____ ions. (7) _____ or (8) _____ responses can damage glomeruli and interfere with (9) _____ formation. In cases of persons with nonfunctioning kidneys, proper bloodstream solute balances may be restored by use of (10) _____ _____ machines. Bony fish in seawater live in a(n) (11) _____ environment and tend to (12) _____ water by osmosis; they continually (13) _____ and pump out excess (14) _____ through their gills. Freshwater fish tend to (15) _____ water by osmosis from their (16) _____ environment; their kidneys pump out excess (17) _____ in dilute (18) _____ .

Kangaroo rats in the desert gain most of their water through (19) _____ oxidation of seeds; their urine and feces are very (20) _____ .

CHAPTER TEST

UNDERSTANDING AND INTERPRETING KEY CONCEPTS

___ (1) The most toxic waste product of metabolism is _____ .

    (a) water
    (b) uric acid
    (c) urea
    (d) ammonia
    (e) carbon dioxide

___ (2) An entire subunit of a kidney that purifies blood and restores solute and water balance is called a _____ .

    (a) glomerulus
    (b) loop of Henle
    (c) nephron
    (d) ureter
    (e) none of the above

___ (3) In humans, the thirst center is located in the _____ .

    (a) adrenal cortex
    (b) thymus
    (c) heart
    (d) adrenal medulla
    (e) hypothalamus

___ (4) Beginning with the kidney, which is the correct order of structures through which urine passes when eliminated from the body?

    (a) kidney $\longrightarrow$ urethra $\longrightarrow$ urinary bladder $\longrightarrow$ ureter
    (b) kidney $\longrightarrow$ urinary bladder $\longrightarrow$ urethra $\longrightarrow$ ureter
    (c) kidney $\longrightarrow$ ureter $\longrightarrow$ urinary bladder $\longrightarrow$ urethra
    (d) kidney $\longrightarrow$ urinary bladder $\longrightarrow$ ureter $\longrightarrow$ urethra

___ (5) The process of greatest importance in controlling water loss from the body is _____ .

    (a) evaporation from the respiratory surface
    (b) evaporation through the skin
    (c) sweating
    (d) excretion by way of the urinary system
    (e) elimination by way of the gastrointestinal tract

___ (6) Filtration of the blood in the kidney begins in the _____ .

    (a) loop of Henle
    (b) proximal tubule
    (c) distal tubule
    (d) glomerulus
    (e) all of the above

___ (7) When the hypothalamus secretes ADH, _____ is being controlled.

   (a) sodium reabsorption
   (b) secretion
   (c) aldosterone
   (d) water reabsorption
   (e) filtration

___ (8) Secretion of excess ions, foreign substances, uric acid, and other wastes occurs when these substances move out of the _____ and into the _____ .

   (a) glomerulus; tubules
   (b) capillaries around the nephron; nephron
   (c) tubules; glomerulus
   (d) nephron; capillaries around the nephron
   (e) loop of Henle; glomerulus

___ (9) Only the urinary system can eliminate excess amounts of _____ .

   (a) oxygen
   (b) H^+
   (c) carbon dioxide
   (d) water

___ (10) The desert-dwelling kangaroo rat obtains most of its water through _____ .

   (a) evaporation from its skin
   (b) excretion
   (c) evaporation from its lungs
   (d) drinking
   (e) metabolism

INTEGRATING AND APPLYING KEY CONCEPTS

The hemodialysis machine used in hospitals is expensive and time-consuming. Artificial kidneys capable of allowing people who have nonfunctional kidneys to purify their blood by themselves, without having to go to a hospital or clinic, have not been developed. Which aspects of the hemodialysis procedure do you think have presented the most problems in developing a method of home self-care? If *you* had an unlimited budget and were appointed head of a team to develop such a procedure and its instrumentation, what strategy would you pursue?

30
NEURAL CONTROL AND THE SENSES

WHAT NERVOUS SYSTEMS DO
NEURONS
    Structure and Function of Neurons
    Neural Messages
    Synapses
PATHS OF INFORMATION FLOW
VERTEBRATE NERVOUS SYSTEMS
    Peripheral Nervous System
    Central Nervous System
THE HUMAN BRAIN
    The Cerebral Hemispheres
    Memory

Emotional States
Commentary: Drug Action on Integration and Control
States of Consciousness
SENSORY INPUT
    Classes of Receptors
    Somatic Senses and Pain Perception
    Taste and Smell
    Hearing
    Vision
SUMMARY

General Objectives

1. Describe the visible structure of neurons and nerves.
2. Understand how a nerve impulse is received by a neuron, conducted along a neuron, and transmitted across a synapse to a neighboring neuron, muscle, or gland.
3. Outline some of the ways by which information flow occurs and is regulated in the human body.
4. Describe the organization of the peripheral versus the central nervous systems.
5. Tell how the human brain is advanced beyond primitive types.
6. Know what a receptor is, list the various types of receptors, and distinguish them from each other.

30-I
(pp. 426–428)

WHAT NERVOUS SYSTEMS DO
NEURONS
    Structure and Function of Neurons

Summary

Vertebrate nervous systems continually sense, interpret, and issue commands for responses to specific aspects of the environment. Their communication lines are highly organized gridworks of neurons.

    The *neuron* is the basic unit of communication; each neuron acts with many others in a system that keeps all parts of the body in communication with each other and with their environment. Complex animals have three classes of nerve cells (sensory neurons, interneurons, and motor neurons). *Sensory neurons* are receptors for specific sensory stimuli (different forms of energy that can be detected by receptor cells). Each stimulus (such as light or pressure) causes an electrical disturbance at the receptor's surface; a message is then sent to the integrators, the brain and spinal cord. The integrators contain *interneurons* that connect with the motor neurons. *Motor neurons* send information from integrator to muscle or gland cells (effectors).

The cell body of a neuron contains the nucleus and metabolic machinery for protein synthesis; neuronal processes include *dendrites* (with the cell body they are input zones), an *axon*, and finely branched endings (output zones) that terminate next to other neurons—glandular cells or muscle cells.

Key Terms

neuron	integrators	effectors
sensory neurons	interneurons	dendrites
receptors	motor neurons	axon

Objectives

1. Define *neuron;* describe the three classes of neurons and tell how each is related to the other.
2. Draw a neuron and label it according to its three general zones, its specific structures, and the specific function(s) of each part.

Self-Quiz Questions

Fill-in-the-Blanks

The basic unit of communication is the (1) _____ . Complex animals have (2) _____ (number) classes of nerve cells. (3) _____ neurons are receptors for specific sensory stimuli. The integrators are the (4) _____ and the (5) _____ _____ ; they contain (6) _____ that connect with the (7) _____ neurons. (8) _____ neurons send information from integrator to muscle or gland cells. Processes (structures) extending from the cell body include input zones called (9) _____ and a(n) (10) _____ with finely branched endings (output zones) that terminate next to muscle cells.

30-II
(pp. 428–433)

NEURONS (cont.)
    Neural Messages
    Synapses

Summary

The resting plasma membrane of the neuron shows a polarity because there is more Na^+ outside than there is inside. A stimulated neuron permits ions to flow across and reverse the polarity—an *action potential*. This membrane excitability is influenced by (1) membrane lipids that limit ion flow; (2) ions crossing the membrane via ion channels that may be open all the time or gated and only open during action potentials; (3) membrane pumps, like the Na^+/K^+ pump, which actively transport ions and restore membrane potential.

At rest, there are more potassium ions inside the neuron than outside and far fewer sodium ions inside than outside. At rest, channels for K^+ are open, and K^+ tends to leak out until there is no net movement of K^+; Na^+ channels are shut and so do not contribute to the resting potential. Now there is a steady "voltage difference" across the membrane; this amount, the *resting membrane potential*, is about 70 millivolts for many neurons.

An action potential is a pulse of electrical activity. At rest, the inside of the neuron is more negative than the outside. During an action potential, the membrane depolarizes. After an action potential, the membrane is repolarized (the resting condition is restored). An action potential is triggered when sufficient stimuli reach a trigger zone and have at least a *threshold level* of energy; depolarization results from a brief flow of Na^+ into the neuron (caused by some gated Na^+ channels opening). The action potential is an *all-or-nothing event*. Most action potentials last a few milliseconds. Each action potential ends because depolarization causes the Na^+ channels to close and the K^+ channels to open; the flow of K^+ out of the neuron restores the resting potential. Following the occurrence of an action potential in a trigger zone, it propagates itself down the neuron. As action potentials travel away from the stimulation site, the opening of gated channels is repeated. A refractory period following each action potential helps prevent backflow; Na^+ gates are shut and K^+ gates are open. During the refractory period, the membrane is insensitive to stimulation.

Many axons serve as communication lines between body regions; they are covered by Schwann cells that form a myelin sheath (a jellyroll-like wrapping) around the axon. Each cell is separated from adjacent ones by a small, exposed node where the axonal membrane is loaded with gated channels. Action potentials jump rapidly from node to node.

When action potentials reach the axonal endings of a neuron, they trigger release of *transmitter substance* into the junction between the neuron and an adjacent cell. A *chemical synapse* is the junction between two neurons separated by a synaptic cleft. In synapses between motor neurons and muscle cells, the branched axonal endings are positioned at the muscle cell membrane. An action potential traveling down the motor neuron spreads through all the endings and acetylcholine (ACh), a transmitter, is released into the synaptic cleft. Muscle cell membrane receptors bind to ACh; the receptors recognize the transmitter ACh and ion channels open. This causes an action potential in the muscle cell and contraction results. *Synaptic integration* is the combining of excitatory and inhibitory signals acting on adjacent regions of a neuronal membrane.

Foreign substances such as *Clostridium tetani* toxin can interfere with synaptic integration, and tetanus results.

Key Terms

membrane excitability
action potential
resting membrane
  potential
threshold level

all-or-nothing event
refractory period
myelin sheath
transmitter substance

chemical synapse
synaptic integration
Clostridium tetani
tetanus

Objectives

1. Describe the distribution of the invisible array of large proteins, ions, and other molecules in a neuron, both at rest and as a neuron experiences a change in potential.

2. List the three membrane properties on which *excitability* depends.
3. Define *resting membrane potential;* explain what establishes it, and how it is used by the cell neuron.
4. Describe what happens to the gradients associated with the neuronal membrane when it is disturbed (during an action potential).
5. Explain the function of a myelin sheath.
6. Describe the events occurring at chemical synapses and relate them to synaptic integration.
7. Give an example of what may occur when a foreign substance interferes with synaptic integration.

Self-Quiz Questions

Fill-in-the-Blanks

The polarity of the neuronal plasma membrane results from differences in the concentrations of (1) _____ ions, (2) _____ ions, and other charged substances in cytoplasm and extracellular fluid. Upon stimulation, the polarity of charge across the plasma membrane can undergo a sudden reversal called a(n) (3) _____ _____ in which the membrane depolarizes. Following this, the membrane is (4) _____ and the (5) _____ condition is restored. An action potential is triggered when sufficient stimuli reach a(n) (6) _____ zone and have at least a(n) (7) _____ level of energy. Depolarization results from a brief flow of (8) _____ into the neuron. The action potential is a(n) (9) _____ - _____ - _____ event. Each action potential ends because depolarization causes the (10) _____ channels to close and the (11) _____ channels to open; the flow of (12) _____ out of the neuron restores the (13) _____ _____ . After an action potential occurs in a trigger zone, it (14) _____ itself down the neuron away from the stimulation site. A(n) (15) _____ period follows each action potential and helps prevent backflow of ions; (16) _____ gates are shut and (17) _____ gates are open during this time. Many axons are covered by Schwann cells that form a(n) (18) _____ sheath around the axon; each cell is separated from the next by a small, exposed (19) _____ ; (20) _____ _____ jump rapidly from node to node. A(n) (21) _____ _____ is the junction between two neurons separated by a synaptic cleft. An action potential traveling down the axon of a motor neuron spreads through the (22) _____ positioned at the muscle cell membranes. Acetylcholine, a(n) (23) _____ is released into the synaptic cleft. This causes an action potential in the muscle cell and (24) _____ results.

(25) _____ _____ is the combining of excitatory and inhibitory signals acting on adjacent membrane regions of a neuron. Foreign substances like *Clostridium tetani* (26) _____ can interfere with synaptic integration, and tetanus results.

30-III
(pp. 433–435)

PATHS OF INFORMATION FLOW
VERTEBRATE NERVOUS SYSTEMS
   Peripheral Nervous System

Summary

Through operation of synaptic integration, signals arriving at any given neuron in the body can be reinforced, sent on, or suppressed. The direction a given signal will travel depends on the organization of neurons into circuits or pathways. Neuronal activity in the brain is confined to local circuits; cordlike *nerves* form communication lines that carry signals between the brain or spinal cord and other body regions. In a nerve, axons of sensory neurons, motor neurons, or both are bundled together. Such bundles within the brain and spinal cord are called *nerve tracts* or *pathways*.

Sensory and motor neurons of many nerves participate in reflexes. In a *reflex arc*, sensory neurons synapse directly on motor neurons. The stretch reflex is an example; it works to contract a stretched muscle. Stretching activates certain muscle receptors that are input zones of sensory neurons that synapse with motor neurons in the spinal cord. Axons of motor neurons lead back to the stretched muscle, and action potentials that reach axon endings trigger ACh release, which initiates contractions. Continued receptor activity excites the motor neurons, allowing stretching to be maintained. A rapid pulling away from an unpleasant or harmful stimulus is called the *withdrawal reflex*. In these reflexes, sensory neurons make connections with interneurons that activate or suppress the motor neurons necessary for a coordinated response. The reflex action can be completed before one is conscious of the stimulus.

The complex vertebrate nervous system evolved by additions of nervous tissue over more ancient reflex pathways. The newest layerings still deal with reflexes but they also form the basis for memory, learning, reasoning, and initiating new responses. The *central nervous system* includes the brain and spinal cord. The *peripheral nervous system* includes all the nerves carrying signals to and from the brain and spinal cord. Both divisions include many neuroglial cells that protect or assist neurons.

The peripheral nervous system has thirty-one pairs of spinal nerves connecting with the spinal cord and twelve pairs of cranial nerves connecting directly to the brain. The somatic system of nerves relays commands to skeletal muscles; the autonomic system of nerves relays commands to smooth muscles and glands. Autonomic nerves have two roles in the overall function of the body. When there is little stress, *parasympathetic nerves* tend to slow down overall body activity; during time of excitement or danger, *sympathetic nerves* prepare one for fight or flight. The rate of internal organ function is determined by the net outcome of opposing sympathetic and parasympathetic signals. *Biofeedback* refers to conscious efforts to enhance or dampen autonomic and other physiological responses.

Key Terms

nerves	central nervous system	autonomic system
reflex arc	peripheral nervous	parasympathetic nerves
stretch reflex	system	sympathetic nerves
withdrawal reflex	somatic system	biofeedback

Objectives

1. Describe paths of information flow within the brain and between the brain, spinal cord, and other body parts.
2. List the components of a *nerve.*
3. Describe all mechanisms of the *reflex arc.*
4. Define and contrast *central* and *peripheral* nervous systems.
5. Distinguish the somatic and autonomic nervous systems with respect to location and chief activities.
6. Explain how parasympathetic nerve activity balances sympathetic nerve activity; define *biofeedback* in connection with the preceding.

Self-Quiz Questions

Fill-in-the-Blanks

Through operation of (1) _____ _____ , signals arriving at any given neuron in the body can be reinforced, sent on, or suppressed. The direction of travel of a given signal depends on the organization of (2) _____ into circuits or pathways. Neuron activity in the brain is confined to (3) _____ circuits. Axons of sensory neurons, motor neurons, or both are bundled together to form a(n) (4) _____ . In a(n) (5) _____ _____ , sensory neurons synapse directly on motor neurons; the stretch reflex is an example. Stretching activates certain muscle (6) _____ that are input zones of sensory neurons that synapse with (7) _____ neurons in the spinal cord. Motor neuron axons lead back to the stretched muscle and (8) _____ _____ that reach axon endings trigger (9) _____ release, which initiates contractions. A rapid pulling away from an unpleasant or harmful stimulus is called the (10) _____ reflex. The (11) _____ nervous system consists of the brain and spinal cord; the (12) _____ nervous system includes all the nerves carrying signals to and from the brain and spinal cord. The (13) _____ system of nerves relays commands to skeletal muscles; the (14) _____ system of nerves relays commands to smooth muscles and glands; it is subdivided into two parts: (15) _____ nerves, which respond to emergency situations, and (16) _____ nerves, which tend to slow down overall body activity. The rate of internal organ function is determined by the net outcome of opposing

(17) _____ and (18) _____ signals. (19) _____ refers to conscious efforts to enhance or dampen autonomic and other physiological responses.

30-IV
(pp. 436–437)

VERTEBRATE NERVOUS SYSTEMS (cont.)
Central Nervous System

Summary

The spinal cord resides within a central canal formed by stacked vertebrae. Its gray matter contains cell bodies and dendrites concerned with reflexes. The white matter consists of major nerve tracts (bundles of sheathed axons) that ascend to and descend from specific brain centers.

The brain, an expanded extension of the spinal cord, is covered by membranes and protected by bones housing the cranial cavity. The brain has three major divisions: the hindbrain, midbrain, and forebrain.

The *hindbrain* includes the medulla oblongata, cerebellum, and pons. The medulla oblongata has reflex centers for respiration and blood circulation; coordination of motor responses and complex reflexes also occurs here. Other brain centers that help you sleep or wake are influenced by the medulla. Reflex centers of the cerebellum maintain posture and refine limb movements. Signal integration from eyes, muscle spindles, and elsewhere occur here. The cerebellum informs other parts of the brain about body positions, movements, and muscle contractions. Nerve tracts pass through the *pons* on their way between brain centers.

The *midbrain* originally coordinated reflexes associated with visual input. The tectum integrates visual and auditory signals. In fish and amphibians, the tectum exerts major control over the body (removal of the frog's cerebrum hardly affects its behavior). In mammals, sensory information still converges on the tectum, but is rapidly sent on to higher centers.

The most recent layerings of nerve tissues are in the *forebrain*. Early vertebrates evolved two olfactory lobes and a primitive cerebrum to integrate input of smell stimuli and to select proper motor responses to it. The thalamus relayed and coordinated sensory signals. The hypothalamus influenced thirst, hunger, and sex. In time, the cerebral cortex developed into an information processing center in mammals; it is most highly developed in the human brain.

Key Terms

central nervous system	pons	thalamus
hindbrain	midbrain	hypothalamus
medulla oblongata	forebrain	cerebral cortex
cerebellum	cerebrum	

Objectives

1. Describe the basic structural and functional organization of the spinal cord. In your answer, distinguish the spinal cord from the vertebral column.
2. Compare the structures of the spinal cord and brain concerning white matter and gray matter.
3. Describe the ways the brain is protected and list its three principal divisions.
4. List the parts of the brain found in the hindbrain, midbrain, and forebrain and tell the basic functions of each.

5. Tell what happened to the importance of the midbrain during the evolution of the vertebrates.

Self-Quiz Questions

Fill-in-the-Blanks

The (1) _____ _____ resides within a central canal formed by stacked vertebrae. The (2) _____ _____ of the spinal cord has cell bodies and dendrites concerned with reflexes. The (3) _____ _____ of the spinal cord contains major nerve tracts that ascend to and descend from specific brain centers. The (4) _____ , an expanded extension of the spinal cord, is covered by membranes and protected by bones housing the cranial cavity. The (5) _____ includes the medulla oblongata, cerebellum, and pons. The medulla oblongata has reflex centers for (6) _____ and (7) _____ _____ . Other brain centers that help you sleep or wake are influenced by the (8) _____ . Reflex centers of the (9) _____ maintain posture and refine limb movements. Signal integration from eyes, muscle spindles, and elsewhere occur in the (10) _____ . The (11) _____ informs other parts of the brain as to body positions, movements, and muscle contractions. Nerve tracts pass through the (12) _____ on their way between brain centers. The (13) _____ originally coordinated reflexes associated with visual input. The (14) _____ integrates visual and auditory signals. In (15) _____ , sensory information still converges on the tectum but is rapidly sent on to higher centers. The most recent layerings of nerve tissues are in the (16) _____ . Early vertebrates evolved two (17) _____ lobes and a primitive (18) _____ to integrate input of smell stimuli and to select proper motor responses to it. The (19) _____ relayed and coordinated sensory signals. The (20) _____ influenced thirst, hunger, and sex. In time, the (21) _____ _____ developed into an information processing center in mammals; the human brain has the most highly developed (22) _____ _____ .

THE HUMAN BRAIN
>   **The Cerebral Hemispheres**
>   **Memory**
>   **Emotional States**
>   *Commentary:* **Drug Action on Integration and Control**
>   **States of Consciousness**

Summary

The human cerebrum is divided into two parts, the cerebral hemispheres. A thin surface layer, the cerebral cortex, contains much of the gray matter; the surface is highly folded. The white matter of the cerebrum consists of major nerve tracts that allow the hemispheres to communicate with each other and the rest of the body. Some functional regions of the cortex are motor centers (which may deal with control of thumb and tongue muscles); other regions are primary receiving centers that coordinate sensory input. In association centers, stored memory is added to primary sensory information.

Memory is the storage and retrieval of information about previous experiences; information is stored in stages and it exists as memory traces. Chemical and structural changes necessary for information storage occur in many different brain regions. Short-term memory storage lasts a few seconds to a few hours and is limited to seven or eight bits of information. Long-term storage is more or less permanent and perhaps of limitless capacity.

States of consciousness include sleeping, dozing, daydreaming, and total alertness; these are governed by the central nervous system and altered by psychoactive drugs. A brain region, the *reticular activating system,* controls changing consciousness levels. Circuit damage may lead to unconsciousness and coma. Sleep center neurons in the reticular formation release serotonin, a transmitter that inhibits other neurons that arouse the brain and maintain wakefulness. Other brain substances counteract effects of serotonin.

The *limbic system* (parts of the cerebral cortex and several brain regions) governs our emotions; it is still closely associated with the sense of smell. The gatekeeper of the limbic system is the hypothalamus; neurons from the cerebral cortex and lower brain pass through it. Reasoning that involves the cerebral cortex can dampen rage and other gut reactions. Internal organs are also monitored by the hypothalamus; the relationship is made clear when one feels the effects of strong emotions (such as passion) in the heart and stomach.

Drugs have no nutritive value; when introduced to the body they elicit some effect on physiological processes. Many *psychoactive drugs* affect a pleasure center in the hypothalamus. Reinforced artificial stimulation of the pleasure center is called an *addiction.* Psychoactive drugs are societal problems. *Stimulants* include caffeine, nicotine, amphetamines, and cocaine. They initially increase alertness and body activity, then they lead to depression. *Depressants* and *hypnotics* lower activity in nerves and parts of the brain. Responses produced range from emotional relief, sedation, sleep, anesthesia, and coma to death. Combining alcohol with barbiturates amplifies behavioral depression. Long-term alcohol addiction destroys nerve cells and causes permanent brain and liver damage. *Analgesics* produced naturally by the body are endorphins and enkephalins; they affect brain centers concerned with emotions and pain perception. The extremely addictive narcotic analgesics include codeine and heroin; they sedate the body and relieve pain. The "mind-expanding" *psychedelics* and *hallucinogens* alter sensory perception. Some affect acetylcholine or norepinephrine activity; LSD affects serotonin activity. LSD, even in small doses, dramatically warps perceptions. Marijuana is another hallucinogen that in low doses acts as a depressant; the body relaxes in mild euphoria. It can produce disorientation, increased anxiety, delusions, and hallucinations. As with alcohol, marijuana can affect an individual's ability to perform complex tasks. Recent studies have established a link between marijuana smoking and immune system suppression. Drug abuse in the modern world is a major problem.

Key Terms

cerebral hemispheres	serotonin	depressants
memory	emotional states	hypnotics
limbic system	psychoactive drug	analgesics
states of consciousness	pleasure center	psychedelics
reticular activating system	stimulants	hallucinogens

Objectives

1. Describe the structure of the cerebral hemispheres and list the activities of each part.
2. State the current information that explains *memory*; distinguish between short-term and long-term memory.
3. Describe the function of the *reticular activating system*.
4. State the location, structure, and function of the *limbic system*.
5. List, define, and give examples of the different classes of psychoactive drugs.

Self-Quiz Questions

Matching

Match each letter with the most appropriate term.

(1) ____ motor cortex

(2) ____ limbic system

(3) ____ hypothalamus

(4) ____ cerebral white matter

(5) ____ reticular activating system

A. Major nerve tracts that maintain communication with the cerebral hemispheres
B. Controls the changing levels of consciousness
C. Controls our emotions
D. Deals with thumb and tongue muscles
E. Gatekeeper of the limbic system

Fill-in-the-Blanks

The human (6) _____ is divided into two parts called hemispheres; its thin surface layer, the (7) _____ _____ contains much of the gray matter. The (8) _____ matter consists of major nerve tracts that allow the hemispheres to communicate with each other and the rest of the body. (9) _____ centers of the cortex deal with control of thumb and tongue muscles; other regions are primary centers for receiving (10) _____ input. In (11) _____ centers, stored memory is added to primary sensory information. (12) _____ is the storage and retrieval of information about previous experiences. (13) _____ - _____ memory storage lasts a few seconds to a few hours; (14) _____ - _____ memory storage is more or less permanent. The (15) _____ _____ system controls consciousness levels. (16) _____ have no nutritive value and have some effect on physiological processes. Many psychoactive drugs affect a pleasure center in the (17) _____ . (18) _____ initially increase alertness and body activity, then they lead to depression. (19) _____ and (20) _____ lower activity in nerves and parts of the brain. Endorphins and enkephalins are natural (21) _____ . Sensory perceptions are altered by (22) _____ and (23) _____ . Recent studies have established a link between (24) _____ _____ and immune system suppression. The (25) _____ system governs our emotions and is still closely associated with the sense of smell; the gatekeeper of this system is the (26) _____ . Rage and other strong emotions can be dampened by reasoning that occurs in the (27) _____ _____ . The (28) _____ monitors internal organs.

Labeling   Identify each indicated part of the accompanying illustration.

(29) _____

(30) _____

(31) _____

(32) _____

(33) _____ _____

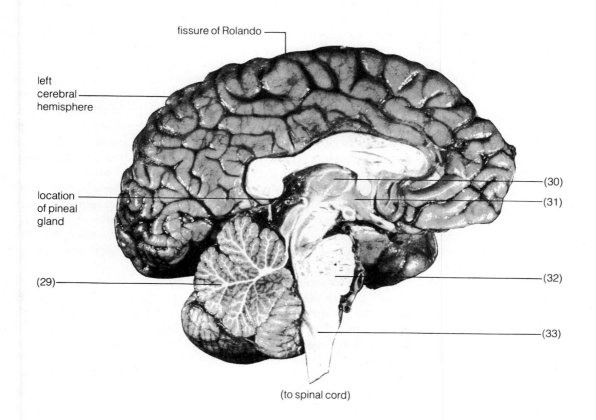

fissure of Rolando

left cerebral hemisphere

location of pineal gland

(29)

(30)

(31)

(32)

(33)

(to spinal cord)

30-VI
(pp. 441–447)

SENSORY INPUT
    Classes of Receptors
    Somatic Senses and Pain Perception
    Taste and Smell
    Hearing
    Vision
SUMMARY

Summary

Receptors are the finely branched endings of sensory neurons. A *stimulus* is any form of energy detected by the receptor. In sensory organs, receptors are organized to amplify or focus the energy of a stimulus. There are four major types of receptors: *chemoreceptors* detect chemical energy including ions or molecules dissolved in body fluids; *mechanoreceptors* detect mechanical energy associated with changes in pressure, position, or acceleration; *photoreceptors* detect light; and *thermoreceptors* detect temperature (including infrared receptors). Different animals have different kinds and numbers of receptors. While all action potentials never vary in magnitude, different sensations are experienced for three reasons: the signals from different receptors arriving from particular nerve pathways end up in specific parts of the brain; action potentials fire more frequently from strong stimulation of a receptor; stronger stimuli may excite more receptors in a given area.

Taste receptors in the mouths of animals enable them to distinguish nutritious from noxious substances; taste receptors on the tongue are often part of taste buds. Olfactory receptors detect odors. Many animals use pheromones as social signals. Male silk moths can detect one molecule of bombykol (a sex attractant) in 10^{15} molecules of air!

Somatic senses refer to sensations of touch, pressure, temperature, and pain near the body surface. Naked nerve endings in skin sense light pressure, temperature, and pain. Pain is the perception of injury to some region of the body. The ability to feel pain seems to require our limbic system. *Referred pain* is pain felt in a tissue some distance from the real stimulation point.

The sense of hearing depends on the bending of *hair cells* under fluid pressure. Sound produces waves, and the peaks are perceived as loudness. Three regions of the mammalian ear receive, amplify, and sort out signals. The outer ear collects sound waves; the external auditory canal channels the waves inward to an eardrum; the middle ear is a series of small bones that amplifies sound as it transfers vibrations from the eardrum to the inner ear; the inner ear contains a coiled part (cochlea) where hair cells under pressure are stimulated to give rise to action potentials that the brain interprets as sound.

Photoreceptors have light-absorbing pigment molecules embedded in their cell membranes. All organisms are sensitive to light. Vision requires a complex set of photoreceptors and a neural program in the brain to interpret signals from them. Vision also requires a lens that is adjustable in more complex animals. *Eyes* are well-developed photoreceptor organs that contribute to image formation.

Many invertebrates cannot see but they do have photosensitive cells in eyespots. Mollusks are the simplest animals with eyes; some have eyes with a cornea and a retina. Squids and octopuses have large paired eyes capable of forming clear images. Their eyes contain a pupil and an iris. Insects and crustaceans have compound eyes; theory has it that each unit samples a portion of the visual field and contributes a small bit to a visual mosaic.

Almost all vertebrates have eyes capable of forming clear images. The eyeball has a lens, sclera, choroid, a receptor-packed retina, and a transparent cornea covering the front of the eye. The photoreceptors are called *rod cells* and *cone cells*. Rods are sensitive to dim light and contribute to visual acuity and coarse perception of movements. Cones respond to high-intensity light and are packed at the fovea where they contribute to visual acuity; three types respond to red, green, or blue light. Each rod has a stack of membranes with rhodopsin molecules; light absorption changes the shape of rhodopsin molecules, and a change in voltage across the membrane results. The change signals the presence of light to neighboring neurons that relay the signals to cells with long axons leading out from the eye; these axons form the optic nerve that brings the signal to the thalamus, then to visual processing centers in the cortex.

Key Terms

receptors	hair cells	pupil
stimulus	outer ear	compound eyes
chemoreceptors	middle ear	sclera
mechanoreceptors	inner ear	choroid
photoreceptors	lens	retina
thermoreceptors	eyes	rod cells
taste receptors	eyespots	cone cells
olfactory receptors	cornea	fovea
pain	retina	rhodopsin
cochlea	iris	

Objectives

1. Define *stimulus* and cite several examples.
2. Define and distinguish among *chemoreceptors, mechanoreceptors, photoreceptors*, and *thermoreceptors*.
3. Explain why, if all action potentials are alike, they convey information that gives rise to different sensations such as smell and taste.
4. Distinguish between the function of *taste receptors* and *olfactory receptors*.
5. List and define what is meant by the *somatic senses*.
6. Contrast the mechanism by which the chemical senses work with that by which the somatic senses work.
7. Define *pain* and distinguish it from *referred pain*.
8. Describe the structures and functions involved with hearing.
9. Explain what is meant by a sound vibration; trace a sound wave through the parts of the ear to the place where the brain interprets sound.
10. Describe how the sense of vision has evolved through time.
11. Describe the major components necessary to achieve vision.
12. Compare and contrast the structure of molluscan eyes, the compound eyes of insects and crustaceans, and the human eye.
13. Draw a medial section of the human eyeball through the optic nerve, identify structures discussed in the text, and tell the function of each.

Self-Quiz Questions

Fill-in-the-Blanks

(1) _____ are the finely branched endings of sensory neurons. A(n)

(2) _____ is any form of energy detected by the receptor. (3) _____

detect chemical energy; mechanical energy is detected by (4) _____ ; light

is detected by (5) _____ ; and temperature is detected by (6) _____ .

Although action potentials never vary in magnitude from neuron to neuron, different sensations are experienced because the signals from different receptors end up in specific parts of the (7) _____ , action potentials fire more frequently from (8) _____ _____ of a receptor, and stronger (9) _____ may excite more receptors in a given area. (10) _____ receptors in animal mouths allow them to distinguish nutritious from noxious substances; they are often part of (11) _____ _____ on the tongue. Many animals use (12) _____ as social signals. (13) _____ senses refer to touch, pressure, temperature, and pain sensations near the body surface. Pain felt in a tissue some distance from the actual stimulation point is (14) _____ pain.

    The sense of hearing depends on the bending of (15) _____ _____ under fluid pressure. The outer ear collects (16) _____ _____ . The external auditory canal channels sound waves inward to a(n) (17) _____ . The middle ear is a series of small (18) _____ that transfers vibrations from the (19) _____ to the (20) _____ _____ . Hair cells in a coiled part of the inner ear are stimulated to give rise to (21) _____ _____ that the (22) _____ interprets as sound. All organisms are sensitive to (23) _____ . Vision requires a complex set of (24) _____ , a neural program in the (25) _____ , and a(n) (26) _____ , which may be adjustable. Well-developed photoreceptor organs that contribute to image formation are called (27) _____ . The simplest animals with eyes are (28) _____ . Insects and crustaceans have (29) _____ eyes. The vertebrate eyeball has a lens, sclera, choroid, receptor-packed (30) _____ , and transparent (31) _____ . Vertebrate photoreceptors are called (32) _____ cells, sensitive to dim light, and (33) _____ cells that respond to high-intensity light. Each rod has a stack of membranes with (34) _____ molecules that absorb light and change their shapes; a change in voltage across the membrane results. The change signals light presence to neighboring neurons that relay signals to the (35) _____ nerve; the signal goes to the thalamus and then to visual processing centers in the cortex.

Labeling   Identify each indicated part of the accompanying illustrations.

(36) _____   (43) _____

(37) _____   (44) _____ _____

(38) _____ _____   (45) _____

(39) _____   (46) _____

(40) _____ _____   (47) _____ _____

(41) _____   (48) _____

(42) _____

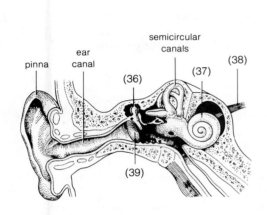

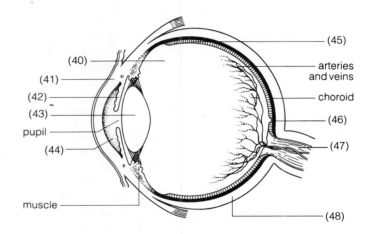

CHAPTER TEST   **UNDERSTANDING AND INTERPRETING KEY CONCEPTS**

___ (1) Which of the following nervous system control schemes is correct?

(a) stimulus → effectors → integrators → receptors → response
(b) response → receptors → integrators → effectors → stimulus
(c) receptors → stimulus → integrators → effectors → response
(d) stimulus → receptors → integrators → effectors → response
(e) effectors → stimulus → integrators → response → receptors

___ (2) Sensory neurons are _____ ; motor neurons are _____ .

(a) effectors; integrators
(b) integrators; receptors
(c) receptors; effectors
(d) effectors; receptors

___ (3) Many neurons have many short, slender extensions of the cell body, the _____ ; with the cell body they are thought of as _____ zones.

(a) axons; output
(b) dendrites; input
(c) axons; input
(d) dendrites; output

___ (4) In response to stimulation, the polarity of charge across the plasma membrane of the neuron can undergo a sudden reversal, called a(n) _____ .

  (a) membrane potential
  (b) excitable membrane
  (c) action potential
  (d) resting potential

___ (5) Which of the following is *not* true of an action potential?

  (a) It is a short-range message that can vary in size.
  (b) It is an all-or-none brief reversal in membrane potential.
  (c) It doesn't decay with distance.
  (d) It is self-propagating.

___ (6) Your *sense* of sound, or hearing, begins when _____ .

  (a) sound waves spread out through the air
  (b) the eardrum vibrates
  (c) specialized hair cells in your ears bend under pressure
  (d) the outer ear collects sound waves

___ (7) The resting membrane potential _____ .

  (a) exists as long as a charge difference sufficient to do work exists across a membrane
  (b) occurs because there are more potassium ions outside the neuron membrane than there are inside
  (c) occurs because of the unique distribution of receptor proteins located on the dendrite exterior
  (d) is brought about by a local change in membrane permeability caused by a greater-than-threshold stimulus

___ (8) The phrase *all-or-nothing event* used in conjunction with discussion about an action potential means that _____ .

  (a) a resting membrane potential has been received by the cell
  (b) if threshold is reached, nothing can stop the full spiking
  (c) the membrane either achieves total equilibrium or remains as far from equilibrium as possible
  (d) refractory periods do not occur

___ (9) _____ nerves generally tend to slow down the body overall and divert energy to basic "housekeeping" tasks.

  (a) Ganglia
  (b) Pacemaker
  (c) Sympathetic
  (d) Parasympathetic
  (e) All of the above

___ (10) The _____ is the information-encoding and information-processing center of the mammalian brain.

  (a) medulla oblongata
  (b) thalamus
  (c) hypothalamus
  (d) cerebellum
  (e) cerebral cortex

___ (11) A brain region called the _____ controls the changing levels of consciousness.

  (a) medulla oblongata
  (b) thalamus
  (c) hypothalamus
  (d) reticular activating system
  (e) cerebral cortex

___ (12) The part of the brain that controls the basic responses necessary to maintain life processes (respiration, blood circulation) is the _____ .

  (a) cerebral cortex
  (b) cerebellum
  (c) tectum
  (d) medulla oblongata
  (e) limbic system

___ (13) Our emotions are governed by the cerebral cortex and by different brain regions collectively called the _____ .

  (a) pons
  (b) thalamus
  (c) cerebellum
  (d) reticular activating system
  (e) limbic system

___ (14) The reflex centers for maintaining posture and refining limb movements in the human brain is the _____ .

  (a) cerebrum
  (b) pons
  (c) cerebellum
  (d) hypothalamus
  (e) thalamus

___ (15) The eardrum vibrates, and the vibrations are picked up and transferred inward by small bones of the _____ .

  (a) outer ear
  (b) ear canal
  (c) middle ear
  (d) inner ear
  (e) none of the above

For questions 16 to 20, choose from these answers:

  (a) fovea
  (b) cornea
  (c) iris
  (d) retina
  (e) sclera

___ (16) The outer coat of the eye is the _____ .

___ (17) Rods and cones are located in the _____ .

___ (18) The highest concentration of cones is in the _____ .

___ (19) The _____ , with its abundant screening pigments and muscle fibers, is used to control the amount of incoming light.

___ (20) The outer transparent protective covering of part of the eyeball is the _____ .

INTEGRATING AND APPLYING KEY CONCEPTS

(1) Suppose that anger eventually is determined to be caused by excessive amounts of specific transmitter substances in the brains of angry people. Suppose that an inexpensive antidote to anger that neutralizes these anger-producing transmitter substances is also readily available. Can violent murderers now argue that they have been wrongfully punished because they were victimized by their brain's transmitter substances and could not have acted in any other way? Suppose an antidote is prescribed to curb violent tempers in a person. Suppose also that the easily angered person forgets to take the pill and subsequently murders a family member. Can the murderer still claim to be "victimized" by transmitter substances?

(2) How might human behavior be changed if human eyes were compound eyes composed of ommatidia and if humans perceived only vibrations—as fish do—rather than sounds?

31
ENDOCRINE CONTROL

"THE ENDOCRINE SYSTEM"
    Discovery of Hormones
    Types of Signaling Molecules
THE HYPOTHALAMUS-PITUITARY CONNECTION
    Posterior Lobe Secretions
    Anterior Lobe Secretions
SELECTED EXAMPLES OF HORMONAL CONTROL
    Adrenal Glands

Thyroid Gland
Parathyroid Glands
Pancreatic Islets
Pineal Gland
SIGNALING MECHANISMS
    Steroid Hormone Action
    Nonsteroid Hormone Action
SUMMARY

General Objectives

1. Know the general mechanisms by which hormones and other signaling molecules integrate and control the various metabolic activities in organisms.
2. Relate hormones to short-term and long-term adjustments in diet and activity levels as well as to growth, development, and reproduction.
3. Explain how stimulation or inhibition of hormone secretions involves negative feedback loops between the neuroendocrine center and many different glands.

31-I
(pp. 449–450)

"THE ENDOCRINE SYSTEM"
    Discovery of Hormones
    Types of Signaling Molecules

Summary

In the early 1900s, Bayliss and Starling first demonstrated that a hormone secreted into the blood triggers the secretion of pancreatic juices. Internal secretions diffused into the bloodstream influence the activities of target tissues and organs. Such internal secretions were named *hormones*. Many vertebrate hormones are now known; hormones are secreted from members of the endocrine system. The boundaries between the nervous system and the *endocrine system* are not complete. Glands once thought to work independently are under neural control and some neurons secrete hormones. For example, the pituitary (master gland) is actually controlled by the hypothalamus.

In complex animals, nervous and endocrine systems work together with *signaling molecules* to integrate countless cell responses. Cells with receptors that fit specific signaling molecules are targets to which these signaling molecules can bind and elicit a response. The four main classes of signaling molecules are

(1) hormones secreted from endocrine glands; (2) transmitter substances secreted from neurons; (3) *local signaling molecules* secreted from cells in different tissues; and (4) *pheromones* secreted from some exocrine glands.

Objectives

1. Explain the origin of the term *hormone*.
2. Describe why there is now said to be no clear-cut division between the nervous system and the endocrine system.
3. List and briefly describe the four main classes of signaling molecules.
4. Understand how neural signals, hormonal signals, local chemical changes, and environmental cues trigger hormone secretions.

**Self-Quiz
Questions**

Fill-in-the-Blanks

Internal (1) _____ released into the bloodstream influence the activities of tissues and organs; such substances were named (2) _____ . We now know the (3) _____ between the nervous system and the endocrine system are not complete. Glands thought to work independently are under (4) _____ control. Some neurons secrete (5) _____ . Integration of cellular responses brought about by the nervous and endocrine systems depend on (6) _____ molecules. Any cell may be a(n) (7) _____ if it has receptors to which specific signaling molecules can bind and elicit a response. The four main classes of signaling molecules are (8) _____ , secreted from endocrine glands; (9) _____ substances secreted from neurons; local (10) _____ molecules secreted from cells in different tissues; and (11) _____ , secreted by some exocrine glands.

31-II
(pp. 450–453)

THE HYPOTHALAMUS-PITUITARY CONNECTION
   Posterior Lobe Secretions
   Anterior Lobe Secretions

Summary

The hypothalamus and pituitary act together as a *neuroendocrine control center*. The hypothalamus monitors internal conditions and emotional states. The lobed pituitary is connected to the hypothalamus by a stalk. Its *posterior lobe* stores and then secretes two hormones actually produced in the hypothalamus. The *anterior lobe* consists of glandular tissue and secretes six hormones and controls the release of others from endocrine glands. Some nonhuman vertebrates have an intermediate lobe that secretes hormones that affect body coloration.

Secretions from the posterior pituitary lobe include antidiuretic hormone (ADH), which helps to reduce fluid loss through the kidneys. Oxytocin stimulates uterine contraction during labor and the release of milk during nursing.

Releasing hormones from the hypothalamus stimulate or inhibit the secretions of hormones from the anterior pituitary lobe. Corticotropin-stimulating hormone (ACTH), thyrotropin-stimulating hormone (TSH), follicle-stimulating hormone (FSH), and luteinizing hormone (LH) are hormones secreted by the anterior lobe; these act on endocrine glands, which in turn produce other hormones. Prolactin and somatotropin have effects on body tissues in general. Prolactin stimulates milk production. Somatotropin influences overall growth, especially of cartilage and bone; it induces protein synthesis and cell divisions in target cells. Too little somatotropin during childhood causes pituitary dwarfism; too much causes gigantism. Too much somatotropin as an adult causes acromegaly—thickened skin and bones.

Key Terms

neuroendocrine control center
hypothalamus
pituitary gland
posterior lobe
anterior lobe
intermediate lobe
antidiuretic hormone, ADH
oxytocin
releasing hormone
corticotropin-stimulating hormone, ACTH

thyrotropin-stimulating hormone, TSH
follicle-stimulating hormone, FSH
luteinizing hormone, LH
prolactin, PRL
somatotropin, STH (= growth hormone, GH)
pituitary dwarfism
gigantism
acromegaly

Objectives

1. Explain why the hypothalamus and pituitary are regarded as the *neuroendocrine control centers*.
2. State how, even though the anterior and posterior lobes of the pituitary are compounded as one gland, the tissues of each part differ in character.
3. Identify the hormones produced by the anterior lobe of the pituitary and tell which target tissues or organs each one influences.
4. Identify the hormones released from the posterior lobe of the pituitary and state their target tissues.
5. Define *pituitary dwarfism, gigantism,* and *acromegaly*; in each case, identify the causative factor.

Self-Quiz Questions

Fill-in-the-Blanks

The (1) _____ and the (2) _____ act together as a neuroendocrine control center. The lobed (3) _____ is connected to the hypothalamus by a stalk. Secretions from the posterior pituitary lobe include (4) _____ hormone, which helps to reduce fluid loss through the kidneys and

(5) _____ , which stimulates labor contractions and milk secretion.

(6) _____ hormones from the hypothalamus stimulate or inhibit hormonal secretions from the (7) _____ pituitary lobe. ACTH, TSH, FSH, and LH are hormones secreted by the (8) _____ pituitary lobe; these act on (9) _____ glands, which in turn produce other (10) _____ .

(11) _____ stimulates milk production; (12) _____ influences overall growth, especially of cartilage and bone; it also induces protein synthesis and cell divisions in (13) _____ cells. A lack of somatotropin during childhood causes (14) _____ _____ ; too much causes (15) _____ . As an adult, too much somatotropin causes (16) _____ , characterized by thickened skin and bones.

31-III
(pp. 454–458)

SELECTED EXAMPLES OF HORMONAL CONTROL
Adrenal Glands
Thyroid Gland
Parathyroid Glands
Pancreatic Islets
Pineal Gland

Summary

Homeostatic feedback loops link many endocrine glands to the neuroendocrine control center. Humans have one adrenal gland above each kidney. The outer portion, or *adrenal cortex*, produces hormones that include glucocorticoids. They function in metabolism and the inflammatory response. Cortisol is an example; it helps to control by negative feedback the concentration of glucose in the blood. Under conditions of stress, higher levels of cortisol are secreted.

The *adrenal medulla* is the inner region of the adrenal gland; it contains hormone-secreting neurons. The adrenal medulla secretes epinephrine and norepinephrine. It controls blood circulation and carbohydrate metabolism. During times of stress, the adrenal medulla enhances the fight or flight response of the sympathetic nervous system; its rate of secretion is controlled by sympathetic nerves.

The *thyroid gland* is at the base of the throat; it secretes thyroxine and triiodothyronine. It influences metabolic rate, growth, and development. Hypothyroidism in adults results in lethargy and weight gain; in infants, retardation and dwarfism can result. People with hyperthyroidism have increased heart rate and blood pressure, weight loss, nervous, agitated behavior, and trouble sleeping. In the absence of iodine, levels of thyroid hormones in the blood decrease; the anterior pituitary responds by secreting thyroid stimulating hormone (TSH). Excessive TSH overstimulates the thyroid gland and it begins to enlarge; goiter can result from a deficiency in iodine.

Parathyroid glands are embedded in tissues in the back of the thyroid gland. The parathyroid glands secrete parathyroid hormone (PTH), which regulates blood levels of Ca^{++} in gene activation and muscle contraction. PTH removes calcium from bone, increases the kidney's reabsorption of Ca^{++}, and activates vitamin D, which enhances calcium absorption. Rickets results from vitamin D deficiency. With a rise in calcium levels in response to PTH-induced events, parathyroid stimulation decreases. At the same time, the thyroid gland secretes calcitonin, which encourages calcium deposition in bones.

The pancreas has about 2 million endocrine cell clusters, the *pancreatic islets;* alpha, beta, and delta cells within the pancreatic islets secrete hormones. Alpha cells secrete *glucagon,* which raises the level of glucose in blood; beta cells secrete *insulin,* which stimulates glucose absorption by cells and protein and fat synthesis and lowers blood glucose levels; delta cells secrete *somatostatin,* which can inhibit secretion of glucagon and insulin. Diabetes mellitus (type 1 diabetes) is caused by insulin deficiency; dehydration and degradation of proteins and fats in body tissues result. Coma and death can follow. Diabetes mellitus is produced by a combination of genetic and environmental factors; insulin injections permit survival. Type 2 diabetics have normal or above-normal levels of insulin but insufficient or abnormal target cells (receptors); this is treated by diet.

Until about 240 million years ago, vertebrates commonly had a third eye. The *pineal gland* remains as a third eye in lampreys; mammals, birds, and reptiles have a modified form of this photosensitive organ, the pineal gland. It secretes melatonin, a hormone that influences reproductive organ development and reproductive cycles. Melatonin is secreted in the absence of light. In winter, increased melatonin suppresses the sexual activity of hamsters. In humans, decreased production of melatonin helps to trigger puberty.

Key Terms

adrenal cortex	thyroid gland	alpha cells
homeostatic feedback loops	thyroxine	glucagon
glucocorticoids	triiodothyronine	beta cells
cortisol	hypothyroidism	insulin
corticotropin-releasing hormone, CRH	hyperthyroidism	delta cells
corticotropin-stimulating hormone, ACTH	goiter	somatostatin
	parathyroid glands	diabetes mellitus
adrenal medulla	parathyroid hormone, PTH	type 1 diabetes
epinephrine	rickets	type 2 diabetes
norepinephrine	calcitonin	pineal gland
	pancreatic islets	melatonin

Objectives

1. List the major endocrine glands and patches of tissue in this section, tell which substance(s) each secretes, tell what tissues or organs each substance affects, and state how the target tissue responds.
2. State the cause of each of the following conditions: fight-flight response, hypothyroidism, hyperthyroidism, goiter, rickets, diabetes, type 1 diabetes, and type 2 diabetes.

Self-Quiz Questions

Fill-in-the-Blanks

Humans have one (1) _____ gland above each kidney. The outer portion is called the (2) _____ _____ , which produces (3) _____ including glucocorticoids. One, cortisol, helps to control the concentration of (4) _____ in the blood by a homeostasis feedback loop. The inner portion of the adrenal gland is the adrenal (5) _____ ; it secretes (6) _____ and (7) _____ . It also controls blood circulation and (8) _____ metabolism. During stress, it enhances the (9) _____ or (10) _____ response. The (11) _____ gland influences metabolic rate, growth, and development. (12) _____ in adults results in lethargy and weight gain; in infants, (13) _____ retardation and (14) _____ can result. In the absence of (15) _____ , levels of thyroid hormone in the blood decrease. (16) _____ can result from an iodine deficiency. The parathyroid glands secrete parathyroid hormone (PTH), which regulates blood levels of (17) _____ used in gene activation and muscle contraction. (18) _____ results from vitamin D deficiency. Alpha, beta, and delta cells are found in the (19) _____ _____ . Alpha cells secrete (20) _____ ; beta cells secrete (21) _____ ; delta cells secrete (22) _____ . Diabetes mellitus is caused by (23) _____ deficiency. Type 2 diabetics have normal or above-normal insulin but insufficient or abnormal (24) _____ _____ . The (25) _____ gland is a modified form of an ancient vertebrate eye in mammals, birds, and reptiles; it secretes (26) _____ ; in humans, decreased production helps to trigger (27) _____ .

Labeling

Identify each indicated part of the accompanying illustration (p. 353).

(28) _____ gland

(29) _____ gland

(30) _____ glands

(31) _____ gland

(32) _____ gland

(33) _____

(34) _____ _____

(35) _____ (in female)

(36) _____ (in male)

31-IV
(pp. 458–460)

SIGNALING MECHANISMS
   Steroid Hormone Action
   Nonsteroid Hormone Action
SUMMARY

Summary

Signaling mechanisms only affect cells with receptors, and they affect different cells in different ways. *Steroid hormones* pass easily through the plasma membrane of a target cell. Inside they bind to a receptor molecule in the nucleus of a target cell and form a hormone-receptor complex that "turns on" specific genes in the nucleus to make specific mRNA transcripts that alter the cell's activity. Testosterone is an example; defective receptors cause testicular feminization syndrome. None of the target cells can respond to testosterone produced in

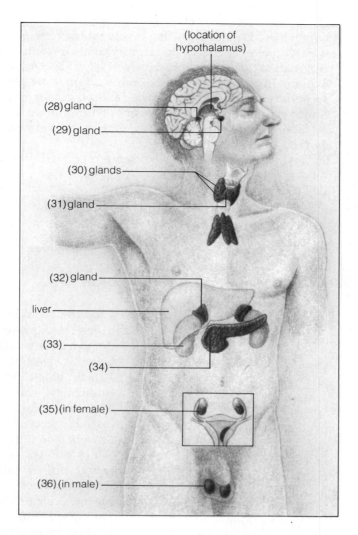

genetic males with functional testes; female secondary sexual traits develop in the individual.

Nonsteroid hormones do not pass through the plasma membrane, but they do bind to the receptor sites on the membrane, which activates *second messengers*. An example is adenylate cyclase, an enzyme that converts ATP to *cyclic AMP*. cAMP activates different enzymes that alter cellular activity. Each of the activated enzymes then activates many enzymes; each enzyme may then form many different enzymes. Thus, second messengers amplify a response to a signaling molecule.

Steroid hormones activate genes and turn on protein synthesis; nonsteroid hormones alter the activity of proteins already present in target cells. These cellular responses help maintain the internal environment and influence developmental and reproductive programs.

Key Terms

steroid hormones
testosterone
hormone-receptor
   complex

testicular feminization
   syndrome
nonsteroid hormones
second messengers

cyclic AMP, cAMP
adenylate cyclase
amplify

Objectives

1. Contrast the behavior of chemical signals that involve intracellular receptors with the behavior of those that involve surface membrane receptors. Identify which types of chemical signals participate in the two different mechanisms.
2. Summarize the distinctions between *steroid* and *nonsteroid* hormones.
3. Explain, in terms of steroid hormone action, how genetic males develop the secondary sexual traits of females in *the testicular feminization syndrome*.
4. Contrast the proposed mechanisms of hormone action on target cell activities by (a) steroid hormones and (b) nonsteroid hormones that are proteins or derived from proteins.
5. Summarize the role of *second messengers* and cite an example.

Self-Quiz Questions

Matching

Match each letter with the most appropriate term.

(1) ___ ACTH, TSH, FSH, LH

(2) ___ ADH

(3) ___ calcitonin

(4) ___ cortisol

(5) ___ epinephrine

(6) ___ somatotropin

(7) ___ insulin

(8) ___ melatonin

(9) ___ parathyroid hormone

(10) ___ testosterone

(11) ___ thyroxine

(12) ___ TSH

A. Growth hormone
B. Influences development of male reproductive organs and sexual traits
C. Governs onset of puberty in humans
D. Affects blood circulation and carbohydrate metabolism
E. The water conservation hormone; released from posterior pituitary
F. Lowers blood glucose level
G. Secreted by the anterior pituitary; cause(s) other endocrine glands to produce hormones
H. Elevates calcium levels in blood by stimulating calcium removal from bone and its movement into extracellular fluid
I. Influences overall metabolic rate, growth, and development
J. Promotes calcium deposition in bone
K. Helps maintain blood levels of glucose between meals; involved in the inflammatory response
L. Secreted by anterior pituitary; stimulates release of thyroid hormones

Fill-in-the-Blanks

Only cells with (13) _____ are affected by signaling mechanisms that affect different cells in (14) _____ ways. (15) _____ hormones pass easily through the plasma membrane of a target cell. In nuclei of target cells steroid hormones form (16) _____ - _____ complexes that turn on specific (17) _____ in the nucleus to make specific mRNA transcripts that (18) _____ the cell's activity. (19) _____ _____ syndrome is a failure of target cells to respond to testosterone produced in genetic males with functional testes. (20) _____ hormones do not pass through the plasma membrane, but they do bind to receptor sites on the membrane, which activates (21) _____ _____ . (22) _____ is one kind of second messenger that activates different enzymes that alter cellular activity. Second messengers (23) _____ a response to a signaling molecule.

CHAPTER TEST

UNDERSTANDING AND INTERPRETING KEY CONCEPTS

___ (1) Which of the following is *not* one of the four main classes of signaling molecules?

  (a) local signaling molecules
  (b) pheromones
  (c) transmitter substances
  (d) secondary messengers
  (e) hormones

___ (2) The pituitary is controlled by the _____ .

  (a) thalamus
  (b) pancreas
  (c) hypothalamus
  (d) parathyroid glands
  (e) thyroid

___ (3) The neuroendocrine control center involves the _____ .

  (a) thalamus and pituitary
  (b) adrenal cortex and thyroid
  (c) hypothalamus and pituitary
  (d) parathyroids and hypothalamus
  (e) medulla oblongata and the limbic system

___ (4) If you were lost in the desert and had no freshwater to drink, the level of _____ would increase in your blood as a means to conserve water.

  (a) insulin
  (b) calcitonin
  (c) oxytocin
  (d) antidiuretic hormone
  (e) salt

For questions 5 to 7, choose from these answers:

  (a) Oxytocin
  (b) Somatotropin
  (c) FSH
  (d) LH
  (e) Prolactin

___ (5) _____ is the growth hormone.

___ (6) _____ triggers muscle contractions during labor.

___ (7) _____ is a hormone produced by the anterior lobe of the pituitary that stimulates follicles to mature in the ovary.

For questions 8 to 10, choose from these answers:

    (a) adrenal medulla
    (b) adrenal cortex
    (c) thyroid
    (d) anterior pituitary
    (e) posterior pituitary

___ (8) The _____ produces glucocorticoid hormones that take part in metabolism and the inflammatory response.

___ (9) The gland that is most closely associated with emergency situations is the _____ .

___ (10) The _____ gland regulates the overall metabolic rate.

___ (11) If all sources of calcium were eliminated from your diet, your body would secrete more _____ in an effort to release calcium stored in your body to the tissues that require it.

    (a) parathyroid hormone
    (b) somatostatin
    (c) calcitonin
    (d) melatonin
    (e) none of the above

INTEGRATING AND APPLYING KEY CONCEPTS

Suppose that you suddenly quadruple your already high daily consumption of calcium. State which body organs would be affected and tell how they would be affected. Name two hormones, the levels of which would most probably be affected, and tell whether your body's production of them would increase or decrease. Suppose that you continue this high rate of calcium consumption for ten years. Can you predict the organs that would be stressed the most?

32

REPRODUCTION AND DEVELOPMENT

THE BEGINNING: REPRODUCTIVE MODES
BASIC PATTERNS OF DEVELOPMENT
    Stages in Development
    Mechanisms of Development
HUMAN REPRODUCTIVE SYSTEM
    Male Reproductive Organs
    Female Reproductive Organs
    Menstrual Cycle
    Sexual Union
FROM FERTILIZATION TO BIRTH
    Fertilization
    Implantation

Membranes Around the Embryo
The Placenta
Embryonic and Fetal Development
Commentary: Mother as Protector, Provider, Potential Threat
Birth and Lactation
POSTNATAL DEVELOPMENT, AGING, AND DEATH
CONTROL OF HUMAN FERTILITY
Commentary: Sexually Transmitted Diseases
SUMMARY

General Objectives

1. Understand how asexual reproduction differs from sexual reproduction. Know the advantages and problems associated with having separate sexes.
2. Describe the major stages of animal development by name.
3. Describe the structure and function of the male and female reproductive tracts.
4. Outline the principal events of prenatal development.
5. Know the principal means of controlling human fertility.

32-I
(pp. 463–467)

THE BEGINNING: REPRODUCTIVE MODES
BASIC PATTERNS OF DEVELOPMENT
    Stages in Development
    Mechanisms of Development

Summary

Asexual reproduction in sponges is by budding; asexual reproduction in flatworms is by fission. In these cases, the offspring are genetically identical copies of the parent and are well-adapted life forms in stable environments. However, most animals live under changing, unpredictable conditions and rely mostly on sexual reproduction even though it requires large energy investments. Sexual reproduction usually involves the fusion of gametes from a male and female parent. Male and female reproductive cycles must be timed so that gametes are released at the same time; energy outlays are required for males and females to meet and recognize each other. External fertilization in water, as in bony fishes, requires the production of large numbers of gametes. Nearly all land-dwelling animals depend on internal fertilization; energy is invested in elaborate reproductive organs such as a penis for depositing sperm inside the female. *Yolk* or attachment to the mother is needed to nourish the embryo. Animals show great diversity in their reproduction and development.

The first stage of animal development is *gamete formation;* sperms and eggs are formed. Oocytes (immature eggs) are larger and more complex than sperms; they contain organelles and stockpiles of proteins, RNA, and other cytoplasmic components for development. The second stage of development, *fertilization,* begins with sperm penetration of an egg; it ends with fusion of egg and sperm nuclei to form the zygote. *Cleavage* now follows as mitotic cell divisions create a ball of cells, the *blastula. Gastrulation,* a time of major cell rearrangements, occurs next. The entire framework of the body is laid out as three germ layers form— *endoderm, mesoderm,* and *ectoderm. Organ formation* occurs as germ layers split into cell lines unique in structure and function; these are forerunners of distinct tissues and organs. During *growth and tissue specialization,* which continues to adulthood, organs acquire special chemical and physical properties.

Cell differentiation and morphogenesis are central to development. *Cell differentiation* occurs through controls over gene expression; a differentiated cell has the same genes as the zygote. The fate of an embryonic cell is partly determined when cleavage partitions different zygote regions into daughter cells. Differentiated cells become organized into tissues and organs through *morphogenesis.* The mechanisms underlying morphogenesis are basically physical and chemical interactions: cells can produce chemical substances that move to other cell types to alter their behavior or a cell may recognize specific proteins on other cell surfaces and stick to them or migrate to them. This produces development patterns. Morphogenesis also depends on localized growth and cell death. Controlled (genetic) cell death transforms paddlelike appendages into hands and feet.

Key Terms

asexual reproduction	gamete formation	endoderm
sexual reproduction	oocyte	mesoderm
budding	fertilization	ectoderm
fission	cleavage	organ formation
reproductive timing	blastula	growth and tissue
external fertilization	gastrula	specialization
internal fertilization	gastrulation	cell differentiation
yolk	germ layers	morphogenesis

Objectives

1. Distinguish between asexual and sexual reproduction; cite an example of each.
2. Discuss some structural, behavioral, and ecological aspects of asexual and sexual reproduction.
3. Describe the stages of development; define *endoderm, mesoderm,* and *ectoderm.*
4. Explain the two mechanisms central to development.

Fill-in-the-Blanks

Asexual reproduction in sponges is by (1) _____ ; asexual reproduction in flatworms is by (2) _____ . In asexual reproduction, offspring are (3) _____ _____ copies of the parent. Sexual reproduction usually involves the fusion of (4) _____ from a male and female parent. Male and female reproductive (5) _____ must be timed for simultaneous (6) _____ release. The cost of males and females meeting and recognizing each other is a huge outlay of (7) _____ . Sexual reproduction in bony fishes involves (8) _____ fertilization. Nearly all land-dwelling animals depend on (9) _____ fertilization. Embryo nourishment in sexually reproducing organisms requires (10) _____ or an attachment to the mother. The first stage of animal development is (11) _____ formation. The second stage of development is (12) _____ , which begins with sperm penetration of an egg; it ends with fusion of egg and sperm nuclei to form the (13) _____ . A hollow ball of cells, the (14) _____ , is created by the process of (15) _____ . (16) _____ is a time of major cell rearrangements. Endoderm, mesoderm, and ectoderm are called the three (17) _____ layers; these layers split into cell lines unique in structure and function to accomplish (18) _____ formation. Cell (19) _____ occurs through controls over gene expression. Differentiated cells become organized into tissues and organs through (20) _____ .

True-False

If false, explain why.

___ (21) Organisms that reproduce sexually produce offspring that are genetically identical to their parents.

___ (22) Endoderm gives rise to the inner lining of gut and the organs derived from it; mesoderm gives rise to nervous system tissues and to the outer layer of the body covering; ectoderm gives rise to muscle, circulatory organs, reproductive and excretory organs, and much of the internal skeleton.

___ (23) In cell differentiation, the fertilized egg gives rise to diverse types of specialized cells.

HUMAN REPRODUCTIVE SYSTEM
Male Reproductive Organs

Summary

The reproductive system is composed of a pair of gonads (*testes* or *ovaries*) with accessory glands and ducts; it produces gametes and sex hormones, which affect reproduction and the development of sexual traits. Sperm development requires a temperature cooler than the body core; hence, the testes are suspended in the scrotum. The testes are divided into as many as 300 wedge-shaped lobes, each containing two or three *seminiferous tubules* where sperm form. Millions are developing on any given day. Each sperm has a head (nucleus with DNA and acrosome with enzymes), midpiece (mitochondria), and tail.

Sperm move from a testis into the epididymis where they mature and are stored until their release from the body. When sperm are to be released, they pass through the two vasa deferentia, ejaculatory ducts, and then through the urethra to the body's surface. As sperm move along this route, glandular secretions become mixed with sperm; this mixture is *semen*. Fluids are added to sperm from the seminal vesicles (fructose and prostaglandins), prostate gland (buffers vaginal pH), and bulbourethral glands (mucus to lubricate penis).

Testosterone, produced by endocrine cells in the testes, stimulates the formation of sperm, reproductive organs, and sexual traits and helps to develop and maintain normal sexual behavior. Other hormones produced by the anterior pituitary that control male reproductive functions are luteinizing hormone (LH) and follicle-stimulating hormone (FSH). The hypothalamus controls testosterone secretion. With low testosterone levels, the hypothalamus signals LH secretion by the anterior pituitary; this stimulates testosterone secretion by the testes. Through negative feedback loops, high levels of testosterone inhibit hypothalamic signals for LH secretion and testosterone production slows. FSH in males stimulates sperm development beginning at puberty; its secretion is also governed by the hypothalamus.

Key Terms

testis, testes	semen	follicle-stimulating
ovary, ovaries	testosterone	hormone, FSH
seminiferous tubules	luteinizing hormone, LH	

Objectives

1. Distinguish between primary and secondary sexual traits and between gonads and accessory reproductive organs.
2. Tell where sperm is formed, list the factors that control its production, and explain how the factors interact.
3. Follow the path of a mature sperm from the seminiferous tubules to the urethral exit. List every structure encountered along the path and state the contribution to the nurture of the sperm.
4. Describe the negative feedback mechanisms that link the hypothalamus, anterior pituitary, and the testes in controlling gonadal function.

Fill-in-the-Blanks

The reproductive system is composed of a pair of (1) _____ , the testes or ovaries with (2) _____ glands and ducts; it produces (3) _____ and sex hormones. Testes are suspended in a scrotum because (4) _____ _____ requires a temperature cooler than body core. Within the testes, sperm form in (5) _____ tubules. Sperm move from a testis into the (6) _____ where they mature and are stored until their release from the body. When sperm are to be released, they pass through the (7) _____ _____ , (8) _____ ducts, and then through the (9) _____ to the body's surface. The mixture of sperm and glandular secretions is called (10) _____ . Fluids are added to sperm from the seminal vesicles, (11) _____ gland, and (12) _____ glands. Endocrine cells in the testes produce (13) _____ , which stimulates sperm formation, reproductive organ formation, and sexual traits and helps to develop and maintain normal sexual behavior. Hormones produced by the anterior pituitary that control male reproductive functions are (14) _____ hormone and (15) _____ - _____ hormone. The (16) _____ controls testosterone secretion. When testosterone levels are low, the (17) _____ signals LH secretion by the anterior pituitary; this stimulates testosterone secretion by the (18) _____ . Through (19) _____ _____ loops, high levels of testosterone inhibit signals from the (20) _____ for LH secretion and testosterone production slows.

Labeling

Identify each indicated part of the accompanying illustration (p. 362).

(21) _____ _____

(22) _____ _____

(23) _____ _____

(24) _____ _____

(25) _____

(26) _____ _____

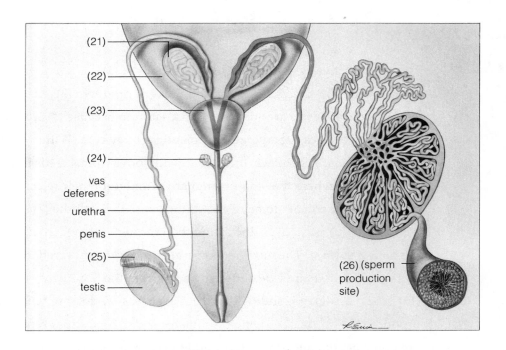

HUMAN REPRODUCTIVE SYSTEM (cont.)
    Female Reproductive Organs
    Menstrual Cycle
    Sexual Union

Summary

In the female reproductive system, two ovaries produce and release eggs monthly and secrete *estrogen* and *progesterone*. Even before the birth of a female, about 2 million immature eggs (oocytes) have begun to form in the ovaries. Division is arrested in meiosis I. About 400 oocytes eventually will mature and be expelled from the body monthly, one at a time. Meiosis II is not completed unless fertilization occurs. *Oviducts* channel oocytes into the *uterus*. The pear-shaped uterus houses the developing embryo during pregnancy; it is mostly muscle with a lining, the endometrium. The lower portion of the uterus is the cervix. The vagina is a muscular tube that extends from the cervix to the body surface; it receives sperm and functions as part of the birth canal. External genitalia (vulva) include the clitoris and other organs for sexual stimulation.

Most female mammals come into "heat," or *estrus*. Humans and other primates have a *menstrual cycle* (in humans there is no relationship between heat and fertility) that takes about twenty-eight days to complete. At about age thirteen human menstrual cycles begin; they continue until menopause in the late forties or early fifties. In the ovary, a *follicle* is a primary oocyte surrounded by a layer of cells. During a menstrual cycle usually only one follicle matures. Meiosis I within the follicle produces a secondary oocyte and a tiny polar body; neither complete meiosis II until fertilization. The developing follicle secretes an estrogen-containing fluid that accumulates and eventually bursts to release the secondary oocyte in *ovulation*. The *corpus luteum* develops from remaining follicle parts; it secretes progesterone and some estrogen. Events in the ovary are controlled by feedback loops involving the hypothalamus, pituitary, and ovaries. At the onset of the menstrual cycle, signals from the hypothalamus cause release of LH and FSH from the anterior pituitary; in turn they signal the ovary to secrete estrogen. Midway through the cycle, the increased estrogen blood level causes a brief LH outpouring from the pituitary; this triggers ovulation. Without fertilization, the corpus luteum persists about twelve days; during this time hypothalamic signals to the anterior pituitary decrease FSH secretions. Thus, development of other follicles is blocked until termination of the menstrual cycle. Lacking fertilization, the corpus luteum disintegrates by secreting prostaglandins near the end of the cycle; progesterone and estrogen levels drop rapidly. Another follicle matures to begin a new cycle as FSH levels increase.

Estrogen and *progesterone* cause the endometrium to develop and prepare for pregnancy. Without fertilization occurring, progesterone and estrogen levels fall, and the *endometrium* disintegrates. Menstrual flow for three to six days consists of blood and sloughed endometrial tissues; its appearance marks the first day of a new cycle. Endometriosis, the spread and growth of endometrial tissue outside the uterus, affects between 4 and 10 million American women. This may lead to pain during menstruation, sexual relations, or urination. Endometrial scar tissue on ovaries or oviducts can cause infertility.

Quickly following sexual arousal, penis erection occurs by blood flow into three spongy tissue cylinders; the penis lengthens and stiffens enabling vaginal penetration. Like the clitoris, the glans penis is loaded with friction-activated sensory receptors. During coitus, pelvic thrusts stimulate the penis and female vaginal walls and clitoral region. Involuntary contractions in the male reproductive tract force secretions from the seminal vesicles and prostate into the urethra; semen is then ejaculated into the vagina. Male orgasm involves muscular contractions, ejaculation, and allied sensations of release, warmth, and relaxation; female orgasm is similar with an intense vaginal awareness, involuntary uterine and vaginal contractions, and sensations of relaxation and warmth. Pregnancy may occur without female orgasm.

Key Terms

estrogen	estrus	menstrual flow
progesterone	menstrual cycle	endometriosis
oocytes	ovary	sexual arousal
oviducts	primary oocyte	penis
uterus	follicle	spongy tissue
endometrium	secondary oocyte	glans penis
cervix	polar body	coitus
vagina	ovulation	ejaculation
clitoris	corpus luteum	orgasm

Objectives

1. Describe the structures and functions of the female reproductive organs.
2. Describe events leading to the menstrual cycle.
3. Describe the events associated with sexual union.

Self-Quiz Questions

Fill-in-the-Blanks

In the female reproductive system, two (1) _____ produce and release eggs monthly and secrete (2) _____ and (3) _____ . About (4) _____ (number) oocytes eventually will mature and be expelled from the body, one at a time. Meiosis II is not completed unless (5) _____ occurs. (6) _____ channel oocytes into the uterus. The (7) _____ is the lining of the uterus. The lower portion of the uterus is the (8) _____ . The (9) _____ is a muscular tube that extends from the cervix to the body surface; it receives sperm and functions as part of the birth canal. Humans and other primates have a(n) (10) _____ cycle that takes about twenty-eight days to complete. In the ovary, a(n) (11) _____ is a primary oocyte surrounded by a layer of cells. In (12) _____ the developing follicle secretes an estrogen-containing fluid that accumulates and eventually bursts to release the secondary oocyte. The (13) _____ _____ develops from remaining follicle parts; it secretes progesterone and some estrogen. Midway through the cycle, the increased (14) _____ level in the blood causes a brief LH outpouring from the pituitary; this causes (15) _____ . Estrogen and progesterone secretions cause the (16) _____ to develop and prepare for pregnancy. Without fertilization occurring, progesterone and estrogen levels fall and the (17) _____ disintegrates. Menstrual flow continues for three to six days; its appearance marks the first day of a new (18) _____ . (19) _____ , the spread and growth of the tissues of the uterine lining, affects between 4 and 10 million American women. Immediately after sexual arousal in males, (20) _____ _____ occurs by blood flow into three spongy tissue cylinders. During (21) _____ , pelvic thrusts stimulate the penis and female vaginal walls and clitoris region. Pregnancy may occur without female (22) _____ .

Labeling

Identify each indicated part of the accompanying illustration.

(23) _____
(24) _____
(25) _____
(26) _____ _____
(27) _____
(28) _____
(29) _____ _____
(30) _____ _____
(31) _____
(32) _____

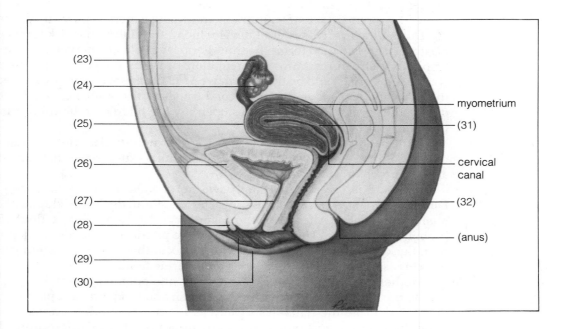

(23)	
(24)	
(25)	myometrium
(26)	(31)
(27)	cervical canal
(28)	(32)
(29)	(anus)
(30)	

32-IV
(pp. 474–481)

FROM FERTILIZATION TO BIRTH
Fertilization
Implantation
Membranes Around the Embryo
The Placenta
Embryonic and Fetal Development
Commentary: Mother as Protector, Provider, Potential Threat
Birth and Lactation

Summary

If sperm ejaculation (150 to 300 million sperm) occurs with ovulation, a few hundred reach the upper regions of oviducts where fertilization commonly occurs. The secondary oocyte is stimulated into completing meiosis II when sperm reach its surface; this forms a mature *ovum*. A sperm nucleus enters the egg cytoplasm and fuses with the egg nucleus; this restores the diploid chromosome number for the zygote.

For three or four days following fertilization, the zygote travels down the oviduct where it receives nutrients and divides. By the time it reaches the uterus, the structure produced by cleavage stages is a blastocyst with a surface layer and an inner cell mass. Before the first week ends, the blastocyst contacts and adheres to the uterine lining through projections that invade maternal tissues (implantation). The inner cell mass becomes transformed into an *embryonic disk* that becomes the embryo proper in which the three germ layers form.

Shelled eggs contain membranes that function in nutrition, respiration, and excretion—the *yolk sac*, the *allantois*, the *amnion*, and the *chorion*. Humans also have a yolk sac (which forms part of the GI tract); the allantois contributes blood vessels, the amnion protects the embryo with fluid, and the chorion contributes to the placenta. Humans have an *umbilical* cord, which connects to these extraembryonic membranes.

The *placenta* is composed of the endometrium and extraembryonic membranes (especially chorion). Projections from the tiny blastocyst develop into many chorionic villi. Across this tissue, the embryo receives nutrients and oxygen and disposes of wastes. In the placenta, the blood vessels of the mother and embryo remain separate; nutrient and waste materials move by diffusion.

During the first trimester of development: the embryonic disk forms in the blastocyst; gastrulation forms the three germ layers; heart, nervous system, and respiratory structures form; segmentation becomes noticeable. By the ninth week a fetus exists, and by the twelfth week all the major organs are formed and the arms and legs move. By the second trimester the lanugo covers the body; during the seventh month the eyes open. By the middle of the third trimester the fetus may be ready for birth.

Birth begins with contractions of the uterine muscles; the cervical canal dilates, and the amniotic sac ruptures. The fetus is expelled before the placenta, and the umbilical cord is severed.

Lactation is the secretion and yielding of milk by the mammary gland. During pregnancy, estrogen and progesterone stimulate the development of the mammary glands. After birth the mammary glands produce a fluid rich in proteins and lactose. Then prolactin from the pituitary stimulates milk production. Suckling stimulates the release of oxytocin that causes the contraction of muscles associated with milk glands and uterine tissue.

Mother is viewed as protector, provider, and potential threat. During pregnancy, the mother requires increased vitamin and food intake; a poor maternal diet can damage the fetus—especially its brain. The mother's antibodies can cross the placenta and protect the fetus from most bacterial infections. Some viral infections, such as German measles, can cause fetal malformations. Some prescription drugs taken by the mother during pregnancy may be harmful; thalidomide, a tranquilizer, caused many birth defects. Alcohol intake by the mother during pregnancy can cause fetal alcohol syndrome; effects include mental retardation. Smoking by the mother during the second half of pregnancy can result in low birth weight and other problems.

Key Terms

ovum	chorion	fetus
embryonic disk	umbilical cord	German measles
yolk sac	placenta	(rubella)
allantois	chorionic villi	thalidomide
amnion	blastocyst	fetal alcohol syndrome

Objectives

1. Compare the events of sperm ejaculation and ovulation; tell where and under what conditions an ovum is formed prior to fertilization.
2. Describe events occurring in the human reproductive system for three or four days following fertilization until implantation is complete.
3. Name and describe the functions of the extraembryonic membranes; explain functions of the human yolk sac and the umbilical cord.
4. Explain how a spherical zygote becomes a multicellular adult with arms and legs.
5. Describe human embryonic and fetal development in terms of trimester developmental periods; tell when the term *fetus* is used.
6. Explain the events that occur during birth and lactation.
7. Cite examples that dramatically demonstrate that the developing individual is at the mercy of the mother's diet, health habits, and life-style.

Self-Quiz Questions

Fill-in-the-Blanks

If sperm ejaculation occurs with (1) _____ , a few hundred sperms reach the upper regions of the (2) _____ where fertilization commonly occurs. The secondary oocyte is stimulated into completing meiosis II when (3) _____ reach its surface; this forms a mature (4) _____ . When a sperm nucleus fuses with an egg nucleus the (5) _____ chromosome number is restored in the zygote. By the time it reaches the uterus, the zygote has divided into a structure called the (6) _____ , which has a surface layer and an inner cell mass. The inner cell mass becomes transformed into a(n) (7) _____ _____ , which becomes the (8) _____ proper in which the three germ layers form. The yolk sac, allantois, amnion, and chorion are termed the (9) _____ membranes. The (10) _____ encloses and bathes the embryo with fluid and the (11) _____ contributes greatly to the placenta. The placenta is composed of the (12) _____ and the (13) _____ membranes. Projections from the tiny blastocyst develop into many (14) _____ _____ through which the embryo receives nutrients and oxygen and disposes of wastes.

During the first trimester of development, the (15) _____ disk forms in the blastocyst; (16) _____ results in germ layers and the formation of heart, nervous system, and respiratory structures. By the ninth week, the embryo is termed a(n) (17) _____ , and by the twelfth week, all the major (18) _____ are formed. By the second trimester, (19) _____ covers the body and the eyes open. By the middle of the third trimester, the fetus may be ready for (20) _____ , which begins by contractions of the (21) _____ muscles; the cervical canal (22) _____ and the (23) _____ sac ruptures. The fetus is expelled before the placenta and the (24) _____ _____ is severed. The secretion and yielding of milk by the mammary gland is called (25) _____ . (26) _____ from the pituitary stimulates milk production; suckling stimulates the release of (27) _____ that causes contraction of muscles associated with milk glands and uterine tissue. A poor maternal diet can damage the fetus—especially the (28) _____ . Viral infections such as German measles can cause (29) _____ . Alcohol intake by the mother during pregnancy can cause (30) _____ _____ in the fetus.

POSTNATAL DEVELOPMENT, AGING, AND DEATH
CONTROL OF HUMAN FERTILITY
Commentary: Sexually Transmitted Diseases
SUMMARY

Summary

After birth, the individual follows a prescribed course of growth and development that leads to the sexually mature *adult*. Later in life, the body gradually deteriorates through *aging* processes. Aging in humans involves loss of hair and teeth, an increase in skin wrinkles and fat deposits, a decrease in muscle mass, the faltering of kidney function, and so on. Gradual physiological changes occur. Collagen, forty percent of the body's protein, undergoes structural changes with aging. Researchers have found cell lines that divide fifty times and die, which implies that cells have a limited division potential. DNA may gradually lose the capacity for repair. The genes that code for self-markers on the cell membranes may deteriorate, and autoimmune responses may intensify.

The control of human fertility has some ethical considerations. When does development begin? Key developmental events occur even before fertilization. When does life begin? The sperms and eggs that fuse to form zygotes are as much alive as any other form of life. Thus, it is difficult to argue that life begins when the zygote is formed. Consider that 9,900 infants are born each hour— how can human fertility be controlled? How can we reconcile the marvel of birth with the confusion surrounding unwanted pregnancies? Each year in the United States alone, there are more than 100,000 marriages based on premarital pregnancy, about 200,000 unwed teenage mothers, and perhaps 1.5 million abortions. Behavioral interventions include abstention, the rhythm method, withdrawal, and douching. Other methods include spermicides, diaphragm, condoms, IUD, the Pill, vasectomy, tubal ligation, and abortion. The termination of pregnancy may be induced with RU-486, the "morning-after pill." With in vitro fertilization, conception outside the body is possible; about fifteen percent of American couples are infertile. In such people, two- to four-day-old embryos implant less than twenty percent of the time.

Sexually transmitted diseases (STDs) have reached epidemic proportions, even in countries with the highest medical standards. *Acquired immune deficiency syndrome* (AIDS) is a set of chronic disorders that can follow infection by the human immunodeficiency virus (HIV). The virus cripples the immune system and is transmitted through body fluids (semen, blood, and vaginal fluid). AIDS is mainly a sexually transmitted disease with most infections occurring through bodily fluid transfer during vaginal or anal intercourse. There is no cure; behavioral controls can limit its spread. During the next decade, as many as 50 million to 100 million could be infected worldwide. *Gonorrhea* is an STD caused by a bacterium, *Neisseria gonorrhoeae*. Unlike AIDS, gonorrhea is an STD that can be cured by prompt diagnosis and treatment. The bacterium infects epithelial cells of the genital tract, eye membranes, and the throat. Multiple reinfections can and do occur. *Syphilis* is caused by a motile bacterium, *Treponema pallidum*. This spirochete is transmitted by sexual contact. Untreated, syphilis can produce lesions of skin and internal organs, including the liver, bones, and aorta; scars form. The brain and spinal cord are damaged in ways that lead to forms of insanity and paralysis. Women may have miscarriages, stillbirths, or sickly and syphilitic infants.

The most prevalent nonspecific sexually transmitted diseases are the *chlamydial infections*. A bacterium, *Chlamydia trachomatis*, causes a variety of diseases. It infects cells of the genitals and urinary tract. Chlamydial infections can be effectively treated. *Pelvic inflammatory disease* (PID) affects about 1.75

million women each year. It is one of the serious complications of gonorrhea, chlamydial infections, and other STDs, but it also can arise when normal vaginal microbes ascend into the pelvic region. The pain may be severe and oviducts may be scarred, leading to abnormal pregnancies and sterility. *Genital herpes* is an extremely contagious viral infection of the genitals. Transmission seems to require intimate sexual contact; the virus does not survive long outside the human body. An estimated 5 to 20 million persons in the United States have genital herpes. Newborns of affected mothers are among these cases. Herpes viruses are classed as type I, infecting mainly lips, tongue, mouth, and eyes, and type II, which causes most genital infections. At present, there is no cure for genital herpes although treatments are helpful.

Key Terms

adult	diaphragm	pelvic inflammatory
aging	condom	disease (PID)
collagen	the Pill	genital herpes
abstention	acquired immune deficiency	vasectomy
withdrawal	syndrome (AIDS)	tubal ligation
douching	gonorrhea	abortion
spermicidal foam	syphilis	RU-486
spermicidal jelly	chlamydia	in vitro fertilization

Objectives

1. Define *aging* and describe cellular and body changes occurring in the human aging process.
2. Identify the factors that encourage and discourage methods of human birth control.
3. Use Figure 32.22 in the main text to identify the three *most effective* birth control methods and the four *least effective* birth control methods.
4. From the Commentary in the main text (p. 485), determine which birth control methods help to control STDs.
5. Describe two different types of sterilization.
6. State the physiological circumstances that would prompt a couple to try in vitro fertilization.
7. For each STD described in the Commentary, know (a) the common name of the causative organism and (b) the symptoms of the disease.

Self-Quiz Questions

Fill-in-the-Blanks

Following birth, each individual follows a prescribed course of growth and development that leads to the sexually mature (1) _____ . The (2) _____ process involves a gradual deterioration of the body. Researchers have discovered that cells have a limited (3) _____ potential. Key events in human development occur even before (4) _____ . Because of previous developmental events and the fact that sperms and eggs are alive, it is difficult to argue that life begins with the formation of the (5) _____ . The most effective method of birth control is (6) _____ .

Males may be sterilized by a surgical technique known as (7) _____ ;
females may be sterilized by a surgical technique known as (8) _____
_____ . The letters STD stand for (9) _____ _____ _____ .

CHAPTER TEST   **UNDERSTANDING AND INTERPRETING KEY CONCEPTS**

___ (1) Examples of organisms that reproduce asexually are _____ .

    (a) frogs
    (b) sponges
    (c) bony fishes
    (d) flatworms
    (e) both b and d

___ (2) Which of the following represents the most correct sequence of development?

    (a) cleavage → gamete formation → fertilization → blastula → gastrulation → organ formation
    (b) gamete formation → fertilization → cleavage → blastula → gastrulation → organ formation
    (c) gamete formation → cleavage → fertilization → blastula → gastrulation → organ formation
    (d) fertilization → cleavage → gamete formation → blastula → gastrulation → organ formation

___ (3) In development, differentiated cells become organized into tissues and organs through _____ .

    (a) gastrulation
    (b) cell differentiation
    (c) morphogenesis
    (d) cleavage

For questions 4 to 8, choose from these answers:

    (a) AIDS
    (b) Chlamydial infections
    (c) Genital herpes
    (d) Gonorrhea
    (e) Syphilis

___ (4) _____ is a disease caused by a bacterium (*Neisseria*); it infects epithelial cells of the genital tract, eye membranes, and the throat.

___ (5) _____ is a disease caused by a motile spirochete bacterium (*Treponema*) that produces lesions of the skin and internal organs; the brain and spinal cord are damaged in ways that lead to forms of insanity and paralysis.

___ (6) _____ is an incurable disease caused by the HIV virus; it seriously cripples the immune system.

—— (7) _____ are the most nonspecific sexually transmitted diseases; cells of the genitals and urinary tract are infected.

—— (8) _____ is an extremely contagious and incurable viral infection of the genitals; one form causes sores on the facial area.

For questions 9 to 11, choose from these answers:

    (a) blastocyst
    (b) allantois
    (c) yolk sac
    (d) oviduct
    (e) cervix

—— (9) The _____ lies between the uterus and the vagina.

—— (10) The _____ is a pathway from the ovary to the uterus.

—— (11) The _____ results from the process known as cleavage.

For questions 12 to 15, choose from these answers:

    (a) endocrine cells in the testes
    (b) seminiferous tubules
    (c) vas deferens
    (d) epididymis
    (e) prostate

—— (12) The _____ is a structure where sperms mature.

—— (13) Testosterone is produced by the _____ .

—— (14) Meiosis to form sperms occurs in the _____ .

—— (15) The _____ is the structure sperms pass through while moving from a testis to the ejaculatory ducts.

INTEGRATING AND APPLYING KEY CONCEPTS

What rewards do you think a society should give a woman who has at most two children during her lifetime? In the absence of rewards or punishments, how else can a society encourage women not to have abortions and yet ensure that the human birth rate does not continue to increase?

Crossword
Number Seven

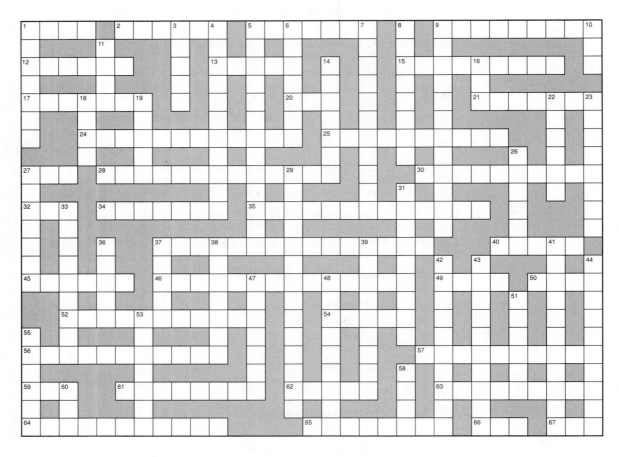

Across

1. nerve cell process that may conduct an action potential from a cell body to a neighboring cell
2. the _____ system is part of the forebrain and influences learning and emotional behavior and, perhaps, helps to perceive pain
5. fatty tissue = _____ tissue
9. small vessel that carries blood away from heart
12. a simple, stereotyped, and repeatable motor action
13. storms; displays of fierce anger
15. blood vessel smallest in diameter
17. system with message-conducting and information-processing pathways
20. rim of a vessel, bell, crater, or basin
21. organs that purify blood
24. localized heat, swelling, or redness caused by infection or irritation
25. a noncellular infectious agent with RNA for its nucleic acid core; a member of this group causes AIDS
27. a pouch or pouchlike structure, sometimes containing fluid, in an organism
28. union of an egg and sperm
30. in this response, lymphocytes attack tissues in organism of which they are a part
32. flat representation of area; relief _____ shows land elevations
34. any substance that stimulates production of an antibody

35. system that protects an animal body from injury, dehydration, and microbial invasion; also involved in controlling body temperature, excretion of some wastes, and reception of external stimuli
37. one of these detects mechanical energy associated with changes in pressure, position, or acceleration
40. circulatory system pump
45. this selection theory suggests how lymphocytes can produce millions of different receptors that distinguish vast numbers of different antibody types in vertebrates
46. one of these detects radiant energy associated with temperature
49. Scottish island where St. Columba's abbey was founded in A.D. 563
50. capital and principal shipping and industrial city of Norway
52. hormone produced by adrenal cortex that stimulates sodium reabsorption and water retention in human body
54. the night hunter of the sky; a constellation containing Rigel and Betelgeuse
56. red blood cell
57. large white blood cell that alerts helper T cells to presence of specific foreign agents; also, cells of this type are phagocytes descended from monocytes in tissues
59. _____ markers are proteins on cell surfaces that identify whether a cell is a "self" cell or a "nonself" cell
61. glands associated with oral region; they secrete substances that prepare food for digestion
62. incensed or outraged
63. connects stomach to anus
64. this nerve action prepares an animal for fight, flight, or intense frolic
65. one of a class of steroid hormones produced in human females by ovarian follicles and in human males by the adrenal cortex
66. the _____ sends excitatory messages to the thalamus that arouse the brain and maintain wakefulness
67. _____ protein utilization is a measure used by nutritionists to compare amino acid composition in proteins from different sources

Down

1. the _____ medulla secretes epinephrine and norepinephrine
3. circulatory fluid pumped by heart
4. most animals have a closed _____ system, in which the heart and blood vessels are continuously connected
5. a process in which antibodies act against "foreign" red blood cells and cause them to clump together
6. a hormone secreted by pancreatic beta cells
7. inner lining of the uterus
8. finely branched peripheral endings of a sensory neuron that respond to a specific kind of stimulus
9. skeleton that is joined to the axial portion at shoulders and hips in humans
10. organ sensitive to light
11. the power to reject a bill passed by a legislative body
14. functional unit of a kidney
16. a small mating territory of grouse species
18. vessel that returns blood to heart
19. relating to a sun
22. a state of reproductive "heat"
23. basic frame consisting of hard components
26. not vulnerable to one or more specific disease-causing agents

27. this nervous system contains efferent nerves connecting the central nervous system to skeletal muscles
29. between hard parts of a spine
31. cells or glands that produce and/or secrete specific hormones
33. these arteries carry deoxygenated blood in humans
36. an organ that extracts specific materials from blood and alters or concentrates them for later secretion
37. a polyphonic composition based on a sacred text and generally sung without accompaniment
38. chemical messenger secreted from an endocrine cell or gland
39. "master gland"
41. in most animals, the overall exchange of oxygen from the environment and carbon dioxide wastes from cells by way of circulating blood
42. occurring at the glomerulus, this is the bulk flow of water and solutes into Bowman's capsule
43. a control point where different bits of information are pulled together and analyzed and a response is selected
44. in this system, about twenty proteins contribute to the inflammatory response directed against bacterial or fungal invaders of the human body
47. condition of being extremely fat
48. connects throat area to stomach
51. _____ luteum
53. part of alimentary canal located between esophagus and small intestine
55. part of skin that cushions body from everyday stresses and strains
58. organ of gas exchange in many terrestrial animals
60. a curved wheel on a rotating shaft producing variable or reciprocating motion in another connecting part

33
POPULATION ECOLOGY

ECOLOGY DEFINED
POPULATION DYNAMICS
    Population Size and Patterns of Growth
    Checks on Population Growth
HUMAN POPULATION GROWTH
    How We Began Sidestepping Controls

Present and Future Growth
Controlling Population Growth
Questions About Zero Population Growth
SUMMARY

General Objectives

1. Learn the major terms associated with the study of population ecology.
2. Understand the factors that affect population density, distribution, and dynamics.
3. Understand the meaning of the equation that describes the rate of increase (r) for the population size (N).
4. Explain the meanings of population curves on graphs that take the shape of J and S.
5. Fully understand the significance of *carrying capacity*.
6. List possible checks on population growth.
7. Explain the concept of *survivorship*.
8. Cite figures relating to the phenomenal human population growth on our planet.
9. Discuss the applications of the *demographic transition model* of population control.
10. Suggest possible solutions for the human population problem.

33-I
(pp. 493–496)

ECOLOGY DEFINED
POPULATION DYNAMICS
    Population Size and Patterns of Growth

Summary

For humans, social, economic, and political considerations may influence the short-term distribution of resources on which any population depends. Biological principles predict the consequences of competition for scarce natural resources.

Ecology involves the interactions of organisms with one another and with the physical and chemical environment. A *population* is a group of the same species that live in a specific *habitat*. The *community* encompasses all species in a habitat; the term is also used for groups of organisms with similar life-styles in the habitat. An *ecosystem* contains a community and its chemical and physical environment; an ecosystem has a biotic component and an abiotic component. The biosphere is the entire realm in which organisms exist; this includes the global physical and chemical environment.

Population dynamics deals with characteristics such as size, density, dispersion, and age structure. The population size is determined by the number of individuals making up its gene pool. Population density is the number of individuals per unit area or volume. Dispersion is the general pattern in which the population's members are dispersed through its habitat; this may vary with time. The age structure of a population is the relative proportion of individuals of each age (prereproductive, reproductive, and postreproductive).

Variables such as birth, death, immigration, and emigration affect population size. When birth rate over the long term is balanced by death rate, there is zero population growth. Populations increase in size when the number of births exceeds the number of deaths. The equation for the rate of increase (r) for the population size (N) is given as:

$$r = \frac{\text{births - deaths (in a specified amount of time)}}{N}$$

As long as the value of r remains positive, population size will increase by ever larger amounts. Size increases of *exponential growth* are plotted on graphs as *J-shaped curves*. This means the populations are increasing in size by ever larger amounts per unit of time.

The *biotic potential* of a population is its maximum rate of increase, per individual, under ideal conditions. Biotic potential may vary for different species. A population may grow exponentially even when it is not expressing its full biotic potential. When any essential environmental resource is in short supply, it becomes a *limiting factor* on population growth. Examples of limiting factors are predators, competition for living space, and pollution. The total of the limiting factors acting on a population collectively represents environmental resistance. *Carrying capacity* is the number of individuals of a given species that can be sustained indefinitely in a given area. *Logistic growth* plots on a graph as an S-shaped curve.

Key Terms

ecology	population size	exponential growth
population	population density	biotic potential
habitat	population dispersion	carrying capacity
community	age structure	limiting factor
ecosystem	reproductive base	environmental resistance
biotic	zero population growth	logistic growth
abiotic	r	S-shaped curve
biosphere	J-shaped curve	

Objectives

1. Understand how the principles of ecology can influence human social, economic, and political considerations.
2. Explain how the kinds of interactions among species can shape the structure of a biological community.
3. Define *size, density, dispersion,* and *age structure* in terms of the population.
4. Explain how birth, immigration, death, and emigration affect the numerical size of a population over a given span of time; define *zero population growth*.
5. Contrast the conditions that promote J-shaped population growth curves with those that promote S-shaped curves.

6. Define *biotic potential, carrying capacity, limiting factor, logistic growth,* and *environmental resistance.*

Self-Quiz Questions

Matching

(1) __ population

(2) __ community

(3) __ ecosystem

(4) __ ecology

(5) __ biosphere

A. The study of the interactions of organisms with one another and with the physical and chemical environment

B. The entire realm in which organisms exist

C. A group of the same species living in a specific habitat at a specific time

D. The unit that encompasses all species that interact in a habitat

E. A biological community and its chemical and physical environment

Fill-in-the-Blanks

Population (6) _____ deals with characteristics such as size, density, dispersion, and age structure. Population (7) _____ is determined by the number of individuals making up its gene pool. Population (8) _____ is the number of individuals per unit area or volume. The general pattern in which the population's members are dispersed through its habitat is (9) _____ . The (10) _____ _____ of a population is the relative proportions of individuals of each age. (11) _____ is the symbol for the rate of increase for population size. J-shaped curves are indicative of population size increases of (12) _____ growth. The maximum rate of increase of a population, per individual, under ideal conditions is its (13) _____ potential. A(n) (14) _____ factor on population growth is any essential environmental resource in short supply. Environmental (15) _____ is the total of all limiting factors acting collectively on a population. The number of individuals of a given species that can be sustained indefinitely in a given area is its (16) _____ _____ . (17) _____ growth appears on a graph as an S-shaped curve.

33-II
(pp. 496–498)

POPULATION DYNAMICS (cont.)
Checks on Population Growth

Summary

Density-dependent controls include competition for resources; predation, parasitism, and disease are the main density-dependent factors. Effects on population growth are generally self-adjusting; when density decreases, pressure eases and population size again increases. *Density-independent controls* are events that tend to increase the death rate more or less independently of population density. Examples include the effects of freak snowstorms on butterfly populations and human poachers on animal populations.

Survivorship curves reflect the age-specific patterns of death for a particular population in a particular environment. Data are gathered by following the fate of a group of newborn individuals until the last one dies and then graphing the number of survivors at each age. Birth schedules document the average number of offspring produced by individuals at each age. The Type I curve is typical of large mammals that provide their offspring with extended parental care (humans with good health-care services). The Type II curve is typical among organisms that are just as likely to be killed or die of disease at any age (songbirds, lizards, and small mammals). The Type III curve reflects a high death rate early in life among organisms that produce large numbers of new individuals (marine invertebrates, most insects, many fish, plants, and fungi). Survivorship and birth schedules may be put together in a life table, which shows the survival rate at each age group in a population.

Key Terms

density-dependent controls

density-independent controls

survivorship curves (Types I, II, and III)

life table

Objectives

1. Define *density-dependent controls*, give two examples, and indicate how density-dependent factors act on populations.
2. Define *density-independent controls*, give two examples, and indicate how they affect populations.
3. Discuss the factors that impart different shapes to Type I, Type II, and Type III survivorship curves; present an example of each.

Self-Quiz Questions

Fill-in-the-Blanks

(1) _____ - _____ controls include competition for resources.

(2) _____ - _____ controls include events that tend to increase the death rate more or less independently of population density. The effects of a freak snowstorm on butterfly populations would be an example of (3) _____ - _____ population controls. A predator killing its prey would be an example of a(n) (4) _____ - _____ population control.

(5) _____ curves reflect the age-specific patterns of death for a particular population in a particular environment. A Type (6) _____ curve is typical of large mammals that provide their offspring with extended parental care. A Type (7) _____ curve is typical among organisms that are just as likely to be killed or die of disease at any age. A Type (8) _____ curve reflects high

death rate early in life among organisms that produce large numbers of new individuals. Most insects would fit the Type (9) _____ survivorship curve. The human survivorship curve is the Type (10) _____ . Songbirds and lizards fit the Type (11) _____ survivorship curve. Survivorship and birth schedules may be integrated into a life (12) _____ , which shows the survival rate at each age group in a population.

33-III
(pp. 498–500)

HUMAN POPULATION GROWTH
  How We Began Sidestepping Controls
  Present and Future Growth

Summary

In 1990, the human population reached 5.3 billion individuals. In 1988 alone, 1.7 million people (238,000 per day, or 9,900 per hour) were added to the human population each week.

Human population growth has increased astoundingly for the past two centuries. Reasons for this are traced to our sidestepping of natural controls: (1) we steadily developed the capacity to expand into new habitats and new climatic zones; (2) carrying capacities for humans increased in the environments we already occupied (agriculture); and (3) we removed several limiting factors (medicine).

It took 2 million years for the human population to reach the first billion but only 12 years to reach the fifth billion. We can expect the population to go beyond 6 billion within the next 10 years. It is not reasonable to expect that food, drinkable water, energy reserves, wood, and steel can be produced to accommodate the predicted population growth of humans. Pollutants are rendering the environment less capable of supporting increased populations. It is realistic to expect an imminent crash in our numbers; exponential growth continues but is not sustainable.

Key Terms

This section is supported by terminology the reader should be familiar with.

Objectives

1. State the total of the human population in 1990.
2. List three reasons that human population growth has increased astoundingly for the past two centuries.
3. State the predicted human population figure for the year 2000.
4. Give the predictable outcome of the present rate of human population growth; list reasons for this expectation.

Self-Quiz Questions

Fill-in-the-Blanks

In 1990, the human population reached (1) _____ billion. By the year 2000, the human population will exceed (2) _____ billion. The astounding increase in human population growth can be traced to the ways we have sidestepped (3) _____ _____ . We have steadily developed the capacity to expand into new (4) _____ and new (5) _____ zones. In environments humans already occupy, (6) _____ _____ increased. We have also removed several (7) _____ _____ . (8) _____ are rendering the environment less capable of supporting increased human population size. The (9) _____ growth of the human population continues but it is not sustainable.

33-IV
(pp. 500–504)

HUMAN POPULATION GROWTH (cont.)
  Controlling Population Growth
  Questions About Zero Population Growth
SUMMARY

Summary

There is widespread awareness of the links among overpopulation, resource depletion, and increased pollution. Most efforts to decrease population focuses on decreasing the birth rate. Two general approaches to decreasing birth rates are economic development and family planning.

Population control through economics can be illustrated by the *demographic transition model* in which population growth changes are linked with changes unfolding during four stages of economic development. In the preindustrial stage, harsh living conditions and high birth and death rates exist; there is little population growth. In the transitional stage, industrialization begins, food production rises, and health care improves; death rates drop and birth rates remain high with population growth. In the industrial stage, growth slows as urban couples regulate family size. In the postindustrial stage, zero population growth is reached; birth rate falls below death rate with a slow decrease in population size. The United States, the Soviet Union, and most western nations are in the industrial stage. Mexico and other less-developed countries are in the transition stage. If Mexico's population growth keeps outpacing economic growth, the death rate may increase. Unless birth rates are controlled, countries stuck in the transitional stage may well return to the harsh conditions of the preceding stage.

Population control through family planning may bring about a faster decline in birth rates than economic development alone. The average "replacement level for fertility" is slightly higher than two children to replace every reproducing couple because some female children die before reaching reproductive age. If replacement level for fertility were achieved on a global scale, the population would still continue to grow for about sixty years because children already existing will be reproducing. Encouraging delayed reproduction slows birth rate by lengthening generation time and lowering the average number of children in each family. China has established the most extensive family planning program in the world; the fertility rate has dropped from a previous high of 5.7 to 2.4 children per woman in 1987.

The biological implications of exponential growth for the human population are staggering as are the social implications of achieving and maintaining zero population growth. With *zero population growth*, far more people will fall in the older age brackets. Younger, productive persons must bear a greater and greater share of the burden for the more numerous older persons. Decisions on this matter must soon be made. All species, including our own, face limits to growth. A global effort must soon be made to limit population growth in accordance with environmental carrying capacity. If we wait, the environment will limit population for us.

Key Terms

demographic transition model
preindustrial stage
transitional stage
industrial stage

postindustrial stage
family planning
zero population growth

Objectives

1. State the basic premise of the *demographic transition model.*
2. Summarize the conditions of each of the four economic development stages.
3. Discuss implications of population control through family planning.
4. List biological and social implications of *zero population growth.*
5. Relate global efforts for limiting human population growth to environmental carrying capacity; share your insights with others you believe might not understand these principles.

Self-Quiz Questions

Fill-in-the-Blanks

Population control through economics can be illustrated by the (1) _____ _____ model in which population growth changes are linked with changes in four stages of economic development. Harsh living conditions along with high birth and death rates are typical of the (2) _____ stage. When industrialization begins, food production increases, and health care improves, it is the (3) _____ stage. Urban couples regulate family size and growth slows in the (4) _____ stage. Zero population growth is achieved and birth rate falls below death rate with a slow decrease in population size in the (5) _____ stage. The United States and the Soviet Union are in the (6) _____ stage. Mexico and other less-developed countries are in the (7) _____ stage. Population control through (8) _____ planning may bring about a faster decline in birth rates than economic development alone. (9) _____ has established the most extensive family planning program in the world. With zero population growth, far more people will fall into the (10) _____ age brackets. All species, including the human species, face limits to (11) _____ . A worldwide effort to limit population growth must be in accordance with environmental (12) _____ _____ .

___ (1) The term that encompasses all species in a habitat is the _____ .

    (a) population
    (b) ecosystem
    (c) community
    (d) biosphere

___ (2) The total number of individuals making up its gene pool defines the population _____ .

    (a) density
    (b) growth
    (c) birth rate
    (d) size

___ (3) The average number of individuals of the same species per unit area or volume at a given time is the population _____ .

    (a) density
    (b) growth
    (c) birth rate
    (d) size

___ (4) A population that is growing exponentially may be plotted on a graph as a(n) _____ .

    (a) S-shaped curve
    (b) J-shaped curve
    (c) curve that terminates in a plateau phase
    (d) Type III survivorship curve

___ (5) A(n) _____ is an essential environmental resource that, when in short supply, adversely affects population growth.

    (a) carrying capacity
    (b) biotic potential
    (c) exponential growth
    (d) limiting factor

___ (6) A situation in which the birth rate over the long term is balanced by the the death rate is called _____ .

    (a) a limiting factor
    (b) exponential growth
    (c) logistic growth
    (d) zero population growth

___ (7) The rate of increase for a population (r) refers to the _____ the birth rate and death rate divided by the number of individuals in the population.

    (a) sum of
    (b) product of
    (c) doubling time between
    (d) difference between

___ (8) Which of the following is *not* an example of a density-independent control?

   (a) Effects of freak snowstorms on butterfly populations
   (b) Competition for living space
   (c) The effects of a killing frost on a population of plants
   (d) A flood on a colony of groundhogs

___ (9) Small mammals are just as likely to be killed or die of disease at any age; therefore, the appropriate survivorship curve is a _____ .

   (a) Type I
   (b) Type II
   (c) Type III
   (d) Type IV

___ (10) _____ is the number of individuals of a given species that can be sustained indefinitely in a given area.

   (a) The carrying capacity of the environment
   (b) Exponential growth
   (c) The doubling time of a population
   (d) Logistic growth

___ (11) In 1990, the human population of our planet reached _____ .

   (a) 5.3 million
   (b) 9,900 million
   (c) 4 billion
   (d) 5.3 billion
   (e) none of the above

___ (12) Which of the following stages of economic development describe Mexico and other undeveloped countries?

   (a) preindustrial
   (b) transitional
   (c) industrial
   (d) postindustrial
   (e) none of the above

INTEGRATING AND APPLYING KEY CONCEPTS

Assume that the world has reached *zero population growth*. The year is 2110, and there are 10.5 billion individuals of *Homo pollutans* on Earth. You have seen stories on the community television screen about how people used to live 120 years ago. List the ways that life has changed and comment on the events that no longer happen because of the enormous human population.

34
COMMUNITY INTERACTIONS

CHARACTERISTICS OF COMMUNITIES
    The Concepts of Niche and Habitat
    Types of Species Interactions
MUTUALLY BENEFICIAL INTERACTIONS
COMPETITIVE INTERACTIONS
    Categories of Competition
    Competitive Exclusion
CONSUMER-VICTIM INTERACTIONS
    On "Predator" Versus "Parasite"
    Dynamics of Predator-Prey Interactions
    Prey Defenses

    Parasitic Interactions
**COMMUNITY ORGANIZATION,
DEVELOPMENT, AND DIVERSITY**
    Resource Partitioning
    Effects of Predation on Competition
    Species Introductions
    Commentary: Hello Lake Victoria,
    Goodbye Cichlids
    Succession
    Patterns of Species Diversity
SUMMARY

**General
Objectives**

1. List and distinguish among the several types of species interactions.
2. Understand mutually beneficial interactions.
3. Describe the different forms of competitive interactions.
4. Discuss the positive aspects and the negative aspects of predation on prey populations.
5. Describe how communities are organized, how they develop, and how they diversify; relate the consequences of introducing species into new geographic areas.

34-I
(pp. 506–509)

CHARACTERISTICS OF COMMUNITIES
    The Concepts of Niche and Habitat
    Types of Species Interactions

Summary

In the rain forests of New Guinea, there are nine species of fruit-eating pigeons; each species has its own unique role in the forest. The *niche* of a species is defined by its relations with other organisms and with its physical surroundings (one can think of the niche as the job an organism has in the environment). Each species of pigeon can eat fruit of different sizes, and their seed dispersal affects where different plants grow. All of the organisms of the forest interact (but not necessarily directly) as a community. A *community* is an association of populations, tied together directly or indirectly by way of competition for resources, predation, and other interactions. The *habitat* of an organism is the type of place where it normally lives (one can think of habitat as the address of the organism);

it is characterized by physical and chemical features as well as general vegetation structure. The number of species in a habitat, the numbers of individuals in each species, and dispersion of those individuals depend on: (1) geochemical and climatic processes, (2) the kinds and amounts of food available through the year, (3) adaptive traits that allow individuals of a species to exploit specific resources, and (4) interactions that occur among different populations in the habitat. As a result of the above factors, several community properties emerge: diversity (number of species), the number of species at different "feeding levels," relative abundances (the number of individuals of each kind of organism), and dispersion (how individuals of each population are spaced in the habitat).

Several types of species interactions occur in a community. In a *neutral interaction*, neither population directly affects the other (for example, eagles and grass). In *commensalism*, one population benefits and the other is not affected (for example, bird's nest in tree). In *mutualism*, both populations directly benefit by interacting. In *interspecific competition*, some individuals of both species are harmed by the interaction. In *predation* and *parasitism*, one species benefits while the other is harmed.

Key Terms

niche	relative abundances	mutualism
community	dispersion	interspecific competition
habitat	neutral interaction	predation
diversity	commensalism	parasitism
feeding levels		

Objectives

1. Describe the characteristics of a *community*.
2. Define and distinguish between *habitat* and *niche*.
3. Discuss factors that determine the number of populations in a habitat, the numbers of individuals in each population, and dispersion of those individuals.
4. Discuss the meanings of the following community properties: diversity, feeding levels, relative abundances, and dispersion.
5. Define *neutral interaction*, *commensalism*, and *mutualism*.
6. Distinguish among *interspecific competition*, *predation*, and *parasitism*.

Self-Quiz Questions

Fill-in-the-Blanks

The (1) _____ of a species is defined by its relations with other organisms and with its physical surroundings. A(n) (2) _____ is an association of populations, tied together directly or indirectly by way of competition for resources, predation, and other interactions. The (3) _____ of an organism is the type of place where it normally lives. Geochemical and climatic processes, kinds and amounts of food available through the year, adaptive traits, and species interactions determine the number of (4) _____ in a habitat, the numbers of individuals in each (5) _____ , and (6) _____ of those individuals. The example of eagles and meadow grass describes a(n) (7) _____ interaction. The interaction between a bird's nest and a tree is called (8) _____ . When both species benefit directly by an interaction, it is known as (9) _____ . In (10) _____ competition, both species are harmed by the interaction. In (11) _____ and (12) _____ , one species benefits while the other is harmed.

34-II
(pp. 509–510)

MUTUALLY BENEFICIAL INTERACTIONS
COMPETITIVE INTERACTIONS
    Categories of Competition
    Competitive Exclusion

Summary

Some community interactions are mutually beneficial. The interactions between flowering plants and their pollinators are classic examples of *mutualism*. A case of *symbiosis* is the interaction between the yucca moth and the yucca plant; neither species can live long without the other.

Interspecific competition occurs when different species compete for a limited resource. There are two types of competitive interactions, regardless of whether they occur within or between species. In *exploitative competition*, all individuals have equal access to a resource but differ in their ability to exploit that resource. In *interference competition*, certain individuals limit or prevent others from using the resource and so control access to it. In pure exploitation competition, one species hampers the growth, survival, or reproduction of the other only indirectly, by reducing the common supply of resources.

When two species exploit the same resource they may continue to coexist, or one species may exclude the other from the habitat by interfering with resource access or exploiting resources so effectively as to starve the other species. A test by G. Gause using two species of *Paramecium* exploiting similar food resulted in strong competition between them. This suggested that complete competitors cannot coexist indefinitely, a concept known as *competitive exclusion*. When two *Paramecium* species feeding on different food sources (bacteria and yeast cells) were grown together, they continued to coexist although the growth rate decreased. Another competitive interaction is the example of four chipmunk species that occupy four different habitats on the eastern slope of the Sierra Nevada in California.

Key Terms

mutualism	exploitation competition	G. Gause
symbiosis	interference competition	competitive exclusion
interspecific competition		

Objectives

1. Define and give an example of *mutualism* and *symbiosis*.
2. Describe the conditions of two types of *interspecific competition*.
3. Relate how the study of two species of *Paramecium* exploiting similar food sources illustrates *competitive exclusion*.

Fill-in-the-Blanks

Interactions between flowering plants and their pollinators provide examples of (1) _____ . When different species compete for a limited resource, (2) _____ competition occurs. When all individuals have equal access to a resource but differ in their ability to exploit that resource, it is (3) _____ competition. In (4) _____ competition, certain individuals limit or prevent others from using the resource and so control access to it. In a test by G. Gause, two species of *Paramecium* exploiting similar food resulted in strong competition between them; this describes a concept known as (5) _____ _____ . When two species of *Paramecium* were grown together while feeding on different food sources, they continued to coexist although the (6) _____ rate decreased.

34-III
(pp. 510–515)

CONSUMER-VICTIM INTERACTIONS
On "Predator" Versus "Parasite"
Dynamics of Predator-Prey Interactions
Prey Defenses
Parasitic Interactions

Summary

Predators get their food from *prey* they may or may not kill, but they do not live on or in the prey. *Parasites* get their food from hosts, and they live on or in the host for a good part of their life cycle; they may or may not kill the host.

The dynamics of predator-prey interactions range from stable coexistence to erratic oscillations depending upon: (1) the carrying capacity for prey in the absence of predation, (2) the reproductive rates of the predator and prey, and (3) the adaptive capacity of individual predators to respond to increases in prey density. Stable coexistence results when predators prevent prey from overshooting the carrying capacity. Fluctuations in population density tend to occur when predators do not reproduce as fast as their prey. Long-term studies of the Canadian lynx and snowshoe hare indicate that cycling of predator and prey abundance may be caused by more than just predation; apparently wildfires, floods, and insect outbreaks destroy mature trees. Hares browse extensively on harmless parts of the newly established plants growing on the disturbed habitats; peaks in prey density correspond to the new plant growth. When hare population density increases, hares are forced to feed on toxic portions of the newly established plants; they then become stressed and have increased vulnerability to predation. This may cause a decline in the hare population.

When a new prey defense evolves during the *coevolution* of predator and prey, only those predators able to counter it will live to reproduce.

A *moment-of-truth defense* occurs when a cornered prey startles or intimidates a predator; for example, a bombardier beetle under attack sprays a noxious chemical. Chemicals produced by many plant and animal species serve as warning odors, repellents, and outright poisons. For example, earwigs, skunks, and skunk cabbages produce awful odors.

Warning coloration occurs when toxic prey organisms present bright colors or bold patterns to warn predators. Other prey not equipped with defenses may resemble the toxic prey by *mimicry.*

Camouflage allows prey and predators to hide in the open; an example is a desert plant that looks like a small rock. It flowers only during a brief rainy season, when other plants and water are available for the plant eaters.

True parasites tend to coevolve with their host; they only kill hosts without coevolved defenses. Parasites that coevolve with their hosts produce less-than-fatal effects. *Parasitoids* are insect larvae whose target hosts are young stages of other insect species; they serve as natural controls over other insects in a community. Use of parasitoids is a promising way to avoid use of environmentally harmful insecticides. Animal species that depend upon the social behavior of another to complete its life cycle are *social parasites.* An example of a social parasite is the cowbird that lays eggs in the nests of other birds.

Key Terms

predator	moment-of-truth defense	true parasite
prey	warning coloration	parasitoids
parasite	mimicry	social parasites
coevolution	camouflage	

Objectives

1. Distinguish *predators* from *parasites.*
2. List factors that determine the nature of predator-prey interactions.
3. Describe factors affecting the predator-prey cycling in the case of the Canadian lynx and snowshoe hare.
4. Define *coevolution* and suggest why it might serve the interests of populations to coevolve.
5. Define and give an example of a *moment-of-truth defense.*
6. Distinguish *mimicry* from *camouflage.*
7. List criteria that distinguish *true parasites* from *parasitoids.*
8. Define and give an example of *social parasitism.*

Self-Quiz Questions

Fill-in-the-Blanks

(1) _____ get their food from prey they may or may not kill, but they do not live on or in the prey. (2) _____ get their food from hosts, and they live on or in the host for a good part of their life cycle; they may or may not kill the host. The dynamics of predator-prey interactions range from stable coexistence to erratic oscillations depending upon: the (3) _____ _____ for prey in the absence of predation; the (4) _____ rates of predator and prey, and the (5) _____ _____ of the individual predators to respond to increases in prey density. Long-term studies of the Canadian lynx and snowshoe hare indicate that (6) _____ of predator and prey abundance may be caused by more than just predation. New prey

defenses come about during the (7) _____ of predator and prey. When a cornered skunk produces an awful odor, it is an example of a(n) (8) _____-_____-_____ defense. The presentation of bright colors or bold patterns to predators is called (9) _____ coloration. Other prey not equipped with defenses may resemble the toxic prey by using (10) _____ . A desert plant that looks like a small rock illustrates the principle of (11) _____ . True (12) _____ tend to coevolve with their hosts; they only kill hosts lacking coevolved defenses. (13) _____ are insect larvae whose target hosts are young stages of other insect species. The cowbird lays its eggs in the nests of other bird species; as such the cowbird is a(n) (14) _____ parasite.

Matching

The same letter may be used more than once. More than one letter may be appropriate for the same blank.

(15) ___ bird nest in a tree

(16) ___ yucca moth and yucca plant

(17) ___ two species of *Paramecium,* similar food

(18) ___ tapeworms

(19) ___ mistletoe

(20) ___ Canadian lynx and snowshoe hare

(21) ___ bombardier beetle and grasshopper mouse

(22) ___ skunk

(23) ___ desert plant resembling a small rock

(24) ___ cowbird

A. Parasitism
B. Commensalism
C. Mutualism
D. Predator-prey relationship
E. Camouflage
F. Competitive exclusion
G. Moment-of-truth defense
H. Warning coloration
I. Social parasite

34-IV
(pp. 515–517)

COMMUNITY ORGANIZATION, DEVELOPMENT, AND DIVERSITY
   Resource Partitioning
   Effects of Predation on Competition
   Species Introductions
   Commentary: **Hello Lake Victoria, Goodbye Cichlids**

Summary

A stable community is one whose forces have come into balance although sometimes this is an uneasy balance. In any community, similar species share the same resource in different ways. This community pattern, called *resource partitioning*, develops in two ways: (1) natural selection increases the differences between competing populations, and (2) to have success, species joining an existing community must be dissimilar to established ones. An example of resource partitioning is given by three species of annual plants in an abandoned field; each exploited different parts of the habitat. Foxtail grasses have shallow, fibrous root systems to absorb rainwater quickly in areas where soil moisture varies from day to day; mallow plants have deeper taproot systems and live where deeper soil areas are moist early in the growing season but drier later on. Smartweeds prevail in areas of continuously moist topsoil with taproots that branch in topsoil.

By reducing prey population densities, predators decrease competition among prey and promote their coexistence; removing predators increases competition among prey species. When sea stars were kept out of experimental plots of the rocky intertidal zone, the number of prey species was reduced from sixteen to eight. Mussels, main prey of sea stars, were the strongest competitors for space.

Few introductions of species into new geographic areas are without ecological consequences. The introduction of the South American water hyacinth into the United States in the 1880s went unchecked by natural controls; they clog ponds, streams, rivers, and canals. Imported honeybees have become part of existing communities in the United States, but in many areas they have displaced native bees.

Cichlids (fishes that eat aquatic plants and detritus) were a favorite food for people dwelling near Lake Victoria in East Africa. Unfortunately, fishermen had overfished the cichlid populations, and the Nile perch, a larger, carnivorous fish, was introduced in an effort to increase the food supply for humans. The Nile perch ate its way through the cichlids, wiping out its own food source. People disliked the oily taste of the Nile perch, which was preserved by wood smoke. To get the necessary wood, the natives destroyed the local forests, which have proved difficult to replant. Knowledge and simple experimentation could have prevented this disaster.

Key Terms

| community stability | species introduction | cichlids |
| resource partitioning | | |

Objectives

1. Discuss the two ways that resource partitioning develops in a community.
2. Explain (by giving an example) the effects of resource partitioning and the effect of predators on prey population densities.
3. Describe how the introduction of nonnative species can lead to undesirable ecological consequences; cite three examples.

Self-Quiz Questions

Fill-in-the-Blanks

A(n) (1) _____ community is one whose forces have come into balance. Two different species with similar niche requirements can occupy that niche if they are able to (2) _____ that resource—that is, use it at different times, in different areas, or in different ways. When similar species share the same resource in different ways, it is called (3) _____ _____ ; this develops in the following two ways. (4) _____ _____ increases differences between competing populations and to be successful, species joining an existing community must be (5) _____ to established ones. Foxtail grasses, mallow plants, and smartweeds provide an example of (6) _____ _____ by exploiting different parts of the habitat. When predators reduce prey population densities, they decrease (7) _____ among prey and promote their coexistence. When sea stars were kept out of experimental plots of the rocky intertidal zone, the number of prey species was (8) _____ . Few introductions of species into new geographic areas are without (9) _____ consequences. The introduction of the South American (10) _____ _____ into the United States in the 1880s went unchecked by natural controls. Imported (11) _____ have become part of existing communities in the United States, but in many areas they have displaced their native counterparts. The introduction of Nile perch into Lake Victoria caused the (12) _____ of cichlids; the new environment lacked (13) _____ controls for the introduced species.

34-V
(pp. 518–521)

COMMUNITY ORGANIZATION, DEVELOPMENT, AND DIVERSITY (cont.)
    Succession
    Patterns of Species Diversity
SUMMARY

Summary

When an environment lacks life, a predictable pattern of community change occurs called *succession*. The first species will first flourish and then be replaced by other species, which are replaced by others in an orderly progression until a more or less stable array of species, the *climax community*, is reached.

Primary succession occurs in a barren habitat, one devoid of life. *Pioneer species* help build soil; these are usually small, low-growing plants with a short life cycle. Pioneers set the stage for their own replacement; perennials then appear and replace the pioneers.

In *secondary succession*, a community progresses back toward the climax state after parts of the habitat have been disturbed. Secondary succession occurs in abandoned fields and in forests where disturbances allow sunlight to reach the forest floor. Secondary succession is similar to primary succession, but many plants grow from seeds or seedlings already present when the process begins.

Winds, fires, insects, and overgrazing can disturb the climax community so as to modify and shape the direction of succession by encouraging some species and eliminating others in different parts of the habitat. Community stability over a broad area may require episodes of local instability that permit cyclic replacement of dominant species. An example is the sequoia tree in the Sierra Nevada; modest brush fires remove leaf litter and eliminate small trees and shrubs that compete with sequoias—this allows some time for germination of sequoia seeds, which establishes sequoia seedlings. Thus, fire promotes the conditions necessary for cyclic replacements in this climax community.

There are patterns of species diversity. Species numbers increase on new islands and reach a stable number. An example of the *island pattern* was furnished when a volcanic eruption formed a new island, Surtsey, southwest of Iceland. Bacteria, fungi, seeds, flies, and some seabirds were established there within six months. After two years, the first vascular plant appeared; two years later, the first moss plant appeared. Soil conditions improved, and the number of plant species increased. All species were colonists from Iceland. Species numbers on Surtsey will not increase indefinitely; islands distant from source areas receive fewer colonizing species, and larger islands tend to support more species than smaller islands at equivalent distances from source areas. Extinctions probably keep species diversity lower on small islands.

There are also *mainland and marine patterns* of species diversity. The most striking patterns of species diversity on land and in the seas relate to distance from the equator. Species diversity is highest near the equator in tropical areas; it then decreases toward the poles. This is so because rainfall and sunlight are available throughout the year, species diversity is self-reinforcing, and the overall rate of speciation in the tropics has long exceeded the rate of extinction from natural causes. Mass extinctions at higher latitudes have helped maintain low species diversity there.

Key Terms

succession	secondary succession	island patterns
climax community	cyclic replacements	species diversity
primary succession	local instability	mainland and marine
pioneer species		patterns

Objectives

1. Define *succession* and generally describe events leading to a stable *climax community*.
2. Distinguish between *primary succession* and *secondary succession*.
3. Generally describe the sequence of communities that might occur if the climax community nearest to you were burned to the ground.
4. Discuss species diversity in terms of the island pattern, the mainland pattern, and the marine pattern.

Self-Quiz Questions

Fill-in-the-Blanks

In an environment lacking life, (1) _____ , a predictable pattern of community change, occurs; species replacements occur until a more or less stable array of species, the (2) _____ community, is reached. (3) _____ succession occurs in a barren habitat, one devoid of life. In (4) _____ succession, a community progresses back toward the climax state after parts of the habitat have been disturbed. Community stability over a broad area may require episodes of (5) _____ instability that permits (6) _____ replacement of dominant species. In the case of sequoia trees in the Sierra Nevada, (7) _____ promotes the conditions necessary for cyclic replacements of older trees in this climax community. In an island pattern of species diversity, species numbers (8) [choose one] ( ) increase ( ) decrease, on new islands and reach a stable number. Islands distant from source areas receive (9) [choose one] ( ) more ( ) fewer, colonizing species and larger islands tend to support (10) [choose one] ( ) more ( ) fewer, species than smaller islands at equivalent distances from source areas. Extinctions probably keep species diversity (11) [choose one] ( ) higher ( ) lower, on small islands. The most striking patterns of species diversity on land and in the seas relate to distance from the (12) _____ . Species diversity is highest in tropical regions because (13) _____ and (14) _____ are available throughout the year. The rate of (15) _____ in the tropics has long exceeded the rate of extinction from natural causes.

CHAPTER TEST

UNDERSTANDING AND INTERPRETING KEY CONCEPTS

___ (1) All of the populations of different species that occupy and are adapted to a given area are referred to as a(n) _____ .

(a) biosphere
(b) community
(c) ecosystem
(d) niche

___ (2) In exploitative competition, _____ .

(a) different species with equal abilities compete to exploit a particular resource
(b) all individuals have equal access to a resource but differ in their ability to exploit that resource
(c) certain individuals limit or prevent others from using the resource and so control access to it
(d) one species hampers the growth, survival, or reproduction of the other only indirectly, by reducing the common supply of resources

___ (3) A one-way relationship in which one species benefits while the other is harmed is called _____ .

(a) commensalism
(b) competitive exclusion
(c) parasitism
(d) an obligatory relationship

___ (4) The weakest symbiotic attachment, in which one species simply lives in the presence of another species (for example, bird's nest in tree), is _____ .

(a) commensalism
(b) competitive exclusion
(c) mutualism
(d) an obligate relationship

___ (5) One type of community interaction in which both species benefit is best described as _____ .

(a) commensalism
(b) mutualism
(c) predation
(d) parasitism

___ (6) Prey not equipped with other defenses may be protected by their resemblance to a toxic prey; this is known as _____ .

(a) moment-of-truth defense
(b) camouflage
(c) mimicry
(d) warning coloration

___ (7) During the process of community succession, _____ .

(a) the initial community is called the climax community
(b) an orderly progression of species replacements occur until a more or less stable array of species results
(c) secondary succession always follows primary succession
(d) pioneer species always remain through all successional stages

___ (8) Island patterns, mainland patterns, and marine patterns refer to _____ .

(a) community stages of succession
(b) species diversity
(c) cyclic replacements in the community
(d) resource partitioning

___ (9) The relationship of Nile perch and cichlids illustrated _____ .

(a) succession in a lake community
(b) parasitism
(c) the effects of an introduced species into a new geographic area
(d) commensalism

___ (10) The relationship between an insect and the plants it pollinates (for example, apple blossoms, dandelions, and honeysuckle) is best described as _____ .

(a) mutualism
(b) competitive exclusion
(c) parasitism
(d) commensalism

___ (11) The relationship between the yucca plant and the yucca moth that pollinates it is best described as _____ .

   (a) camouflage
   (b) commensalism
   (c) competitive exclusion
   (d) a case of symbiosis in which neither species can live long without the other

___ (12) In 1934, G. Gause utilized two species of *Paramecium* in a study that described _____ .

   (a) interspecific competition and competitive exclusion
   (b) resource partitioning
   (c) the establishment of territories
   (d) coevolved mutualism

INTEGRATING AND APPLYING KEY CONCEPTS

(1) If you were Ruler of All People on Earth, how would you organize industry and human populations in an effort to solve our most pressing pollution problems?

(2) Is there a *fundamental niche* that is occupied by humans? If you believe so, describe the minimal abiotic and biotic conditions that populations of humans require to live and reproduce. (Note that to *thrive* and *be happy* are not criteria.) If you do not think so, state why.

(3) The above minimal niche conditions can be viewed as resource categories that must be protected by populations if they are to survive. Do you believe that the Cold War that used to exist between the United States and the Soviet Union primarily involved protection of minimal niche conditions, or do you believe that the Cold War was based on other, more (or less) important factors?

   (a) If the former, how do you think *minimal* niche conditions might be guaranteed for all humans willing and able to accept certain responsibilities as their contribution toward enforcing this guarantee?

   (b) If the latter, identify what you think those factors are, and explain why you consider them more (or less) important than minimal niche conditions.

35
ECOSYSTEMS

CHARACTERISTICS OF ECOSYSTEMS
STRUCTURE OF ECOSYSTEMS
    Trophic Levels
    Food Webs
ENERGY FLOW THROUGH ECOSYSTEMS
    Primary Productivity
    Major Pathways of Energy Flow
    Ecological Pyramids

BIOGEOCHEMICAL CYCLES
    Hydrologic Cycle
    Carbon Cycle
    Commentary: Greenhouse Gases and a Global Warming Trend
    Nitrogen Cycle
    Transfer of Harmful Compounds Through Ecosystems
SUMMARY

General Objectives

1. Understand how materials and energy enter, pass through, and exit an ecosystem.
2. Understand the various trophic roles and levels.
3. Describe the nature of biogeochemical cycles and be able to discuss examples.
4. Discuss the basis of the greenhouse effect and its significance to life on our planet.
5. Tell the significance of harmful compounds passing through the ecosystem.

35-I
(pp. 523–527)

CHARACTERISTICS OF ECOSYSTEMS
STRUCTURE OF ECOSYSTEMS
    Trophic Levels
    Food Webs

Summary

Each region of the earth is a system where energy is secured by autotrophs serving as *producers;* the energy flows to heterotrophs. Some heterotrophs serve as *consumers;* those called *herbivores* eat plants, *carnivores* eat animals, *parasites* reside in or on living hosts and extract energy from them, and *omnivores* take in a variety of edibles. The *decomposers* are heterotrophs that include fungi and bacteria. *Detritivores,* such as crabs, nematodes, and earthworms, obtain energy from partly decomposed particles of organic matter. Autotrophs secure nutrients as well as energy for the entire system. The components above describe an *ecosystem,* an entire complex of organisms interacting with one another and the physical environment through (1) a flow of energy and (2) a cycling of materials. Ecosystems are *open systems,* not self-sustaining; they require *energy input* (as from the sun) and often *nutrient input.* Energy in the ecosystem cannot be recycled; all ecosystems have *energy output* (much is lost as heat). Nutrients are usually recycled but some may be lost through *nutrient output*.

Feeding relationships in all ecosystems are similar. Members fit somewhere in a hierarchy of energy transfers ("who eats whom") called *trophic levels*. Primary producers are closest to the initial energy source (for example, photosynthetic autotrophs such as aquatic plants in a lake). Organisms feeding directly on producers form the next trophic level (for example, herbivores such as rotifers and snails). Birds and other carnivores preying directly on the herbivores are part of the next trophic level. Decomposers and humans may feed from more than one trophic level.

A general linear sequence of who eats whom is a *food chain* (for example, algae → fish → fisherman → shark). Because the same food resource is often part of more than one food chain, more complex *food webs* are constructed. In the above food chain, crustaceans would also feed on algae, larger fishes feed on smaller ones, sharks eat other fishes, and the fisherman probably ate plants as part of his meal.

Key Terms

producers
photosynthetic autotroph
consumers
herbivores
carnivores
parasites
omnivores

decomposers
detritivores
ecosystem
open system
energy input
nutrient input

mineral
energy output
nutrient output
trophic level
food chain
food webs

Objectives

1. Define the term *ecosystem* and state how autotrophic organisms are related to ecosystems.
2. List the principal trophic levels in an ecosystem of your choice; state the source of energy for each trophic level and give one or two examples of organisms associated with each trophic level.
3. Give an example of a short food chain; using that food chain, construct a small food web.

Self-Quiz Questions

Fill-in-the-Blanks

Photosynthetic (1) _____ capture (2) _____ and concentrate (3) _____ for ecosystems. Some heterotrophs serve as consumers: (4) _____ eat plants; (5) _____ eat animals; (6) _____ reside in or on living hosts and extract energy from them; (7) _____ take in a variety of edibles. Other heterotrophs, the (8) _____ , include fungi and bacteria; (9) _____ are invertebrates that feed on partially decomposed bits of organic matter. The above components describe a(n) (10) _____ , an entire complex of organisms interacting with one another and the physical environment through a flow of (11) _____ , and a cycling of (12) _____ . Ecosystems are (13) _____ systems, not self-sustaining.

They require (14) _____ input and often (15) _____ input. All

ecosystems have (16) _____ output. (17) _____ are usually recycled

in the ecosystem. All organisms in a community that are the same number of

energy transfers away from the initial energy input into their ecosystem

constitute a(n) (18) _____ _____ . A general linear sequence of who

eats whom is a(n) (19) _____ _____ . When the same food resource is

often part of more than one food chain, more complex (20) _____

_____ are constructed to show feeding relationships.

35-II
(pp. 528–530)

ENERGY FLOW THROUGH ECOSYSTEMS
   Primary Productivity
   Major Pathways of Energy Flow
   Ecological Pyramids

Summary

The ecosystems *primary productivity* is the rate at which its producers capture and store a given amount of energy in a given length of time. Amount of energy stored depends on (1) balance between photosynthesis and aerobic respiration in the plants and (2) how many plants escape the attention of consumers and decomposers. *Gross* primary productivity is the total rate of photosynthesis for the ecosystem during a specified interval. *Net* primary productivity is the rate of energy storage in plant tissues in excess of the rate of respiration by the plants themselves. Temperature range and rainfall amounts affect net primary production quantities, their seasonal patterns, distribution through the habitat, and size and forms of primary producers. Productivity tends to be lower in harsher environments.

Only a small portion of sunlight energy becomes fixed in plants; plants metabolize about half of what they store. Other organisms use what fixed energy remains in plant tissues, remains, or wastes; they too lose heat to the environment. Heat losses represent a one-way flow of energy out of the ecosystem. Energy flows from plants to herbivores and then to carnivores in *grazing food webs*. In *detrital food webs*, energy flows from plants through decomposers and detritivores to consumers. Both types of food webs are interconnected in an ecosystem. Energy amounts moving through these food webs vary throughout the year. Detrital food webs handle the largest portion of net primary production in natural ecosystems. When cattle graze heavily, about half of the net primary production passes through grazing food webs but cattle do not use all the energy; undigested residues in feces become available for decomposers and detritivores. When plants die in marshes, stored energy in dead plants becomes available for detrital food webs.

Pyramids are used to represent the trophic structure of an ecosystem. A large number of small producers or a small number of large producers form the base of such pyramids. Pyramids can be based on the numbers of members in an ecosystem, or the weight or *biomass* of all members at each trophic level. The most accurate representation of the ecosystem's trophic structure is the *energy pyramid*; it is based on energy flow at each transfer to a different trophic level. With a pyramid of energy, only 6 to 16 percent of the energy of one trophic level is available to the next level.

Key Terms

primary productivity
gross primary
productivity
net primary
productivity

grazing food web
detrital food web
ecological pyramid
"head count"

pyramid of numbers
biomass
pyramid of biomass
energy pyramid

Objectives

1. Define *primary productivity*.
2. Distinguish between *net* and *gross* primary productivity.
3. Contrast the ways in which energy and nutrients pass through an ecosystem. Explain why nutrients can be completely recycled but energy cannot.
4. Compare *grazing food webs* with *detrital food webs*. Present an example of each.
5. Explain why ecologists use pyramids to represent the trophic structure of an ecosystem; list the types of pyramids used.
6. Tell which type of pyramid most accurately represents the ecosystem's trophic structure; explain why.

**Self-Quiz
Questions**

Fill-in-the-Blanks

The ecosystem's (1) _____ productivity is the rate at which its producers capture and store a given amount of energy in a given length of time.

(2) _____ primary productivity is the total rate of photosynthesis for the ecosystem during a specified interval. (3) _____ primary productivity is the rate of energy storage in plant tissues in excess of that energy plants use for their own respiration. Heat losses represent a one-way flow of (4) _____ out of the ecosystem. In (5) _____ food webs, energy flows from plants to (6) _____ and then to carnivores. In (7) _____ food webs, energy flows from plants through decomposers and detritivores. The (8) _____ shape is used to represent the trophic structure of an ecosystem; the base is formed by large numbers of small (9) _____ or a small number of large (10) _____ . The most accurate representation of the ecosystem's trophic structure is the (11) _____ _____ .

35-III
(pp. 531–537)

BIOGEOCHEMICAL CYCLES
    Hydrologic Cycle
    Carbon Cycle
    Commentary: **Greenhouse Gases and a Global Warming Trend**

Summary

In addition to energy, nutrient availability profoundly affects the trophic structure of ecosystems. Along with carbon, oxygen, and hydrogen, plants require about thirteen mineral elements; this includes nitrogen and phosphorus. Mineral deficiencies affect plant growth and thus ecosystem primary productivity. *Biogeochemical cycles* include the movement of water, nutrients, and other elements and compounds from the reservoir of the physical environment to organisms, and then back to the environment. Elements essential for life move in biogeochemical cycles. There are separate *hydrologic* (oxygen and hydrogen in water), *atmospheric* (for example, carbon and nitrogen), and *sedimentary* (for example, phosphorus from land → sea sediments → land) *cycles.*

In the hydrologic cycle, which is driven by solar energy, water is moved or stored by evaporation, precipitation, and transpiration. Water moves other nutrients in or out of ecosystems. Water released as precipitation remains on land for 10 to 120 days. *Watersheds,* which can be of any size, funnel rain or snow into a single river (for example, the Mississippi River watershed extends across one-third of the United States; Hubbard Brook Forest watersheds average thirty-six acres). Water also percolates through soil to the water table (groundwater) or moves into a stream. Plants hold nutrients (roots "mine" soil) and slow the rates at which those nutrients move through biogeochemical cycles; they are made available to ecosystem food webs. Clearing land of vegetation greatly increases rate of ecosystem nutrient loss (calcium in stream outflow by six times); this has long-term disruptive effects on nutrient availability throughout the ecosystem.

In the *carbon cycle,* carbon moves from reservoirs in the atmosphere and oceans (mostly as CO_2 from aerobic respiration, fossil fuel burning, and volcanic eruptions), through the biomass of organisms, then back to the "holding stations" of the reservoirs. Photosynthesizers incorporate billions of metric tons of the carbon from carbon dioxide into organic compounds each year. The turnover of such captured carbon atoms is rapid in tropical forest regions but slow in bogs and marshes where carbon accumulates in forms such as peat. In aquatic food webs, carbon is incorporated into shells and other hard parts of organisms; carbon can be buried for long periods in deep oceans or be converted to fossil fuel reserves. Human burning of fossil fuels adds carbon to the atmosphere, which intensifies the greenhouse effect.

The atmospheric concentrations of carbon dioxide, water, ozone, methane, nitrous oxide, and chlorofluorocarbons profoundly influence the average temperature near Earth's surface, and that temperature influences global climates. Together, these gas molecules act like a pane of glass in a greenhouse (greenhouse gases). Visible light passes upward through the "pane" of gases, but escape of longer, infrared heat wavelengths are impeded and reradiated back toward Earth; this causes heat to build up in the lower atmosphere, a global warming action called the *greenhouse effect.*

Key Terms

biogeochemical cycles	precipitation	runoff
hydrologic cycle	transpiration	carbon cycle
atmospheric cycles	watersheds	chlorofluorocarbons, CFCs
sedimentary cycle	nutrients	greenhouse effect
phosphorus cycle	Hubbard Brook Forest	infrared wavelengths
evaporation	soil infiltration	

Objectives

1. Describe completely what is meant by *biogeochemical cycles;* list two examples.
2. Distinguish among *hydrologic, atmospheric,* and *sedimentary cycles.*
3. Describe the hydrologic cycle, correctly using words such as *solar energy, evaporation, precipitation, transportation, watershed,* and *transpiration.*

4. Describe the carbon cycle and explain how your life is affected by carbon reservoirs, carbon dioxide, organic compounds, and the greenhouse effect.

Self-Quiz Questions

Fill-in-the-Blanks

In addition to energy, (1) _____ availability profoundly affects the trophic structure of ecosystems. Along with carbon, oxygen, and hydrogen, plants require about (2) _____ (number) mineral elements; these include nitrogen and (3) _____ . (4) _____ cycles include the movement of water, nutrients, and other elements and compounds from the reservoir of the physical environment to organisms, and then back to the environment. There are three separate types of cycles: (5) _____ (for example, oxygen and hydrogen in water; (6) _____ (for example, carbon and nitrogen); (7) _____ (for example, phosphorus). The (8) _____ cycle is driven by solar energy. (9) _____ , which can be of any size, funnel rain or snow into a single river. Plants hold (10) _____ and slow their rate of movement through biogeochemical cycles; they are made available to ecosystem food webs. Clearing land of vegetation greatly (11) [choose one] ( ) increases ( ) decreases, rate of ecosystem nutrient loss. In the carbon cycle, carbon moves from (12) _____ in the atmosphere and oceans through the biomass of organisms, then back to the "holding stations" of the (13) _____ . Photosynthesizers incorporate billions of metric tons of carbon from carbon dioxide into (14) _____ compounds each year. The turnover of such captured carbon atoms is (15) [choose one] ( ) rapid ( ) slow, in tropical forest regions but (16) [choose one] ( ) rapid ( ) slow, in bogs and marshes where carbon accumulates in forms such as peat. Greenhouse gases together act like a pane of glass in a greenhouse; visible light passes upward through the "pane," but longer infrared heat wavelengths are impeded and reradiated back toward Earth. This causes a global warming action known as the (17) _____ effect.

35-IV
(pp. 538–540)

BIOGEOCHEMICAL CYCLES (cont.)
   Nitrogen Cycle
   Transfer of Harmful Compounds Through Ecosystems
SUMMARY

Summary

Nitrogen is needed for proteins and nucleic acids; it is abundant in the atmosphere but not in the Earth's crust. Of all nutrients needed for plant growth, nitrogen is often in the shortest supply. Nearly all nitrogen in soils has been put there by nitrogen-fixing organisms. The largest nitrogen reservoir is the atmosphere (eighty percent gaseous N_2). Some bacteria and lightning can convert N_2 into forms that can be used in the ecosystem. Nitrogen is lost from ecosystems by metabolic denitrification activities (which "unfix" nitrogen) of bacteria and soil leaching. Nitrogen may be gained from nutrient input of rivers and streams. There are six processes of the nitrogen cycle: nitrogen fixation, assimilation and biosynthesis, decomposition, ammonification, nitrification, and denitrification. In *nitrogen fixation*, bacteria convert N_2 of the atmosphere to NH_3 (dissolves in water to produce NH_4^+), which is then used in the synthesis of amino acids, then of proteins and nucleic acids. Decomposition of dead nitrogen fixers releases nitrogen-containing compounds. *Ammonification* is a process in which bacteria and fungi decompose dead plants and animals and use released amino acids and proteins for growth; released excess ammonia or ammonium ions are picked up by plants. *Nitrification* is a type of chemosynthesis where NH_3 or NH_4^+ is converted to NO_2^-; other nitrifying bacteria use the nitrite for energy and release NO_3^- or nitrate; plants also use this form of nitrogen to synthesize amino acids and then proteins. Soil nitrogen is scarce; it is lost by leaching, bacterial *denitrification* (some bacteria convert nitrate or nitrite to N_2), and farming. Nitrogen losses are great in agricultural regions; modern agriculture depends on costly nitrogen-rich fertilizers. In the race against hunger, soil enrichment with nitrogen-containing fertilizers is essential.

Transfer of harmful compounds occurs through ecosystems. DDT, used to kill mosquitoes, accumulates in fatty tissues and results in *biological magnification* and unexpected results. *Ecosystem analysis* attempts to predict the complex effects of a single change in an ecosystem.

Key Terms

gaseous nitrogen, N_2	nitrification	DDT
cycling processes	nitrite, NO_2^-	chlorinated hydrocarbon
nitrogen fixation	nitrate, NO_3^-	biological magnification
ammonia, NH_3	chemosynthesis	malaria
ammonium, NH_4^+	leaching	sylvatic plague
ammonification	denitrification	ecosystem analysis

Objectives

1. Define the chemical events that occur during *nitrogen fixation, nitrification, ammonification,* and *denitrification.*
2. Explain why agricultural methods in the United States tend to put more energy into the soil in the form of nitrogen-rich fertilizers than do simpler methods of agriculture.
3. Describe how DDT damages ecosystems and define *biological magnification.*
4. Explain the goal of *ecosystem analysis.*

Self-Quiz Questions

Fill-in-the-Blanks

(1) _____ can assimilate nitrogen from the air in the process known as

(2) _____ _____ . (3) _____ is a process in which bacteria and

fungi decompose dead plants and animals and release (4) _____ or (5) _____ ions. (6) _____ is a type of chemosynthesis where ammonia or ammonium ions are converted to (7) _____ . Other nitrifying bacteria use nitrite for energy and release (8) _____ , which can also be used by plants to synthesize amino acids and then proteins. Soil nitrogen is scarce; it is lost by leaching and bacterial (9) _____ . Modern agriculture depends on costly nitrogen-rich (10) _____ . Transfer of harmful compounds occurs through ecosystems; when DDT accumulates in fatty tissues, it is referred to as (11) _____ _____ . (12) _____ analysis attempts to predict the complex effects of a single change in an ecosystem.

CHAPTER TEST

UNDERSTANDING AND INTERPRETING KEY CONCEPTS

___ (1) A network of interactions that involve the cycling of materials and the flow of energy between a community and its physical environment is a(n) _____ .

    (a) population
    (b) community
    (c) ecosystem
    (d) biosphere

___ (2) In an ecosystem, fungi and bacteria are classified as _____ .

    (a) herbivores
    (b) carnivores
    (c) decomposers
    (d) detritivores

___ (3) Which of the following is *not* true of an ecosystem?

    (a) cycles materials
    (b) cycles energy
    (c) requires energy
    (d) sometimes needs nutrient input
    (e) sometimes has nutrient output

___ (4) _____ is a process in which nitrogenous waste products or organic remains of organisms are decomposed by soil bacteria and fungi that use the amino acids being released for their own growth and release the excess as ammonia or ammonium ions.

    (a) Nitrification
    (b) Ammonification
    (c) Denitrification
    (d) Nitrogen fixation

___ (5) In a natural community, the primary consumers are _____ .

   (a) herbivores
   (b) carnivores
   (c) scavengers
   (d) decomposers

___ (6) Gross primary productivity is _____ .

   (a) the rate at which ecosystem consumers capture and store a given amount of energy in a given length of time
   (b) total global productivity
   (c) the total rate of photosynthesis for the ecosystem during a specified interval
   (d) the rate of energy storage in plant tissues in excess of the rate of respiration by the plants themselves

___ (7) Which of the following is a primary consumer?

   (a) cow
   (b) dog
   (c) hawk
   (d) all of the above

___ (8) Energy flows from plants to herbivores and then to carnivores in _____ .

   (a) detrital food webs
   (b) some food chains
   (c) grazing food webs
   (d) both b and c

___ (9) The most accurate representation of the ecosystem's trophic structure is the _____ pyramid.

   (a) number
   (b) biomass
   (c) double
   (d) energy

___ (10) The hydrologic cycle is driven by _____ .

   (a) water
   (b) solar energy
   (c) watershed energy
   (d) precipitation

___ (11) Which of the following is *not* true of the *greenhouse effect?*

   (a) atmospheric concentration of gases is involved
   (b) visible light is prevented from passing through a layer of atmospheric gases
   (c) the average temperature near the Earth's surface is affected
   (d) a global warming gradually occurs
   (e) heat builds up in the lower atmosphere

INTEGRATING AND APPLYING KEY CONCEPTS

In 1971 *Diet for a Small Planet* was published. Frances Moore Lappe, the author, felt that people in the United States of America wasted protein and ate too much meat. She said, "We have created a national consumption pattern in which the majority, who can pay, overconsume the most inefficient livestock products [cattle] well beyond their biological needs (even to the point of jeopardizing their health), while the minority, who can not pay, are inadequately fed, even to the point of malnutrition." Cases of *marasmus* (a nutritional disease caused by prolonged lack of food calories) and *kwashiorkor* (caused by severe, long-term protein deficiency) have been found in Nashville, Tennessee, and on an Indian reservation in Arizona, respectively. Lappe's partial solution to the problem was to encourage people to get as much of their protein as possible directly from plants and to supplement that with less meat from the more efficient converters of grain to protein (chickens, turkeys, and hogs) and from seafood and dairy products. Most of us realize that feeding the hungry people of the world is not just a matter of distributing the abundance that exists—that it is being prevented in part by political, economic, and cultural factors. Devise two full days of breakfasts, lunches, and dinners that would enable you to exploit the lowest acceptable trophic levels to sustain yourself healthfully.

36

THE BIOSPHERE

CHARACTERISTICS OF THE BIOSPHERE
  Biosphere Defined
  Global Patterns of Climate
THE WORLD'S BIOMES
  Deserts
  Dry Shrublands and Woodlands
  Grasslands
  Forests

  Tundra
THE WATER PROVINCES
  Lake Ecosystems
  Marine Ecosystems
  Commentary: El Niño and Oscillations in
   the World's Climates
SUMMARY

General Objectives

1. Describe the ways in which climate affects the biomes of Earth and influences how organisms are shaped and how they behave.
2. Contrast life in lake ecosystems with that in oceans and estuaries.

36-I
(pp. 543–547)

CHARACTERISTICS OF THE BIOSPHERE
  Biosphere Defined
  Global Patterns of Climate

Summary

Unrelated species in distant regions often show striking similarities. The distribution of species is related to climate, topography, and species interactions. The term *biosphere* refers to all regions of Earth where organisms live. The *hydrosphere* includes all water on or near the Earth's surface. The *atmosphere* is a region of gases, airborne particles, and water vapor enveloping the Earth.

Climate, or prevailing weather conditions, includes temperature, humidity, wind velocity, cloud cover, and rainfall. Our climate is shaped by the amount of solar radiation, Earth's rotational and orbital movements, the distribution of land and water, and elevation of land masses. Ultraviolet radiation is absorbed by ozone and oxygen in the upper atmosphere. Our atmosphere reflects or absorbs half of the solar radiation. Heat from the sun drives Earth's weather systems.

Incoming rays from the sun differentially heats equatorial and polar regions; the nonuniform distributions of land and water cause pressure differences, which result in air currents. Warm air moving north or south from equatorial regions is deflected because of Earth's rotation. Heated air rising at the equator gives up moisture (rain) when it rises to cooler altitudes; drier air moves away from the equator and descends at about 30° latitudes; deserts occur there. The air warms again, picks up moisture, and ascends at 60° latitudes, then travels poleward where no precipitation accompanies its descent. Regional differences in rainfall result—and rain influences the type of ecosystem that can occur in a specific region.

Ocean currents form because of Earth's rotation, winds, temperature variations, and distribution of land masses. Two immense, circular water movements occur in each ocean and bring warm equatorial waters poleward. Thus, *atmospheric and oceanic circulation patterns influence the distribution of different types of ecosystems.*

Topography, that is, the distribution of mountains, valleys, and other land formations, also influences regional climates. For example, the mountains on the west coast of California cause air to rise and lose moisture. Vegetation belts (semiarid shrubs at the base → deciduous and evergreen trees → subalpine evergreen trees → grasses and sedges above the subalpine belt) occur at different elevations, which correspond to changes in air temperature and moisture. Air flows over mountain crests, descends as it warms, and draws moisture from plants and soil; it is given up as rain. There is a *rain shadow,* a reduction in rainfall on leeward sides of high mountains; only plants adapted to arid or semiarid conditions grow there.

The tilt in Earth's axis causes seasons; throughout the year, the amount of solar radiation reaching Earth's surface changes in the Northern and Southern hemispheres, leading to seasonal variations of climate.

Key Terms

biosphere	climate	leeward
hydrosphere	topography	windward
atmosphere	rain shadow	

Objectives

1. Describe how the biosphere is related to its three components: the narrow zone of water, the lower atmosphere, and the fraction of Earth's crust in which organisms live.
2. Name four factors that interact to shape climate.
3. Explain how certain components of Earth's atmosphere moderate some of the harsh effects of incoming solar radiation.
4. State how air currents, ocean currents, and topography affect the nature of Earth's ecosystems.
5. Discuss the cause of seasonal variations in climate and give examples of organism responses to seasonal variations.

Self-Quiz Questions

Fill-in-the-Blanks

The term (1) _____ refers to all regions of Earth where organisms live. The (2) _____ includes all water on or near Earth's surface. The (3) _____ is a region of gases, airborne particles, and water vapor enveloping Earth. Prevailing weather conditions are referred to as (4) _____ . Due to the mediating effects of the (5) _____ , only about half the solar radiation reaching the atmosphere actually gets to Earth's surface. Earth's rotation, winds, temperature variations, and distribution of land masses causes (6) _____ currents to form. From the base of the coastal mountains of California to their peaks, one generally encounters four zones: (7) _____ shrubs at the base; deciduous and evergreen (8) _____ ; (9) _____ evergreen trees; grasses and sedges above the (10) _____ belt. An area of reduced rainfall on leeward sides of high mountains is called the (11) _____ _____ . A tilt in Earth's axis causes (12) _____ .

36-II
(pp. 547–553)

THE WORLD'S BIOMES
Deserts
Dry Shrublands and Woodlands
Grasslands

Summary

Climatic factors determine patterns of vegetation (grasslands, deserts, forests, and tundra) and provide reasons to explain why unrelated species may have similar adaptations. Each region has a distinct array of species that evolved independently of those elsewhere. Oceans, mountain ranges, deserts, and other barriers have isolated species and restricted dispersal. W. Sclater and then Alfred Wallace were the first to propose six *biogeographic realms*. *Biomes* are broad vegetational subdivisions of the biogeographic realms; each biome is dominated by certain types of plants and other organisms. Biome distribution corresponds to climate, topography, and soil type. *Soil* is a mixture of rock, mineral ions, and organic matter combined with water, air, and a variety of organisms.

Deserts are areas where evaporation exceeds rainfall. More than a third of Earth's land area is arid or semiarid due to drought and overgrazing; these conditions prevail at 30° north and south latitudes. Deserts do not have much vegetation but still there is considerable plant diversity. Each plant type has adaptations for obtaining and holding moisture during stress. There is an alarming trend toward desertification (dry wastelands) in many parts of the world.

Dry shrublands and *dry woodlands* are biomes that occur in semiarid coastal regions of continents between 30° and 40° latitude. Dry shrublands are semiarid regions (annual rainfall less than 60 centimeters) where dominant plants are short, woody, and multibranched and have tough, evergreen leaves. California has about 6 million acres of chaparral. These plants are adapted to survive fire episodes. Dry woodlands are semiarid regions (annual rainfall less than 10 centimeters); dominant trees are sparse but tall. Examples are eucalyptus woodlands of southwestern Australia and oak woodlands of western North America.

Great temperate grasslands occur in South Africa, South America, and midcontinental regions of North America and the Soviet Union. *Grasslands* are lands that are usually flat or rolling and dry, with high evaporation rates. *Shortgrass prairie* is found in areas of strong winds and light rainfall, with infrequent and rapid evaporation. Plant roots above permanently dry subsoil soak up brief, seasonal rainfall. *Tallgrass prairie,* now mostly farmland, once extended west from the temperate deciduous forests of North America. Legumes were abundant. *Tropical grasslands* include the broad belts of the African *savanna.* Here tufted grasses predominate in low rainfall regions; acacia and other shrubs grow in scattered patches where there is slightly more rainfall. In areas of higher rainfall, tall, coarse grasses, shrubs, and low trees grow. Other tropical regions have comparable biomes called *monsoon grasslands*. Here heavy rainfalls alternate with a pronounced dry season.

Key Terms

W. Sclater and A. Wallace	desert	shortgrass prairie
biogeographic realms	desertification	Dust Bowl
biome	dry shrublands	tallgrass prairie
soil	chaparral	tropical grasslands
loam	dry woodlands	savanna
humus	grasslands	monsoon grasslands
topsoil		

Objectives

1. Explain the basis of the six *biogeographic realms*.
2. List the factors that influence the distribution of *biomes* on land.
3. State the relationship between temperature and rainfall (on the one hand) and the abundance and diversity of producers (on the other hand) in the different biomes.
4. Explain how soil characteristics influence the distribution of ecosystems on land.
5. Distinguish between *dry shrublands* and *dry woodlands*.
6. List the factors that encourage tallgrass and shortgrass prairie to form.
7. Describe *tropical grasslands* that include the African *savanna* and *monsoon grasslands*.

Self-Quiz Questions

Fill-in-the-Blanks

Climatic factors determine patterns of (1) _____ . Regions like grasslands, deserts, forests, and tundra have distinct arrays of (2) _____ that evolved independently of those elsewhere. W. Sclater and then Alfred Wallace were the first to propose six (3) _____ _____ . (4) _____ are broad vegetational subdivisions; each is dominated by certain types of plants and other organisms. (5) _____ are areas where evaporation exceeds rainfall. Dry biomes where dominant plants are short, woody, multibranched, and with tough, evergreen leaves are called (6) _____ _____ . Dry biomes with dominant trees that are sparse but tall are called (7) _____ _____ . (8) _____ are lands that are usually flat or rolling and dry, with high evaporation rates. (9) _____ prairie is found in areas of strong winds, light rainfall, with infrequent and rapid evaporation; plant roots above permanently dry subsoil soak up brief, seasonal rainfall. (10) _____ prairie is now mostly farmland; it once extended west from the temperate deciduous forests of North America. (11) _____ grasslands include the broad belts of the African savanna. Some tropical regions have grassland biomes with heavy rainfall alternating with a pronounced dry season called (12) _____ grasslands. (13) _____ is partly decomposed organic matter that helps soil retain water-soluble ions. (14) _____ is a type of rich, well-aerated soil that is a balanced mixture of sand, silt, clay, and humus. (15) _____ is the fertile soil layer.

36-III
(pp. 553–557)

THE WORLD'S BIOMES (cont.)
  Forests
  Tundra

Summary

The world's major forest biomes have tall trees growing close enough together to form a fairly continuous canopy over a broad region. *Evergreen broadleaf forests* include the tropical rain forests with a great diversity of plants and animals. These are areas of regular heavy rainfall and high humidity. Some evergreen trees produce new leaves and shed old ones throughout the year.

Tropical rain forest diversity includes vines, orchids, mosses, lichens, and bromeliads. Entire communities of insects, spiders, and amphibians are associated with leaves of aerial plants. Various herbivores and predators are associated with the rain forest. With continual leaf drop, much litter is produced; decomposition and mineral cycling are rapid in the hot, humid climate. Soil is not a significant nutrient reservoir.

Trees in *deciduous broadleaf forests* drop their leaves during a pronounced dry season. This includes the *monsoon forests* of India and southeastern Asia as well as the *temperate deciduous forests*. At one time, deciduous broadleaf forests stretched across northeastern North America, Europe, and eastern Asia.

Evergreen coniferous forests are composed of cone-bearing trees, typically with needlelike leaves adapted to arid conditions. Conifers are primary producers in a variety of biomes. *Boreal forests,* or *taiga,* include the broad expanses of coniferous trees in northern Europe, Asia, and North America; they occur in glaciated regions with cold lakes and streams. Most rain occurs in summer with low evaporation and cool air; the cold dry winters are more severe in eastern parts of continents. Spruce and balsam dominate boreal forests of North America with abundant deciduous birches and aspens occurring in disturbed areas. *Montane coniferous forests* extend southward through the great mountain ranges; spruce and fir dominate in mountains paralleling the Canadian Pacific coast and the United States, but fir and pine dominate in the southern extensions of the Rockies and Cascades.

Some temperate lowlands also support coniferous forests. A *temperate rain forest* parallels the coast all the way from Alaska to northern California; it includes the tall Sitka spruces and redwoods. Also, the sandy, nutrient-poor soil of New Jersey's coastal plain supports the *pine barrens*, a scrub forest, with grasses and low shrubs growing among open stands of pitch pine and oak trees. Pines recover quickly from frequent fires.

Tundra is a largely flat, desolate, treeless plain with *permafrost* (permanently frozen layer) just beneath the surface; temperatures range from cool in short summer to below freezing in winter. Little water vapor forms in the cool air and so rainfall is sparse. Sunlight is nearly continuous for three months. The vast *arctic tundra* biome lies to the north, between the polar ice cap and huge belts of boreal forests in North America, Europe, and Asia. Decomposition is inhibited in this environment. *Alpine tundra* occurs at high elevations in mountains throughout the world. There is low productivity with dominant plants forming cushions or mats. There is no permafrost layer.

Key Terms

canopy
evergreen broadleaf
  forests
tropical rain forests
bromeliads
deciduous broadleaf
  forest

evergreen coniferous
  forest
boreal forest
taiga
montane coniferous
  forest

temperate rain forest
pine barrens
arctic tundra
permafrost
peat
alpine tundra

Objectives

1. List the factors that encourage the development of *evergreen broadleaf forests.*
2. Describe a *tropical rain forest* and a *deciduous broadleaf forest;* distinguish between *monsoon forests* and *temperate deciduous forests.*
3. List the factors that encourage the development of *evergreen coniferous forests.*
4. Describe the location and dominant trees of the *montane coniferous forests.*
5. Discuss the features and location of *boreal forests;* give examples of coniferous forests found in temperate lowlands.
6. Describe the physical and biotic features of the *tundra* ecosystem. Distinguish between *alpine* and *arctic tundra.*

Self-Quiz Questions

Fill-in-the-Blanks

(1) _____ _____ forests include the tropical rain forests with a great diversity of plants and animals; these are areas of regular heavy (2) _____ and high (3) _____ . With continuous leaf drop and high litter accumulation of the tropical rain forests, decomposition and mineral cycling are (4) [choose one] ( ) rapid ( ) slow, in the hot, humid climate. Trees in (5) _____ _____ forests drop their leaves during a pronounced dry season; this includes (6) _____ forests of Indian and southeastern Asia as well as the (7) _____ _____ forests of North America, Europe, and eastern Asia. (8) _____ _____ forests are composed of cone-bearing trees with needlelike leaves adapted to arid conditions. (9) _____ are primary producers in a variety of biomes including (10) _____ forests of northern Europe, Asia, and North America. (11) _____ _____ forests extend southward covering the great mountain ranges. Some (12) _____ lowlands also support coniferous forests; one temperate rain forest parallels the coast all the way from Alaska to northern California. New Jersey's coastal plain supports the (13) _____ _____ , a scrub forest. (14) _____

_____ is a largely flat, desolate, treeless plain with a permanently frozen layer just beneath the surface, the (15) _____ . (16) _____ tundra occurs at high elevations in mountains throughout the world.

36-IV
(pp. 558–559)

THE WATER PROVINCES
Lake Ecosystems

Summary

Water provinces are extensive regions that encompass diverse ecosystems; these include freshwater lakes, rivers, swamps, marshes, bogs, and similar wetlands, as well as brackish seas. There is no such thing as a "typical" ecosystem.

Zones of a deep lake are the littoral, limnetic, and profundal. The *littoral* zone extends from the shore to where rooted plants stop growing; diversity is greatest here. The open, sunlit waters of the *limnetic* zone go to the maximum depth of photosynthesis; it includes *plankton.* The *profundal* zone is the deep, open water below the depth of effective light penetration; here bacteria decompose materials into detritus.

In temperate regions, seasonal changes occur in deep lakes that bring about changes in water density and temperature. Water is densest at 4° C. In winter, most of the lake is at 4° C. In spring, daylength increases and the air warms; ice melts and gradually warms to 4° C. Temperatures become uniform throughout the lake and surface winds cause a *spring overturn;* dissolved oxygen moves from surface to depths and nutrients released by decomposition move to the surface layer. By midsummer, the surface has warmed and a *thermocline* (region of sudden temperature change) develops on which surface water floats; this prevents mixing. The upper layer cools in autumn, increases in density, and sinks, causing the thermocline to disappear. The lake water then mixes vertically during the *fall overturn* when dissolved oxygen moves down and nutrients move up. Cycles of primary productivity correspond to the seasonal changes.

Geologic processes give rise to lakes as when glaciers form lakes. Interactions among soils, basin shape, and climate produce conditions ranging from oligotrophy to eutrophy. *Oligotrophic lakes* are often deep, poor in nutrients, and low in primary productivity. *Eutrophic lakes* are often shallow, rich in nutrients, and high in primary productivity. As lake sediments accumulate, a lake may progress from oligotrophy to eutrophy and then to the final successional stage—a completely filled-in basin. Human activities disrupt geologic, climatic, and biological forces that determine the trophic nature of lakes.

Key Terms

lake ecosystems	zooplankton	fall overturn
littoral zone	profundal zone	oligotrophic
limnetic zone	spring overturn	eutrophic
plankton	thermocline	basin
phytoplankton		

Objectives

1. Describe the causes of thermal stratification in bodies of water.
2. Discuss how spring and fall overturns can occur in freshwater ecosystems.

Self-Quiz Questions

Fill-in-the-Blanks

The (1) _____ zone extends from the shore to where rooted plants stop growing; diversity is greatest here. The (2) _____ zone goes to the maximum depth of photosynthesis and includes plankton. The (3) _____ zone is the deep, open water below the depth of effective light penetration; here bacteria decompose materials into detritus. Water is densest at 4° C; at this temperature it sinks to the bottom of its basin, displacing the nutrient-rich bottom water upward and giving rise to spring and fall (4) _____ . By midsummer, the surface has warmed and a(n) (5) _____ develops on which the surface water floats; this prevents mixing. (6) _____ lakes are often deep, poor in nutrients, and low in primary productivity. (7) _____ lakes are often shallow, rich in nutrients, and high in primary productivity.

36-V
(pp. 560–567)

THE WATER PROVINCES (cont.)
    Marine Ecosystems
      Commentary: **El Niño and Oscillations in the World's Climates**
SUMMARY

Summary

Like freshwater ecosystems, the oceans and seas vary in their physical and chemical properties. Types of marine environments include estuaries, intertidal zones, and open oceans. An *estuary* is a partly enclosed coastal region where freshwater and saltwater meet. Estuaries do not all look alike. Along the New England and mid-Atlantic coasts, *Spartina*, a salt-tolerant plant, is a major estuary producer. Estuaries are feeding grounds but also nurseries for organisms belonging to food webs in the oceans. Estuaries are highly productive ecosystems that are being rapidly altered by pollution.

Life along the coasts is varied. Coastlines with rocky and sandy shores have an *intertidal zone*. Here residents are battered by waves, changing moisture, and varying temperature conditions. *Rocky shores* often have three vertically arranged zones: the sparsely populated *upper littoral* is only submerged during the highest tide of the lunar cycle; the *mid-littoral* is exposed during the lowest tide of each day (tide pools rich in life are characteristic); the lower littoral is only exposed during the lowest tide of the lunar cycle (diversity is greatest here). Grazing food webs predominate in all three zones. *Sandy and muddy shores* are areas of unstable stretches of loose sediment. Few large plants grow in either area; organic debris from offshore or land ecosystems form the basis of detrital food webs. Not much vertical zonation occurs. Along temperate coasts, blue crabs and sea cucumbers live below low tide mark; burrowing marine worms, isopods, and crabs live between high and low tide marks.

Beyond the intertidal are two vast provinces of the open ocean. The *pelagic province* is divided into the *neritic zone* (shallow waters, some with tropical reefs overlying continental shelves) and the *oceanic zone* (water over the ocean basins that may include tropical reefs). The *benthic province* includes all sediments and rocky formations of the ocean bottom. It begins with the continental shelf and extends down through the deep-sea trenches. In the neritic and oceanic zones, surface photosynthetic activity by phytoplankton forms vast, suspended zooplankton pastures. Surface organic remains and wastes sink to the benthic zone to provide nutrients for most communities. In the deep zones of the benthic province, remarkable communities thrive at *hydrothermal vents*. Producers here are chemosynthetic bacteria. *Upwelling* occurs when currents move nutrient-rich water to the surface.

Periodically, warm surface waters of the western equatorial Pacific move eastward. Prevailing winds are affected by the massive displacement of warm water, which accelerates the eastward movement. Upwelling is prevented by displacement of the cooler waters of the Humboldt Current. This phenomenon, called El Niño (El Niño Southern Oscillation or ENSO) by local fishermen, causes a decline in productivity; there are catastrophic effects on anchovy-eating birds and on the anchovy industry. Interactions among the atmosphere, oceans, and land profoundly influence the world of life.

Key Terms

estuary	lower littoral	benthic zone
Spartina	sandy and muddy shores	continental shelf
intertidal zone	open ocean	hydrothermal vents
rocky shores	pelagic province	upwelling
upper littoral	neritic zone	El Niño Southern
mid-littoral	oceanic zone	Oscillation (ENSO)

Objectives

1. Define *estuaries*, *intertidal zone*, and *open oceans*.
2. List and define the three vertically arranged zones of *rocky shores*.
3. Describe habitat conditions of sandy and muddy shores.
4. Give a detailed description of the *neritic zone* and the *oceanic zone* of the *pelagic province*.
5. Describe habitat conditions of the *benthic province*.
6. Explain how a *hydrothermal vent community* operates; cite the function of *upwelling*.

Self-Quiz Questions

Fill-in-the-Blanks

A region where freshwater mixes with saltwater is a(n) (1) _____ . Rocky and sandy shores of coastlines have a(n) (2) _____ zone. Rocky shores often have three vertically arranged zones: the sparsely populated (3) _____ _____ is only submerged during the highest tide of the lunar cycle; the (4) _____ - _____ is exposed during the lowest tide of each day; the (5) _____ _____ is only exposed during the lowest tide of the lunar cycle. (6) _____ and (7) _____ shores are areas of unstable stretches of loose sediment with little vertical zonation. Here few plants grow but organic debris from offshore or land ecosystems form the basis of (8) _____ food webs. The (9) _____ province is divided into the (10) _____ zone (shallow waters, some with tropical reefs) and the (11) _____ zone (water over the ocean basins that may include tropical reefs). The (12) _____ province includes all sediments and rocky formations of the ocean bottoms. In the deep zones of the benthic province, remarkable communities thrive at (13) _____ _____ ; producers here are chemosynthetic bacteria. (14) _____ occurs when currents move nutrient-rich water to the surface.

CHAPTER TEST   **UNDERSTANDING AND INTERPRETING KEY CONCEPTS**

___ (1) Which of the following statements pertaining to *climate* is *not* correct?

(a) Incoming rays from the sun equally heat equatorial and polar regions.
(b) Warm air moving north or south from equatorial regions is deflected because of Earth's rotation.
(c) Heated air rising at the equator gives up moisture as rain when it rises to cooler altitudes.
(d) Drier air moves away from the equator and descends at about 30° latitudes; deserts occur there.

___ (2) Air moving up mountains loses moisture as it rises and flows over mountain crests; there is a _____ on *leeward* sides of high mountains.

(a) tropical zone
(b) dry shadow
(c) rain shadow
(d) subtropical zone

___ (3) In a(n) _____ , water draining from the land mixes with seawater carried in on tides.

(a) neritic zone
(b) oceanic zone
(c) upwelling
(d) estuary

___ (4) A biome with grasses as primary producers and scattered trees adapted to prolonged dry spells is known as a _____ .

(a) warm desert
(b) savanna
(c) tundra
(d) taiga

___ (5) The eucalyptus woodlands of southwestern Australia and the oak woodlands of western North America are examples of _____ .

    (a) dry shrublands
    (b) grasslands
    (c) a boreal forest
    (d) dry woodlands

___ (6) Areas that include tall Sitka spruces, redwoods, and a scrub forest called the "pine barrens" describes _____ .

    (a) temperate lowland coniferous forests
    (b) evergreen coniferous forests
    (c) the alpine tundra
    (d) boreal forests

___ (7) In tropical rain forests, _____ .

    (a) there is little rainfall and low humidity
    (b) there is a great diversity of organisms
    (c) trees rarely shed leaves
    (d) there are many deciduous broadleaf trees

___ (8) Decomposition of organic materials would be inhibited most in the _____ .

    (a) evergreen broadleaf forest
    (b) tundra
    (c) deciduous broadleaf forest
    (d) boreal forest

___ (9) In a lake, the open sunlit water with its suspended plankton is referred to as its _____ zone.

    (a) epipelagic
    (b) limnetic
    (c) littoral
    (d) profundal

___ (10) Which of the following is appropriate for the *spring overturn* of a lake?

    (a) An area of sudden temperature change develops below warmer waters.
    (b) The upper layer cools, increases in density, and sinks.
    (c) Dissolved oxygen moves from surface to depths and nutrients released by decomposition move to the surface.
    (d) Mixing is prevented by the thermocline.

___ (11) Which of the following describes the eutrophic lake?

    (a) deep
    (b) rich in nutrients
    (c) shallow
    (d) poor in nutrients
    (e) both b and c

___ (12) An area of great diversity in the intertidal zone that is only exposed during the lowest tide of the lunar cycle is the _____ .

   (a) upper littoral
   (b) mid-littoral
   (c) lower littoral
   (d) sandy and muddy shore

___ (13) The most productive soil is _____ .

   (a) humus
   (b) sand
   (c) clay
   (d) loam
   (e) silt

___ (14) Temperature, humidity, wind velocity, cloud cover, and rainfall all describe _____ .

   (a) the biosphere
   (b) climate
   (c) the hydrosphere
   (d) the atmosphere

___ (15) _____ most directly causes seasons of the year.

   (a) Ocean currents
   (b) Air currents
   (c) A tilt in the Earth's axis
   (d) The rotation of the Earth

INTEGRATING AND APPLYING KEY CONCEPTS

Telephone your local lumber supply store and ask the owner to list the six most widely selling types of wood from different kinds of trees. Then determine from where in the world and from which type of forest the different woods originated. Tell which types are cheapest and most expensive.

37

HUMAN IMPACT ON THE BIOSPHERE

ENVIRONMENTAL EFFECTS OF HUMAN
POPULATION GROWTH
CHANGES IN THE ATMOSPHERE
    Local Air Pollution
    Acid Deposition
    Damage to the Ozone Layer
CHANGES IN THE HYDROSPHERE
    Consequences of Large-Scale Irrigation
    Maintaining Water Quality
CHANGES ON LAND
    Solid Wastes

Conversion of Marginal Lands for
Agriculture
Deforestation
Commentary: Tropical
Forests—Disappearing Biomes?
Desertification
A QUESTION OF ENERGY INPUTS
    Fossil Fuels
    Nuclear Energy
    Commentary: Biological Principles and
    the Human Imperative
SUMMARY

**General
Objectives**

1. Understand the magnitude of pollution problems in the United States.
2. Examine the effects modern agriculture has wrought on desert, grassland, and tropical rain forest ecosystems.
3. Describe how our use of fossil fuels and nuclear energy affects ecosystems.

37-I
(pp. 569–574)

**ENVIRONMENTAL EFFECTS OF HUMAN POPULATION GROWTH
CHANGES IN THE ATMOSPHERE**
    Local Air Pollution
    Acid Deposition
    Damage to the Ozone Layer

Summary

Interactions among the atmosphere, oceans, and land are the engines of the biosphere. Humans have been straining these engines without appreciating that they can crack. For example, our CO_2 waste is contributing to a greenhouse effect, or global warming. Population growth and individual demands are stressing the environment.

    Humans have brought about drastic changes in the composition of the atmosphere. Each day 700,000 metric tons of *pollutants* are dumped into the atmosphere in the United States alone. *Thermal inversions* can trap pollutants close to the ground. *Industrial smog* is gray air found in industrial cities that burn fossil fuel. *Photochemical smog* is brown air found in large cities in warm climates; its formation can begin with gases from car exhausts. Nitric oxide is the chief culprit; it reacts with oxygen in the air to form nitrogen dioxide. In sunlight, nitrogen dioxide reacts with *hydrocarbons* to form photochemical oxidants. Other smog components are ozone and PANs.

Acid deposition can occur from chemical reactions of atmospheric pollutants. Burning coal produces sulfur dioxide; burning fossil fuels and fertilizers result in sulfur and nitrogen oxides. These compounds can fall to Earth as *dry acid deposition* or *wet acid deposition*—acid rain (weak solutions of sulfuric acid and nitric acid). Sometimes as acidic as lemon juice, the acids attack marble, metals, mortar, rubber, plastic, even nylon stockings; they are disrupting ecosystems.

Damage can occur to the ozone layer. Ozone in the lower stratosphere absorbs most of the ultraviolet radiation from the sun. The ozone layer has been thinning since 1976 and a hole appears over the Antarctic each spring. The reduction in the ozone layer is allowing more ultraviolet radiation to reach Earth's surface; rates of skin cancer have increased, cataracts may increase, and phytoplankton may be affected. *Chlorofluorocarbons* seem to be the cause of the thinning of the ozone layer; one chlorine atom can convert 10,000 ozone molecules to oxygen.

Key Terms

greenhouse effect	nitric oxide	sulfuric acid
pollutants	nitrogen dioxide	wet acid deposition
carbon dioxide	hydrocarbons	acid rain
thermal inversion	photochemical oxidants	ozone
industrial smog	PANS, peroxyacyl nitrates	chlorofluorocarbons
oxides of sulfur	dry acid deposition	(CFCs)
photochemical smog		

Objectives

1. Identify the principal air pollutants, their sources, their effects, and the possible methods for controlling each pollutant.
2. Define *thermal inversion* and indicate its cause.
3. Distinguish *photochemical smog* from *industrial smog.*
4. Explain what acid rain does to an ecosystem. Contrast those effects with the action of CFCs.

Self-Quiz Questions

Fill-in-the-Blanks

Each day 700,000 metric tons of (1) _____ are dumped into the atmosphere in the United States alone. (2) _____ _____ can trap pollutants close to the ground. (3) _____ _____ is gray air found in industrial cities that burn fossil fuel. (4) _____ _____ is brown air (as from automobile exhausts) found in large cities in warm climates. (5) _____ oxide is the chief culprit; it reacts with oxygen in the air to form (6) _____ _____ . In sunlight, (7) _____ _____ reacts with hydrocarbons to form photochemical oxidants. Other smog components are ozone and (8) _____ . Acid (9) _____ can occur from chemical reactions involving atmospheric pollutants. Burning coal produces

(10) _____ dioxides; burning fossil fuels and fertilizers results in sulfur and (11) _____ oxides. These compounds can fall to Earth as (12) _____ _____ deposition or (13) _____ _____ deposition. (14) _____ _____ consists of weak solutions of sulfuric acid and nitric acid; these acids destroy many manufactured goods and are disrupting (15) _____ . The ozone layer has been thinning since 1976; the reduction in the ozone layer is allowing more (16) _____ radiation to reach the Earth's surface. (17) _____ cancer incidence has increased, cataracts may increase, and phytoplankton may be affected. (18) _____ seem to be the cause of the thinning of the ozone layer.

37-II
(pp. 574–580)

CHANGES IN THE HYDROSPHERE
  Consequences of Large-Scale Irrigation
  Maintaining Water Quality
CHANGES ON LAND
  Solid Wastes
  Conversion of Marginal Lands for Agriculture
  Deforestation
  Commentary: **Tropical Forests—Disappearing Biomes?**
  Desertification

Summary

There is a huge amount of water in the world. Despite this, two of every ten humans do not have enough water or, if they do, it is contaminated.

About half the food being produced today to support the exponential growth of human populations grows on irrigated land. The consequences of such large-scale irrigation are salt buildup (salination) and waterlogging of soil. Farmers draw amounts of water for irrigation from the groundwater of the Ogallala aquifer nearly equal to the annual flow of the Colorado River.

In addition to the serious problem of not having enough water, human waste, insecticides, herbicides, chemicals, radioactive materials, and heat can pollute available water supplies. Pollutants accumulate in lakes, rivers, and bays before reaching the oceans. Most liquid wastes from urban populations in the United States are treated on one or more of three levels of *wastewater treatment.* *Primary treatment* removes sludge from wastewater and then burns it before it is dumped in landfills; chlorine is added to the reclaimed water. *Secondary treatment* uses microbes to degrade organic matter—nitrates, viruses, and toxic substances remain in the water. *Tertiary treatment* involves expensive and largely experimental methods of precipitating suspended solids and phosphate compounds, adsorption of dissolved organic compounds, reverse osmosis, stripping nitrogen from ammonia, and disinfecting water through chlorination or ultrasonic energy vibrations; it is used on only about five percent of the nation's wastewater. Most wastewater is not being properly treated. Drinking water is removed *upstream* from a city and wastes from industry and sewage treatment are discharged *downstream.*

The more affluent countries face a challenge to move from a "throwaway" society to one of conservation and reuse. Billions of metric tons of solid wastes are dumped, burned, and buried annually in the United States alone.

Another type of assault on the land is occurring throughout the world. Cultivation and grazing is expanding into areas that are only marginally suitable for agriculture. Almost twenty-one percent of the land is used for agriculture; another twenty-eight percent is available but may not be worth the cost. The *green revolution* has increased yields 4 times but uses 100 times more energy and mineral resources (the costs of fertilizers and machinery are reflected in market food prices). A growing human population is moving into marginal lands to meet increasing needs.

The world's great forests play major roles in the biosphere. Forests are watersheds; they control erosion, flooding, and sediment buildup in rivers and lakes. Deforestation can reduce fertility, change rainfall patterns, increase temperatures, and increase carbon dioxide levels. Tropical forests contain an enormous variety of organisms. Despite the diversity, they are one of the worst places to grow crops. Because of rapid decomposition in the hot, humid climates, there is little nutrient storage in the subsoil. Minerals released during decomposition are rapidly picked up by roots and mycorrhizae in the topsoil layers, and most become tied up in the standing biomass. Developing countries have been clearing their tropical forests on a massive scale with *slash-and-burn agriculture;* cleared plots quickly become infertile and are abandoned. Clearing tropical forests means extinction for thousands of species; these genetic resources for potential food and medicines are lost forever.

Desertification is the conversion of grasslands and cropland to desertlike conditions. About 9 million square kilometers have become deserts during the past fifty years; at least 200,000 square kilometers are still being transformed each year. Today, large-scale desertification is occurring mainly as a result of cattle and goats overgrazing marginal lands.

Key Terms

saline, salination
waterlogging
Ogallala aquifer
wastewater treatment
primary treatment
sludge

secondary treatment
tertiary treatment
precipitation
reverse osmosis
solid wastes
subsistence agriculture

animal-assisted agriculture
green revolution
slash-and-burn agriculture
alkaloids
desertification

Objectives

1. Define *primary, secondary,* and *tertiary wastewater treatment* and list some of the methods used in each of the three types of treatment.
2. List the disadvantages and dangers of trying to maintain our present throwaway system for handling solid wastes.
3. Distinguish a recycling system for solid wastes from a resource recovery center.
4. Cite individual actions you, as a consumer, can take to assist the transition from a throwaway society to one based on conservation and reuse.
5. Discuss some disadvantages to exploiting lands marginally suitable for agriculture.
6. Explain how deforestation has caused soils, water quality, and genetic diversity to deteriorate.
7. Define *desertification;* state the principal cause of the large-scale desertification occurring today.

Self-Quiz Questions

Fill-in-the-Blanks

The consequences of large-scale irrigation are (1) _____ and (2) _____ of soil. Most liquid wastes from urban populations in the United States are treated on as many as three levels. (3) _____ treatment removes solids and then burns the sludge before it is dumped in landfills; chlorine is added to the reclaimed water. (4) _____ treatment uses microbes to degrade organic matter—nitrates, viruses, and toxic substances remain in the water. (5) _____ treatment uses largely experimental methods and is used on only about five percent of the nation's wastewater. Drinking water is removed (6) _____ from a city and wastes from industry and sewage treatment are discharged (7) _____ . The more affluent countries face a challenge to move from a(n) (8) _____ society to one of conservation and reuse. Another type of assault on the land is occurring; cultivation and grazing are expanding into areas that are only (9) _____ suitable for agriculture. The world's great forests play major roles in the biosphere. (10) _____ can reduce fertility, change rainfall patterns, increase temperatures, and increase carbon dioxide levels. Land once occupied by a tropical rain forest is one of the worst places to grow crops. Because of (11) _____ _____ in the hot, humid climate; there is little nutrient storage in the subsoil. (12) _____ released in decomposition are rapidly picked up by roots and mycorrhizae in the topsoil layers. Developing countries have been clearing their tropical forests on a massive scale with (13) _____ - _____ - _____ agriculture; cleared plots quickly become infertile and are abandoned. Clearing tropical forests means (14) _____ for thousands of potentially valuable species. (15) _____ is the conversion of grassland and croplands to desertlike conditions. Today, large-scale conversion of lands to desert is occurring mainly as a result of (16) _____ of marginal lands.

37-III
(pp. 580–585)

A QUESTION OF ENERGY INPUTS
    Fossil Fuels
    Nuclear Energy
    Commentary: **Biological Principles and the Human Imperative**
SUMMARY

Summary

Increases in human population and extravagant life-styles increase energy consumption. When one hears talk of abundant energy supplies, keep in mind that there is an enormous difference between the total supply and the net amount available. *Net energy* is the energy left over after subtracting the energy used to locate, extract, transport, store, and deliver energy to consumers. In addition, some sources of energy are not renewable. *Fossil fuels* are a limited resource; extraction requires more and more energy and is increasing atmospheric levels of carbon dioxide and sulfur dioxides. Colorado, Utah, and Wyoming probably have more potential oil than the entire Middle East in the form of *oil shale*, which contains the hydrocarbon *kerogen*. However, extraction disfigures the land, increases water and air pollution, and taxes existing short water supplies. There are enough coal reserves to meet the energy needs of the human population for several centuries. Coal is the single largest source of air pollution; low-quality coal releases sulfur dioxide into the air and adds to acid deposition. Burning fossil fuels also releases carbon dioxide and amplifies the greenhouse effect. Strip-mining limits the usefulness of land for agriculture, grazing, and wildlife.

With nuclear energy, the net energy produced is low and the cost high compared with coal-burning plants; *meltdowns* may release large amounts of radioactivity to the environment. In 1986, the potential dangers of nuclear power were brought into sharp focus by a meltdown at the Chernobyl power station in the Soviet Union. Radiation was released into the atmosphere, several people died immediately, and others died of radiation sickness. A breeder reactor is theoretically possible; it uses a rare isotope of uranium and is more energy efficient, but unlike a conventional reactor, could undergo a small nuclear explosion. Fusion power, in which hydrogen atoms fuse to form helium atoms, is a future possible energy source; these reactions are similar to those on the sun producing heat energy. At best, this technology would be available to replace electricity in the last half of the next century.

Waste from nuclear reactors is so radioactive that it must be isolated for 10,000 years. There are plans to bury radioactive waste deep underground in ceramic containers; such facilities will not be available until after the turn of the century. The nuclear exchange of a major war and the resulting *nuclear winter* may cause catastrophic extinctions. The Cenozoic Era may be brought to a close and what would follow is unknown.

Although humans are not the first organisms to modify the nature of living systems, we now have the population size, the technology, and the cultural inclination to use energy and modify the environment at frightening rates. Where will this accelerated change lead us? Feedback controls will not be enough to correct the environmental deviations we have caused. Even feedforward controls are not enough. We can avert ecological disaster if we use our ability to anticipate events before they happen. We can adapt to a future we can partly shape by redesigning and constructing ecosystems that harmonize with human values and biological models. We must come to terms with the principles of energy flow and resource utilization that govern the survival of all systems of life.

Key Terms

net energy	kerogen	Chernobyl incident
fossil fuels	meltdown	nuclear winter
oil shale	breeder reactor	fusion power

Objectives

1. List disadvantages of heavy dependency on fossil fuels even though the Earth still holds large amounts.
2. Explain why exploiting oil shale deposits may not be worth doing.

3. Suggest advantages in using the theoretical breeder reactor over conventional nuclear power plant reactors.
4. Explain what a *meltdown* is.
5. Define *fusion power* and assess the prospects for its use.
6. Describe lessons learned from the Chernobyl incident.
7. List five ways in which you personally could become involved in ensuring that institutions serve the public interest in a long-term, ecologically sound way.
8. Describe the probable path humans must follow to avert global ecological disaster.

Self-Quiz Questions

Fill-in-the-Blanks

Increases in human population and extravagant life-styles increase (1) _____ consumption. (2) _____ energy is the energy left over after subtracting the energy used to locate, extract, transport, store, and deliver energy to consumers. (3) _____ fuels are a limited resource; their extraction requires more and more energy and atmospheric levels of carbon dioxide and sulfur dioxides are rising. Oil shale is buried rock that contains (4) _____ , a hydrocarbon compound. There are enough (5) _____ reserves to meet energy needs of the human population for several centuries; however, that resource is the largest single source of air pollution. Burning fossil fuels also releases carbon dioxide and amplifies the (6) _____ effect. With nuclear energy, the net energy produced is (7) [choose one] ( ) high ( ) low, and the cost is (8) [choose one] ( ) high ( ) low, compared with coal-burning plants. (9) _____ may release large amounts of radioactivity to the environment. (10) _____ reactors theoretically may replace conventional nuclear power reactors in the future. (11) _____ of hydrogen atoms is a process of energy release that may be possible in the distant future. The nuclear exchange of a major war and the resulting (12) _____ _____ may cause catastrophic extinctions. We may avert ecological disaster if we use our ability to (13) _____ events before they happen. We can adapt to a future we can partly shape by redesigning and constructing (14) _____ that harmonize with human values and biological models.

UNDERSTANDING AND INTERPRETING KEY CONCEPTS

___ (1) Which of the following processes is *not* generally considered a component of tertiary wastewater treatment?

    (a) microbial action
    (b) precipitation of suspended solids
    (c) reverse osmosis
    (d) adsorption of dissolved organic compounds

___ (2) Gray air that accumulates in industrial cities that burn fossil fuel is _____ .

    (a) photochemical smog
    (b) industrial smog
    (c) a thermal inversion
    (d) both a and c

___ (3) _____ results when nitrogen dioxide and hydrocarbons react in the presence of sunlight.

    (a) Photochemical smog
    (b) Industrial smog
    (c) A thermal inversion
    (d) Both a and c

___ (4) Nitric oxide is the chief culprit in photochemical smog; it reacts with oxygen to form _____ .

    (a) PANs
    (b) nitrogen dioxide
    (c) sulfur oxide
    (d) CFCs

___ (5) Breeder reactors would be _____ than other types of nuclear reactors.

    (a) more energy efficient
    (b) more widely used
    (c) safer
    (d) all of the above

___ (6) It is believed that thinning of the ozone layer is caused mainly by _____ .

    (a) CFCs
    (b) PANs
    (c) photochemical oxidants
    (d) nitrogen oxides

___ (7) There is a huge amount of water in the world. Despite this, _____ of every ten humans do(es) not have enough water or, if they do, it is contaminated.

    (a) one
    (b) two
    (c) three
    (d) four

___ (8) Secondary treatment uses _____ to degrade organic material—nitrates, viruses, and toxic substances remain.

(a) microbes
(b) aeration with pure oxygen
(c) mechanical screens
(d) chemicals such as aluminum sulfate

___ (9) Almost twenty-one percent of the world's land is used for agriculture; another twenty-eight percent is available but probably not worth the cost and is referred to as _____ .

(a) land for the green revolution
(b) grazing land
(c) marginal land
(d) land to be deforested

___ (10) The term "slash-and-burn agriculture" is associated with _____ .

(a) using marginal lands
(b) the green revolution
(c) tropical forests
(d) desertification

___ (11) Burning which of the following is the single largest source of air pollution?

(a) oil
(b) kerogen
(c) gasoline
(d) coal

___ (12) Which of the following statements is *incorrect?*

(a) Net energy is the amount of energy before subtracting the energy used to locate, extract, transport, store, and deliver energy to consumers.
(b) A nuclear winter may cause catastrophic extinctions.
(c) Fossil fuels are a limited resource that are taking increasing energy to extract.
(d) Burning fossil fuels also releases carbon dioxide and amplifies the greenhouse effect.

___ (13) Each day in the United States, _____ metric tons of pollutants are discharged into the atmosphere.

(a) 1,000
(b) 100,000
(c) 700,000
(d) 5,000,000

___ (14) The most abundant fossil fuel in the United States is _____ .

(a) carbon monoxide
(b) oil
(c) natural gas
(d) coal

INTEGRATING AND APPLYING KEY CONCEPTS

(1) If you were Ruler of All People on Earth, how would you encourage people to depopulate the cities and adopt a way of life by which they could supply their own resources from the land and dispose of their own waste products safely on their own land?

(2) Explain why some biologists believe that the endangered species list now includes all species.

38
ANIMAL BEHAVIOR

GENES, HORMONES, AND BEHAVIOR
    Genetic Basis of Behavior
    Hormonal Effects on Behavior
INSTINCT AND LEARNING
    Instinctive Behavior
    Learned Behavior
    Imprinting
THE ADAPTIVE VALUE OF BEHAVIOR
    Adaptive Feeding Behavior

    Anti-Predator Behavior
    Reproductive Behavior
SOCIAL BEHAVIOR
    Social Communication
    Costs and Benefits of Social Life
    Social Life and Self-Sacrifice
    Evolution of Altruism
    Human Social Behavior
SUMMARY

General Objectives

1. Understand the components of behavior that have a genetic and/or hormonal basis.
2. Distinguish behavior that is primarily instinctive from behavior that is learned.
3. Know the aspects of behavior that have an adaptive value.
4. Describe how forms of communication organize social behavior.
5. List the costs and benefits of social life.
6. Explain the roles of self-sacrifice and altruism in social life.

38-I
(pp. 587–588)

GENES, HORMONES, AND BEHAVIOR
    Genetic Basis of Behavior
    Hormonal Effects on Behavior
INSTINCT AND LEARNING
    Instinctive Behavior

Summary

Animal behavior involves coordinated responses to external and internal stimuli. Behavioral responses are brought about by interactions among nervous, endocrine, and skeletal-muscular systems. *Heredity* provides the animal with its basic response mechanisms. *Learning* has a role in behavior; it allows the animal to modify behavior while interacting with the environment.

    Animal behavioral responses depend on the physical layout of the nervous system; genes contribute to behavior by dictating how the physical layout of the nervous system develops. Young white-throated sparrows and young white-crowned sparrows reared under the same conditions were allowed to listen to tape-recorded songs of adult white throats and white crowns. Male white throats sing, "Sam Peabody, Sam Peabody," and the white crowns sing a different song. Since environmental conditions are similar, the difference in singing behavior must be due to genetic differences between the birds. The number of genes causing the differences is not important.

Behavior is profoundly influenced by hormones, the signaling molecules of the endocrine system. In white throats and other songbirds, melatonin from the pineal gland suppresses growth and function of gonads. Sunlight inhibits melatonin secretion. With spring's increasing amount of daylight, melatonin levels decline and the gonads secrete their own hormones (estrogen and testosterone); this begins mating behaviors in the birds. In very young male songbirds, estrogen *organizes* the development (brain regions and nerve pathways leading to vocal organs) of the song system; then testosterone *activates* the song system and prepares the bird to sing when properly stimulated.

Instinct is the animal capacity to complete fairly complex, stereotyped responses to a number of environmental cues, even without having had prior experience with those cues. Instinctive behavior is set in motion automatically. Cuckoo eggs are laid in the nests of other birds; newly hatched cuckoos instinctively respond to the shape of any eggs touched by pushing them out of the nest. Male stickleback fish have red bellies; in guarding eggs they have fertilized, rival males with red bellies (the cue for aggressive behavior) are chased away. In experiments, male stickleback fish try to strike any red object. Human infants instinctively smile at a flat face mask with two dark spots (one does not work) representing eyes.

Key Terms

animal behavior	learning	instinctive behavior
heredity	hormones	instinct

Objectives

1. Define *animal behavior* and relate heredity and learning to behavior.
2. Explain the genetic basis of behavior; cite an example.
3. Explain how hormones affect behavior; cite an example.
4. Discuss *instinctive behavior*; cite an example.

Self-Quiz Questions

Fill-in-the-Blanks

Animal (1) _____ involves coordinated responses to external and internal stimuli. Behavioral (2) _____ are brought about by interactions among nervous, endocrine, and skeletal-muscular systems. (3) _____ provides the animal with its basic response mechanisms. (4) _____ allows the animal to modify behavior while interacting with the environment. It was determined that (5) _____ differences were the basis for white-throated sparrows and white-crowned sparrows singing different songs despite being raised under similar conditions. In very young songbirds, estrogen (6) _____ the development of the song system; then testosterone (7) _____ the song system and prepares the bird to sing when properly stimulated. (8) _____ is the animal capacity to complete fairly complex, stereotyped responses to a number of environmental cues, even without having had prior experience with those cues. Cuckoo eggs are laid in the nests of other birds; newly hatched cuckoos instinctively respond to the (9) _____ of any eggs touched by pushing them out of the nest. In guarding eggs they have fertilized, male stickleback fish drive rival males away; their cue for aggressive behavior is the (10) _____ belly of rival males.

38-II
(pp. 588–589)

INSTINCT AND LEARNING (cont.)
    Learned Behavior
    Imprinting

Summary

Learning is the adaptive (genetically advantageous) modification of behavior in response to specific experiences. Newly hatched peacocks peck the ground (an instinct) and learn from personal experience that some habitat areas have more food to peck at, and that some times of the day are better to go after food than others. The ability to learn requires a nervous system that is genetically determined. Learning occurs in a variety of ways. *Associative learning* is the capacity to make a connection between a new stimulus and a familiar one; it often occurs by trial-and-error with behavior reward or punishment. Dogs salivate when meat extract is placed on the tongue; if they hear the sound of a bell just before getting meat extract, they later salivate at the sound of the bell alone. Earthworms placed in a T-maze where wrong turns yield an electric shock and correct turns yield a moist, dark chamber eventually make correct responses more frequently. *Insight learning* is problem solving without trial-and-error learning. Chimpanzees could pile scattered boxes to climb and take a banana dangling from the ceiling without prior trial-and-error practice. Pigeons performed a similar feat.

    The capacity to learn specific information is sometimes pronounced during certain early developmental stages. With *imprinting,* the capacity to learn is time dependent. For example, young animals are primed during a short period early in life to form a learned attachment to a moving object, normally their mother. Konrad Lorenz discovered that newly hatched goslings separated from their mother formed an attachment to and followed a moving object, even a human. Hatchlings not offered an object to follow lost imprint readiness. Imprinting has long-term consequences. When the male goslings who had imprinted on a human matured, their sexual behavior was directed not toward females of their species but toward humans. Imprinting also plays a role in migratory behavior, as in bird travel from a summer breeding area to a distant winter refuge. Migrating animals have a compass sense and a navigational sense. Indigo buntings show a compass sense. During a sensitive period young buntings imprint on the visual image of the night sky. When just a few months old they fly at night to their wintering grounds using the position of stars in the night sky as an orientation guide. Young buntings reared in the dark cannot directionally migrate.

Key Terms

learned behavior   insight learning   compass sense
learning   imprinting   navigational sense
associative learning

Objectives

1. Define *learned behavior;* cite an example.
2. Describe associative learning and insight learning; cite examples of each.
3. Define *imprinting* and give examples.

Self-Quiz Questions

Fill-in-the Blanks

(1) _____ is the adaptive modification of behavior in response to specific experiences. The ability to learn requires a nervous system that is (2) _____ determined. Learning occurs in a variety of ways. (3) _____ learning is the capacity to make a connection between a new stimulus and a familiar one; it often occurs by (4) _____ - _____ - _____ with appropriate behavior rewarded and inappropriate behavior being punished. (5) _____ learning is problem solving without trial-and-error learning. Meat extract placed on the tongue of a dog just after hearing the sound of a bell causes salivation; later, when the dog salivates at just the sound of a bell, it is an example of (6) _____ learning. When chimpanzees are able to pile boxes and climb to obtain a dangling banana without trial-and-error experience, it is an example of (7) _____ learning. (8) _____ is the capacity to learn specific information during certain developmental stages. Konrad Lorenz discovered that newly hatched goslings separated from their mother formed an attachment to and followed a moving object, even a(n) (9) _____ . Hatchlings not offered an object to follow lost their readiness to (10) _____ . (11) _____ also plays a role in migratory behavior. Migrating animals have a(n) (12) _____ sense and a(n) (13) _____ sense. During a sensitive period young buntings (14) _____ on the visual image of the night sky. When only a few months old the buntings fly at night to their wintering grounds using the position of (15) _____ in the night sky as an orientation guide.

THE ADAPTIVE VALUE OF BEHAVIOR
    Adaptive Feeding Behavior
    Anti-Predator Behavior
    Reproductive Behavior

Summary

Many forms of behavior are adaptive traits that are subject to evolution by natural selection. Several definitions help to understand the evolutionary mechanisms involved. *Reproductive success* refers to the survival and production of offspring. *Adaptive behavior* promotes reproductive success. *Selfish behavior* occurs when an individual increases its chances of producing offspring regardless of consequences to the group to which it belongs. *Altruistic behavior* is self-sacrificing behavior that helps others and decreases the individual's own chance to reproduce. *Natural selection* refers to differential reproductive success among individual members of a group that varies in heritable traits—including many behavioral traits—that promote survival and reproduction. Animals are not consciously aware that their selfish or altruistic behavior is related to reproductive success. When populations of Norwegian lemmings become dense, the individuals swarm away but are not really programmed to sacrifice themselves.

Adaptive behavior does evolve through individual selection. Arnold's studies of garter snakes provide an example of natural selection for feeding behavior. Two populations of garter snakes show different feeding preferences from birth. It was demonstrated that the populations differ in the genes controlling chemical responses to food odors. Individuals attracted to a particular food for which they have a genetic predisposition leave more descendants and spread acceptance for a particular food. When a black heron stands motionless and holds its wings over its head like an umbrella, it is attracting minnows in the patch of shade it has created over the water. The behavior enhances heron survival and reproduction.

Animals exhibit anti-predator behavior. It does an animal no good to go out in the world and eat if it is attacked and killed in the process. For example, the caterpillar of certain tropical moths eats voraciously during the darkness of night but remains motionless on vines during the day. If the caterpillar is poked during the day, it drops partway from the vine and puffs to simulate a striking snake; small birds hesitate to continue an attack.

Reproductive behavior is also adaptive. Sexual selection is the outcome of individuals of one sex competing for mating with individuals of the opposite sex. A male's reproductive success may be measured by the number of descendants. A female's reproductive success is measured by how many eggs she can produce or how many offspring she can care for—females are concerned with the quality of a mate, not the number of matings. Male white-throated sparrows compete for insect-rich patches of forest that will attract females; control over a resource-based territory is advertised by a song that attracts females and discourages rival males. Female reproductive success depends on feeding nestlings well; this means having huge quantities of insects available. If the resources sought by females are clumped in space, then *resource-defense behavior* should evolve (winners get mates, losers are kept from reproducing). Males often show *female-defense behavior* when females live together but do not concentrate at patches of a useful resource. Male red-winged blackbirds, lions, elk, and bighorn sheep compete fiercely in response to female clustering behavior. Competition favors the fiercest combatants but winners have a ready-made harem. Females benefit by choosing superior genes or superior material benefits or both. For example, female hangingflies mate with mates that offer them the largest fly or moth.

Key Terms

reproductive success   sexual selection
adaptive behavior   territory
selfish behavior   resource-defense behavior
altruistic behavior   female-defense behavior
natural selection

Objectives

1. Contrast *altruistic behavior* with *selfish behavior* and cite examples of each in animal populations.
2. State why the example of lemming suicide demonstrates that individuals do not sacrifice their own chances of reproducing to promote the welfare of their species.
3. Describe Arnold's studies that suggest that feeding behavior in certain California garter snakes has some genetic basis.
4. Explain the value of anti-predator behavior; cite an example.
5. Define *sexual selection*; contrast *resource-defense behavior* with *female-defense behavior*, and cite specific examples.

Self-Quiz Questions

Matching

Match the example or definition with the corresponding term at the left.

(1) ___ altruistic behavior

(2) ___ female-defense behavior

(3) ___ reproductive success

(4) ___ resource-defense behavior

(5) ___ selfish behavior

(6) ___ sexual selection

A. Male bison, lions, elk, or bighorn sheep competing in combat to gain a ready-made harem

B. Resources sought by females are clumped rather than spread out through the habitat

C. A male white-throated sparrow singing his "Sam Peabody" song in his chosen breeding site, trying to attract a female

D. The captain of a ship shouting "Women and children first!" as the boat begins to sink and the lifeboats are launched

E. Survival and reproduction of offspring

F. A male cat eating a female cat's newborn kittens and then mating with her

---

38-IV
(pp. 592–594)

SOCIAL BEHAVIOR
   Social Communication

Summary

Social behavior is the tendency of individual animals to enter into cooperative, interdependent relationships with others of their kind. The ability of animals to *communicate* with each other, using a complex array of signals, is at the heart of social behavior. A colony of Australian nasute termites may have a million members working together, so communication is necessary. Termite soldiers communicate by releasing alarm scents from their nose; the odor attracts more soldiers to the danger site.

Animals use several channels of communication. A *communication signal* is a stimulus produced by one animal that changes the behavior of another individual of the same species; these signals may be chemical, visual, tactile, and acoustical. Examples of *chemical signals* are alarm odor calls of termites, sex pheromones of male hangingflies, and trail markers of ants. Female insects often attract mates by releasing sex pheromones. External genitalia of female baboons swell and they produce sex pheromones that elicit male mating behavior. *Visual signals* are used by animals that are active during the day. Male baboons may threaten a rival with a "yawn." Albatrosses and other birds assume exaggerated and often contorted postures with specific meanings, such as courtship. Animals that are active at night sometimes use *bioluminescent signals*. Male fireflies signal females; they reply with single flashes until precisely located by the males for mating. Females of a predatory species of firefly sometimes similarly reply to male signals; the male is eaten after being lured to the female.

Acoustical signals are distinctive sounds with specific meaning. Worker termites bang their heads specifically to attract soldiers; male birds sing to stake out territories, attract females, and discourage rivals. Male frogs call to communicate with rival males and receptive females. *Tactile signals* are distinctive touch patterns important for communication over short distances. Honeybees "dance" in the dark beehive to tell others the location of food. A round dance means food is close to the hive; a waggle dance is used when food is far away. The more waggles and the faster the dance, the closer the food source. The angle of the straight run up the comb surface indicates position of food relative to the sun and the beehive.

Key Terms

social communication	visual signals
social behavior	bioluminescent signals
channels of communication	acoustical signals
communication signal	tactile signals
chemical signals	waggle dance

Objectives

1. Define *social behavior*; relate the example of termite communication.
2. List four kinds of signals with which animals convey information; cite an example of each.

Self-Quiz Questions

Matching

Select the best linkage.

(1) ___ acoustical signal

(2) ___ chemical signal

(3) ___ tactile signal

(4) ___ visual signal

A. Pheromone
B. Sam Peabody, Peabody, Peabody
C. Bioluminescent messages
D. Dance of the foraging honeybee

Fill-in-the-Blanks

(5) _____ behavior is the tendency of individual animals to enter into cooperative, interdependent relationships with others of their kind. The ability of animals to (6) _____ with each other, using a complex array of signals, is at the heart of (7) _____ behavior. Termite soldiers (8) _____ by releasing alarm scents from their nose; the odor attracts more soldiers to the danger site. A(n) (9) _____ _____ is a stimulus produced by one animal that changes the behavior of another individual of the same species. Examples of (10) _____ signals are sex pheromones of male hangingflies and trail markers of ants. A male baboon "yawns" to threaten a rival; this is an example of a(n) (11) _____ signal. Male fireflies signal female fireflies for mating with a special type of visual signal called a(n) (12) _____ signal. When worker termites bang their heads specifically to attract soldiers they are using a(n) (13) _____ signal. Honeybees use (14) _____ signals when they "dance" in the dark to tell others the location of food.

38-V
(pp. 594–597)

SOCIAL BEHAVIOR (cont.)
    Costs and Benefits of Social Life
    Social Life and Self-Sacrifice

Summary

Social life has costs and benefits. Social behavior (or its absence) can be explained in terms of how it affects *individual reproductive success* in different environments. Hundreds of individuals pair and nest closely together in the large rookeries of egrets, gulls, or terns. The harmony is an illusion; they may eat unguarded chicks or eggs of neighbors and the crowded birds are highly vulnerable to contagious diseases. Predation has a relationship to sociality. The cooperative hunting of a lion pride can bring down a prey that perhaps a single lion would not dare attack. Predation also is the main selection pressure that favors social behavior among prey species; in a large group, a prey animal is less likely to become a victim because of the dilution effect. Australian fairy penguins improve survival odds by waddling along beaches in groups under patrolling hawks and eagles. Social animals collectively have many eyes to spot predators, and in large groups, each animal has more time to feed because less time is required for vigilance. Prey animals can often repel predators through group defense; nasute termites and musk oxen are examples. Some animals apparently live in groups (*selfish herds*) simply to "use" others as living shields against predators. For example, wildebeests flee from hyenas or hunting dogs in a mass as if they were a single animal; each tries to maneuver itself to the safety of the center of the herd.

    Social life sometimes shows self-sacrifice. Self-sacrifice exists if it enhances the continuity of the genetic line of the altruistic individual. Parental behavior is a familiar form of self-sacrificing behavior. A pair of Caspian terns devote full time in rearing young to ensure at least some offspring survive to reproductive age. Even though the individual sacrifices some of its reproductive capacity, its genes still spread through the population. There are cooperative societies in which self-sacrificing individuals direct friendly, helpful behavior to more than just offspring. Subordinates in a group may ultimately move up to a higher position in the social hierarchy when dominant members die, become injured, old, or feeble. With patience, a subordinate may eventually reproduce. Remaining a subordinate in a group may have the advantage of retaining the protection of the group and the possibility of reproduction. Striking out on its own means the group advantages of protection are lost.

Key Terms

individual reproductive
   success
dilution effect

selfish herd
parental behavior

cooperative societies
dominance hierarchy

Objectives

1. Explain how being preyed upon improves a population's sociality.
2. Explain the role of self-sacrifice in parental behavior and cooperative societies.

**Self-Quiz
Questions**

Fill-in-the-Blanks

Social life has (1) _____ and (2) _____ . Social behavior (or its absence) can be explained in terms of how it affects individual (3) _____ _____ in different environments. (4) _____ is the main selection pressure that encourages social behavior among prey species. In a large group, a prey animal is less likely to become the (5) _____ due to (6) _____ . As wildebeests flee from predators, they mass as if they were a single animal as each tries to move to the center of the herd. This group is called the (7) _____ herd. A pair of Caspian terns rearing their young illustrate (8) _____ - _____ behavior. Remaining a subordinate in a group may have the advantage of retaining the (9) _____ of the group and the possibility of (10) _____ . An individual that leaves the group to strike out on its own means the group advantages of (11) _____ are lost.

38-VI
(pp. 597–598)

SOCIAL BEHAVIOR (cont.)
    Evolution of Altruism
    Human Social Behavior
SUMMARY

Summary

With *altruistic behavior*, the "helper" reduces its own reproductive potential while the "helped" has increased its reproductive success. Altruistic behavior directed toward kin may be favored by natural selection. In *kin selection*, a nonreproducing, subordinate member of a group may help pass on copies of its genes, as long as the members benefiting the altruistic behavior are relatives. An example would be a flock of jays in which there is a breeding pair and as many as six nest helpers (older offspring of the pair). Helpers may never reproduce but instead feed and protect younger brothers and sisters and in that way perpetuate some shared genes when their siblings reproduce. In insect societies like bees, sterile guards may protect the queen by stinging an intruder and committing suicide. Sterile worker honeybees care for the queen and accomplish necessary tasks for the general well-being of the hive. Suicidal behavior and sterility among these insects are adaptive traits, for close relatives are the beneficiaries. They sacrifice their reproductive chances to increase sibling numbers; siblings have many genes in common with them.

Can individual selection theory be applied to analysis of human behavior? This question is controversial. Humans have had an extraordinary cultural evolution and are unique in many ways. But beneath the elaborate, diverse layerings of culture, there still is a biological core for human behavior. Experiments strongly support the view that human behavioral development has a genetic basis and is subject to individual selection. Some human social behavior is universal; for example, smiling is innate and helps to form strong emotional ties between infants and parents. Many human emotions are linked to specific facial movements—pleasure, anger, surprise, and rage are examples. Genes code for proteins, not behavior—genes give us a capacity to behave. For example, baby egrets sometimes batter younger siblings to death. If the baby egret could talk, it wouldn't say "my genes made me do it." Rather, the baby egret has a genetically based capacity to promote its individual survival and reproduction. "Adaptive" does not mean the same thing as "moral." It means more frequent transmission of an individual's genes. Research into the relationships among evolution, genes, and behavioral capacity is in its infancy. Knowledge of human behavior may help us understand how to alleviate negative aspects of the human condition.

Key Terms

altruism	kin selection	social insects
kin	self-sacrificing behavior	

Objectives

1. Cite examples where altruistic behavior may be favored by natural selection.
2. Describe examples of human social behavior.
3. State the value of research that reveals the relationships among evolution, genes, and human behavior.

Self-Quiz Questions

Fill-in-the-Blanks

With (1) _____ behavior, the "helper" reduces its own reproductive potential while the "helped" has increased its reproductive success.

When this type of behavior is directed toward kin, it may be favored by

(2) _____ _____ . In (3) _____ _____ , a nonreproducing, subordinate member of a group may help pass on copies of its genes, as long as the members benefiting the altruistic behavior are relatives. Helper jays perpetuate some shared (4) _____ when they feed and protect younger brothers and sisters that will one day reproduce. In bee societies, sterile guards may protect the queen by stinging an intruder and committing (5) _____ . Sterile worker honeybees care for the (6) _____ and accomplish necessary tasks for the well-being of the hive. Suicidal behavior and sterility among these insects are (7) _____ traits, for close relatives are the beneficiaries. These insects sacrifice their (8) _____ chances to increase sibling numbers and siblings have many genes in common with them. Humans have had an extraordinary (9) _____ evolution. Beneath the elaborate, diverse layerings of culture, there is still a(n) (10) _____ core for human behavior. Experiments strongly support the view that human (11) _____ development has a(n) (12) _____ basis and is subject to individual selection. Many human (13) _____ are linked to specific facial movements. Genes code for (14) _____ , not behavior—genes give us the capacity to (15) _____ . One cannot say that genes make baby egrets batter their siblings to death; rather, the baby egret has a genetically based capacity to promote its individual (16) _____ and (17) _____ . "Adaptive" does not mean the same thing as (18) "_____ ." It means more frequent (19) _____ of an individual's genes.

CHAPTER TEST **UNDERSTANDING AND INTERPRETING KEY CONCEPTS**

___ (1) In experiments, male stickleback fish try to strike any red object; this illustrates _____ .

  (a) learning
  (b) organization
  (c) instinct
  (d) a cue for aggressive behavior
  (e) both c and d

___ (2) _____ provides the animal with its basic response mechanisms; _____ allows the animal to modify behavior while interacting with the environment.

  (a) Behavior; instinct
  (b) Heredity; learning
  (c) Instinct, heredity
  (d) Learning; heredity

___ (3) Newly hatched goslings follow any large moving objects to which they are exposed shortly after hatching; this is an example of _____ .

   (a) homing behavior
   (b) imprinting
   (c) piloting
   (d) migration

___ (4) A chimpanzee piling scattered boxes to remove a dangling banana without benefit of trial-and-error learning illustrates _____ .

   (a) imprinting
   (b) insight learning
   (c) associative learning
   (d) instinct

___ (5) When an individual increases its chances of producing offspring regardless of the consequences to the group to which it belongs, it illustrates _____ .

   (a) natural selection
   (b) altruistic behavior
   (c) sexual selection
   (d) selfish behavior

___ (6) Male red-winged blackbirds, lions, elk, and bighorn sheep compete fiercely in response to female clustering. This is an example of _____ behavior.

   (a) resource-defense
   (b) selfish
   (c) female-defense
   (d) altruistic

___ (7) Male white-throated sparrows compete for insect-rich patches of forest that will attract females. Control over this resource-based territory is advertised by a male song that attracts females and discourages rival males. This is an example of _____ behavior.

   (a) resource-defense
   (b) selfish
   (c) female-defense
   (d) altruistic

___ (8) Female insects often attract mates by releasing sex pheromones. This is an example of a(n) _____ signal.

   (a) chemical
   (b) visual
   (c) acoustical
   (d) tactile

___ (9) Worker termites bang their heads specifically to attract soldiers; male birds sing to stake out territories, attract females, and discourage males. These are examples of _____ signals.

   (a) chemical
   (b) visual
   (c) acoustical
   (d) tactile

___ (10) The "selfish herd" is illustrated by _____ .

    (a) a lion pride bringing down a prey animal that perhaps a single lion would not dare attack

    (b) wildebeests fleeing from hunting dogs in a mass; each tries to maneuver inside the mass of wildebeests

    (c) musk oxen

    (d) nasute termites

___ (11) The example used to demonstrate that individual selection explained some behavioral traits better than "good-of-the-group" selection was _____ .

    (a) the dilution effect in wildebeest populations

    (b) baby egrets battering younger siblings

    (c) courtship behavior in albatrosses

    (d) the dispersal of Norwegian lemmings when population densities become extremely high

___ (12) A breeding pair of jays with as many as six related nest helpers that may not reproduce illustrates _____ .

    (a) kin selection

    (b) predation

    (c) group defense

    (d) altruistic behavior

    (e) both a and d

___ (13) The behavior of a pair of Caspian terns devoting full time in rearing young to ensure at least some offspring survive to reproductive age shows _____ .

    (a) selfish behavior

    (b) female-defense behavior

    (c) self-sacrifice behavior

    (d) learning behavior

    (e) behavior with no sacrifice

___ (14) In highly integrated insect societies, _____ .

    (a) natural selection favors individual behaviors that lead to greater diversity among members of the society

    (b) there is scarcely any division of labor

    (c) cooperative behavior predominates

    (d) patterns of behavior are flexible, and learned behavior predominates

    (e) all of the above

___ (15) Social behavior among insects depends on _____ .

    (a) diversity

    (b) altruism

    (c) acoustical signals

    (d) genetic similarity

    (e) communication

INTEGRATING AND APPLYING KEY CONCEPTS

Explain whether you think humans have any critical periods for establishing the ability to learn certain kinds of knowledge. State whether you think humans undergo imprinting. Do you think humans employ resource-defense behavior? If so, can you cite an example? Female-defense behavior? Example?

If you were Ruler of All People on Earth, what mechanisms would you employ to distribute essential resources to minimize deadly competition and wars? Would a person who believed in biological determinism be likely to advocate the equal distribution of essential resources among all members of a social group?

Crossword
Number Eight

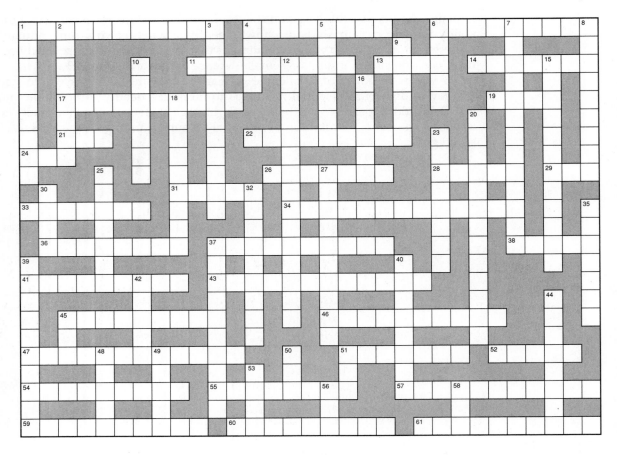

Across

1. transition model that describes how human societies are affected by reproductive rates and patterns of childbearing
4. organism that preys upon another organism
6. narrow zone on Earth that harbors life
11. unable to function without the aid or use of another
13. El _____ : an irregular, but episodic warming of surface waters in the eastern equatorial Pacific
14. level containing organisms that are the same number of energy transfers away from sunlight that enters the ecosystem
17. the distribution pattern within boundaries of a population
19. nutritive material
21. a food _____ shows feeding relationships among many organisms
22. pertaining to a creature that lives in or on a host organism
24. a silvery metal used to coat other metals; forms part of many alloys; Sn
26. situation in which one species bears deceptive resemblance in color, form, and/or behavior to another species that enjoys some survival advantage
28. Jewish holiday celebrating the deliverance of the Jews from massacre by Haman
29. female chicken
31. socketed, hard structure used in chewing or tearing food
33. place where an organism typically dwells

34. occurring between two different species
36. nitrogen _____ : gaseous nitrogen from the air forms ammonia or ammonium ions by means of reduction reactions
37. permanently frozen subsoil of polar regions
38. to gain knowledge or mastery of by experience or study
41. a community and its physical and chemical environment
43. resource _____ allows similar species to coexist in the same habitat
45. biotic _____ : maximum rate of increase in a population living under ideal conditions
46. a change that enables an organism to live under changed environmental conditions
47. primary _____ : rate at which a given amount of energy is captured and stored during a given interval
51. one of the first organisms to become established in a particular habitat
52. a food _____ shows the general sequence of who eats whom in an ecosystem
54. a mutually beneficial relationship involving continuous, intimate contact between interacting populations
55. the fall _____ thoroughly mixes water temperatures, densities, and dissolved oxygen in a lake
57. the middle layer of a lake in which temperature drops rapidly with depth
59. a reduction in rainfall on the leeward side of high mountains, where only plants adapted to arid or semiarid conditions grow
60. the populations of all species interacting in a given habitat
61. the _____ cycle involves the properties, distribution, and effects of water on Earth's surface, in the soil and underlying rocks, and in the atmosphere

Down

1. variety of different organisms
2. overheated reactor core could percolate through its concrete containment slab and contaminate groundwater
3. the process of two different interdependent species changing body structure and/or behavior in a complementary fashion over time
4. that which predators pursue
5. plants that live for only one year, produce seeds, and die
6. a vast expanse of a particular type of vegetation shaped by climate, topography, and soil composition
7. _____ smog; air pollution caused by sunlight interacting with a variety of different airborne chemicals
8. competitive _____ : two species with identical niche requirements cannot coexist
9. _____ potential: maximum rate of increase in a population living under ideal conditions
10. province that includes the open sea over the ocean basins
12. the conversion of land with its characteristic trees, shrubs, and grasses into a desert
15. _____ vent ecosystems generally include chemosynthetic bacteria, tube worms, clams, sea anemones, crabs, and fishes
16. a limiting _____ controls growth rate of a population
18. competition for mates and discrimination among potential mates = sexual _____
20. process by which soil bacteria convert ammonia (or ammonium ions) to nitrite and other soil bacteria convert nitrite to nitrate

23. carrying _____ : maximum number or biomass of a specific population that a particular environment can maintain
25. where freshwater mixes with saltwater
27. zone between highest high tide and lowest low tide marks
30. a forsaken or orphaned child or young animal
32. a dominance _____ establishes a pecking order that determines which sequence of individuals has access to specific resources
35. number of individuals per unit of area
37. all members of one species that live in a given area at a given time
39. generally, a heterotrophic bacterium or fungus that obtains organic nutrients by breaking down remains or products of other organisms
40. innate aspect of behavior that is unlearned, complex, and normally adaptive
42. belief, idea, value, or dogma held by a person or organization
44. factor that helps keep population growth controlled
45. a professional, especially in sports
48. basic structural or functional constituents of a whole
49. Homer's story of the Trojan War
50. what remains after all necessary deductions have been made or all losses accounted for
51. Peter _____ ; also, *peroxyacyl nitrate*
53. _____ population growth: no increase in the rate of population growth
56. to move on foot fast enough that both feet leave the ground during each stride
58. a unit of energy absorbed from ionizing radiation

ANSWERS

CHAPTER 1 Introduction

1-I
(pp. 2–3)

1.	F	11.	F	20.	Photosynthesis
2.	T	12.	C	21.	respiration
3.	T	13.	E	22.	Metabolism
4.	F	14.	G	23.	egg
5.	T	15.	D	24.	larva
6.	T	16.	B	25.	pupa
7.	F	17.	H	26.	adult
8.	F	18.	A	27.	Homeostasis
9.	T	19.	I	28.	Mutations
10.	T				

1-II
(pp. 4–5)

1.	Darwin	7.	D	13.	G
2.	Animalia	8.	F	14.	F
3.	Protista	9.	A	15.	T
4.	Fungi	10.	E	16.	F
5.	Monera	11.	B	17.	F
6.	Plantae	12.	C	18.	F

1-III
(pp. 6–7)

1.	F	6.	D	11.	O
2.	T	7.	F	12.	O
3.	T	8.	E	13.	C
4.	F	9.	A	14.	O
5.	B	10.	C	15.	C

Understanding and Interpreting Key Concepts

(pp. 7–8)

1.	d	5.	d	9.	e
2.	a	6.	b	10.	b
3.	b	7.	d	11.	e
4.	c	8.	d		

CHAPTER 2 Chemical Foundations for Cells

2-I
(p. 15)

1.	T	7.	B	13.	D
2.	T	8.	M	14.	E
3.	T	9.	J	15.	A
4.	F	10.	K	16.	G
5.	F	11.	F	17.	H
6.	C	12.	L		

2-II
(p. 18)

1.	F	5.	T	9.	F
2.	T	6.	F	10.	F
3.	F	7.	T	11.	F
4.	T	8.	T	12.	T

2-III
(pp. 19–20)

1.	F	8.	pH scale	15.	E
2.	T	9.	hydrogen ion	16.	B
3.	F	10.	7	17.	C
4.	F	11.	4	18.	A
5.	F	12.	Buffers	19.	G
6.	hydrogen	13.	salt	20.	D
7.	hydroxide	14.	F		

2-IV (p. 21)	1. F	2. F	3. F	4. T	5. F	6. F

2-V (p. 22)			
1. F	7. T	12. A, I
2. F	8. A, H	13. C, E
3. F	9. C, D	14. B, G
4. T	10. C, E	15. A, H
5. T	11. A, I	16. B, F
6. T		

2-VI (pp. 23–24)			
1. F	5. T	9. A
2. T	6. T	10. B
3. T	7. E	11. C
4. T	8. D	

2-VII (p. 25)	1. F	2. T	3. F	4. F	5. T	6. F

2-VIII (p. 26)			
1. T	5. T	8. B, D
2. F	6. T	9. C, E
3. T	7. T	10. A, F
4. F		

Understanding and Interpreting Key Concepts

(pp. 26–28)			
1. a	6. e	11. b
2. c	7. d	12. c
3. b	8. b	13. d
4. c	9. d	14. a
5. a	10. c	

CHAPTER 3   Cell Structure and Function

3-I (p. 32)			
1. T	6. D	11. Resolution
2. F	7. C	12. nanometer
3. T	8. B	13. transmission
4. E	9. micrograph	14. scanning
5. A	10. light	15. nucleus

3-II (p. 34)			
1. fluid mosaic	8. channels	15. low
2. lipid bilayer	9. water-soluble	16. differentially permeable
3. proteins	10. metabolism	17. hypertonic
4. mosaic	11. gradient	18. F
5. Phospholipids	12. concentration	19. T
6. phosphate	13. movement	20. T
7. Surface proteins	14. high	21. T

3-III (pp. 35–36)			
1. T	5. F	8. carbon dioxide
2. F	6. Water (Oxygen)	9. ions
3. T	7. oxygen (water)	10. active transport
4. T		

3-IV (p. 37)			
1. D	5. C	8. prokaryotes
2. D	6. A	9. Animal
3. B	7. Eukaryotes	10. cell wall
4. E		

3-V (pp. 38–39)			
1. F	6. T	11. E
2. T	7. F	12. B
3. F	8. nucleolus	13. A
4. T	9. chromatin	14. C
5. T	10. D	

3-VI	1. F	10. F	17. chloroplast
(pp. 41–42)	2. T	11. G	18. ATP
	3. F	12. I	19. ATP
	4. F	13. H	20. microtubules
	5. F	14. J	21. Microfilaments
	6. B	15. D	22. cytoskeleton
	7. A	16. infolded plasma	23. tubulins
	8. E	membrane	
	9. H		

3-VII	1. T	9. nucleolus	17. Golgi complex
(pp. 42–43)	2. T	10. mitochondrion	18. nucleus
	3. T	11. chloroplast	19. nuclear envelope
	4. F	12. central vacuole	20. centrioles
	5. ribosomes	13. cell (plasma)	21. mitochondrion
	6. Golgi complex	membrane	22. lysosome (or vacuole)
	7. rough endoplasmic	14. cell wall	23. cytoplasm
	reticulum	15. flagellum	24. microfilaments
	8. nuclear pore	16. ribosomes	

Understanding and Interpreting Key Concepts

(pp. 43–45)	1. d	5. c	8. d
	2. c	6. a	9. b
	3. d	7. a	10. c
	4. d		

CHAPTER 4   Ground Rules of Metabolism

4-I	1. T	2. F	3. T	4. T	5. F
(p. 47)					

4-II	1. D	2. F	3. C	4. A	5. B	6. E
(p. 48)						

4-III	1. Enzymes	7. activation energy	13. feedback inhibition
(pp. 50–51)	2. catalysts	8. pH (temperature)	14. T
	3. dynamic equilibrium	9. temperature (pH)	15. F
	4. substrate	10. enzymes	16. T
	5. active site	11. weak bonds	17. F
	6. induced-fit model	12. inhibitors	18. F

4-IV	1. cofactors	8. ATP	16. electron
(pp. 52–53)	2. NAD$^+$ (or NADP$^+$)	9. ATP	17. electron carrier
	3. directly	10. phosphorylation	18. energy
	4. adenosine	11. metabolic	19. ions
	triphosphate	12. electron transport	20. F
	5. adenine (ribose)	13. enzymes (cofactors)	21. T
	6. ribose (adenine)	14. cofactors (enzymes)	22. T
	7. phosphate	15. oxidation-reduction	

Understanding and Interpreting Key Concepts

(pp. 53–54)	1. d	4. a	6. d
	2. c	5. c	7. a
	3. d		

CHAPTER 5   Energy-Acquiring Pathways

5-I	1. Photosynthetic	6. glycolysis	10. Carbon dioxide
(p. 56)	autotrophs	7. aerobic respiration	11. water
	2. Chemosynthetic	8. light-dependent	12. glucose
	autotrophs	(light-independent)	13. thylakoids
	3. bacteria	9. light-independent	14. grana
	4. bacteria	(light-dependent)	15. stroma
	5. Heterotrophic		

5-II (pp. 57–58)	1. thylakoid membranes 2. grana 3. stroma 4. photon 5. pigments	6. chlorophylls 7. red (blue) 8. blue(red) 9. Carotenoids	10. photosystem 11. light 12. electron 13. acceptor

5-III (pp. 59–60)	1. electron transport 2. photophosphorylation 3. cyclic (noncyclic) 4. noncyclic (cyclic) 5. P700 6. cyclic	7. noncyclic 8. P680 9. P700 10. P700 11. transport system	12. ion 13. $NADP^+$ 14. Photolysis 15. P680 16. Oxygen

5-IV
(p. 61)
	1. ATP (NADPH)
	2. NADPH (ATP)
	3. carbon dioxide
	  (ribulose biphosphate)
	4. ribulose biphosphate
	  (carbon dioxide)
	5. PGA
	6. fixation
	7. PGA
		8. PGAL
		9. six
		10. RuBP
		11. carbon
		12. PGAL
		13. sugar phosphate
		14. fixation
		15. light-dependent
			16. NADPH (ATP)
			17. ATP (NADPH)
			18. Sugar phosphate
			19. photorespiration
			20. C4
			21. food production
			22. oxygen
			23. oxaloacetate

Understanding and Interpreting Key Concepts

(pp. 62–63)	1. a 2. b 3. c	4. a 5. c 6. d	7. a 8. b 9. c

CHAPTER 6   Energy-Releasing Pathways

6-I (p. 65)	1. ATP 2. phosphate group	3. photosynthesis 4. aerobic respiration	5. fermentation

6-II (p. 66)	1. Aerobic respiration 2. thirty-six 3. Glycolysis	4. two 5. oxygen 6. water	7. glycolysis 8. electron transport 9. oxygen

6-III
(p. 68)
	1. aerobic
	2. Krebs
	3. electron transport
	  phosphorylation
	4. 34
	5. ATP
	6. aerobic respiration
	7. carbon dioxide
	8. electrons
		9. NAD^+ (FAD)
		10. FAD (NAD^+)
		11. transport
		12. ATP
		13. oxygen
		14. mitochondria
		15. inner compartment
		16. inner membrane
		17. outer compartment
			18. outer membrane
			19. cytoplasm
			20. ATP
			21. oxygen, O_2
			22. $FADH_2$
			23. NADH
			24. electron transport
			  system

6-IV (pp. 69–70)	1. Autotrophic 2. Glucose 3. ATP 4. ATP	5. NADH 6. oxygen, O_2 7. fermentation (anaerobic)	8. lactate 9. ethanol 10. carbon dioxide

6-V (pp. 70–71)	1. F 2. T 3. T 4. T	5. fatty acids 6. glycerol 7. glycolysis 8. amino acids	9. Krebs cycle 10. acetyl-coA 11. pyruvate

6-VI (p. 72)	1. F 2. F 3. T 4. T

Understanding and Interpreting Key Concepts

(pp. 72–73)	1. c 2. c 3. d	4. b 5. d 6. d	7. a 8. d

CHAPTER 7 Cell Division and Mitosis

7-I
(pp. 81–82)

1. reproduction
2. DNA
3. Mitosis (Meiosis)
4. meiosis (mitosis)
5. Cytokinesis
6. body
7. germ
8. gametes

9. zygote
10. chromosome
11. sister chromatids
12. centromere
13. homologous
14. body
15. mitosis
16. germ

17. meiosis
18. haploid
19. F
20. F
21. T
22. T
23. F
24. F

7-II
(pp. 82–83)

1. interphase
2. mitosis
3. G_1
4. S
5. G_2

6. prophase
7. metaphase
8. anaphase
9. telophase

A. 5
B. 2
C. 4
D. 3
E. 1

7-III
(p. 84)

1. F
2. F
3. F
4. F
5. T

6. I
7. F
8. H
9. B
10. A

11. G
12. J
13. C
14. D
15. E

Understanding and Interpreting Key Concepts

(pp. 85–86)

1. a
2. d
3. d

4. a
5. d
6. c

7. e
8. e

CHAPTER 8 Meiosis

8-I
(pp. 89–90)

1. B
2. E
3. C
4. H

5. D
6. G
7. A
8. F

9. F
10. T
11. F
12. F

8-II
(p. 91)

1. 2
2. 5
3. 4
4. 1

5. 3
6. F
7. T

8. F
9. T
10. F

Understanding and Interpreting Key Concepts

(pp. 91–92)

1. a
2. a
3. d

4. b
5. c

6. b
7. d

CHAPTER 9 Observable Patterns of Inheritance

9-I
(pp. 96–97)

1. traits
2. strains (varieties)
3. pea
4. cross-fertilization

5. true-breeding (they are homozygous)
6. hybrids
7. homozygous recessive

8. blending
9. monohybrid
10. segregation

9-II
(pp. 98–99)

1. F   2. T   3. T
4. albino = pp
   normal pigmentation = pp or Pp
   Woman of normal pigmentation with an albino mother → Pp; received her recessive gene from her mother and her dominant gene (P) from her father. It is likely that half of the couple's children will be albinos (pp) and half will have normal pigmentation but be heterozygous (Pp).

5. d	6. a	7. d	8. c		

9-III
(pp. 100–101)

1. F	2. F	3. F	4. F	5. T	6. F

9-IV
(p. 101)

1. T	2. F

Understanding and Interpreting Key Concepts

(pp. 102–103)

1. d	5. c	8. a
2. b	6. b	9. c
3. a	7. a	10. b
4. c		

Integrating and Applying Key Concepts

(p. 103)

1. 100% of offspring will be spotted because there are no dominant genes included.

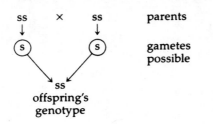

s = spotted
S = solid color

parents

gametes possible

ss
offspring's genotype

2. black = B   red = b
    solid color = S   white spots = s

    Solid Red ♀   Black and White ♂

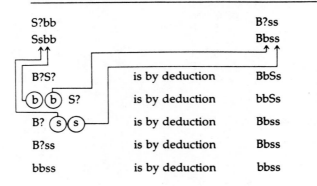

S?bb		B?ss
Ssbb		Bbss
B?S?	is by deduction	BbSs
b b S?	is by deduction	bbSs
B? s s	is by deduction	Bbss
B?ss	is by deduction	Bbss
bbss	is by deduction	bbss

CHAPTER 10   Chromosome Variations and Human Genetics

10-I
(pp. 106–107)

1. Flemming	7. C	13. a
2. Weismann	8. G	14. d
3. meiosis	9. A	15. a
4. Homologous	10. B	16. b
5. E	11. F	17. b
6. D.	12. c	18. c

10-II
(p. 109)

1. T	5. F	8. dominant
2. F	6. Humans	9. autosomal recessive
3. F	7. Huntington's disorder	10. autosomal dominant
4. F		

| 10-III
(pp. 111-112) | 1. C
2. A
3. B
4. D
5. E
6. E
7. F
8. D
9. E
10. B | 11. E
12. E
13. C
14. A
15. A
16. E
17. B
18. D
19. G
20. deletion | 21. translocation
22. inversion
23. duplication
24. Nondisjunction
25. trisomy
26. monosomy
27. crossing over
28. sex chromosome
    abnormalities
29. XYY |

Understanding and Interpreting Key Concepts

| (pp. 112–113) | 1. d
2. d
3. c
4. c | 5. b
6. a
7. a
8. d | 9. b
10. b
11. c |

Integrating and Applying Key Concepts

(p. 114) Answer to genetics problem:
Yes; the husband could not have supplied either of his daughter's recessive genes because his only X chromosome bears the N for normal iris.

$X^N Y$
father

$X^N X^n$
mother

N = normal iris
n = fissured iris

$X^n X^n$
daughter's
genotype

The mother must also carry the recessive gene in order to be her daughter's mother.

CHAPTER 11   DNA Structure and Function

| 11-I
(p. 117) | 1. T
2. F
3. T
4. F | 5. F
6. F
7. F
8. T | 9. T
10. F
11. T
12. T |

| 11-II
(pp. 118–119) | 1. F
2. F
3. T
4. T
5. deoxyribose
6. phosphate group
7. purine (guanine because it has 3 H-bonds) | 8. pyrimidine (thymine because it has 2 H-bonds)
9. purine (adenine because it has 2 H-bonds) | 10. pyrimidine (cytosine because it has 3 H-bonds)
11. nucleotide (thymidine monophosphate) |

| 11-III
(p. 120) | 1. DNA
2. histones | 3. histone
4. DNA | 5. nucleosome
6. genes |

Understanding and Interpreting Key Concepts

| (pp. 120–121) | 1. d
2. d
3. a | 4. d
5. b
6. d | 7. c
8. d |

CHAPTER 12   From DNA to Proteins

| 12-I
(pp. 124–125) | 1. RNA
2. ribose
3. uracil
4. adenine
5. transcription | 6. DNA
7. RNA polymerase
8. one of the strands
9. complementary
10. mRNA | 11. cytoplasm
12. introns
13. exons
14. nucleotide cap
15. nucleotide tail |

12-II (pp. 126–127)	1. amino acids 2. three 3. one 4. mRNA 5. codon	6. mRNA 7. translation (assembly or synthesis) 8. Transfer	9. amino acid 10. protein (polypeptide) 11. codon 12. anticodon

Problems   13. AUG-UUC-UAU-UGU-AAU-AAA-GGA-UGG-CAG-UAG
14. met-phe-tyr-cys-asn-lys-gly-try-gln-stop (start)

12-III (pp. 127–128)	1. High-energy radiation 2. mutagenic chemicals (substances)	3. Insertions 4. deletions 5. Sickle cell anemia	6. amino acid 7. hemoglobin

12-IV (p. 129)	1. *Escherichia coli* 2. operon 3. transcription controls 4. regulator gene 5. promoter	6. negative control 7. low 8. RNA polymerase (mRNA transcription)	9. blocks 10. repressor protein 11. operator 12. needed (required)

12-V (p. 131)	1. DNA (genes) 2. differentiation 3. Cell differentiation 4. selective gene 5. mosaics	6. anhidrotic ectodermal dysplasia 7. selective gene 8. growth (division) 9. division (growth)	10. F 11. T 12. T 13. F 14. F 15. T

Understanding and Interpreting Key Concepts

(pp. 131–134)	1. c 2. b 3. c 4. a 5. c 6. a 7. a	8. d 9. d 10. b 11. d 12. d 13. b 14. d	15. b 16. b 17. c 18. c 19. a 20. b

CHAPTER 13   Recombinant DNA and Genetic Engineering

13-I (p. 137)	1. F 2. T 3. F 4. F 5. T	6. B 7. D 8. F 9. H 10. I	11. C 12. J 13. E 14. G 15. A

13-II (p. 138)	1. Restriction fragment length 2. Polymorphism	3. genetic fingerprint 4. splicing enzymes	5. introns 6. sequencing

13-III (pp. 139–140)	1. genetic engineering 2. frost damage	3. ice-minus 4. body cells	5. gene therapy 6. eugenic engineering

Understanding and Interpreting Key Concepts

(pp. 140–141)	1. a 2. c 3. b 4. c	5. d 6. a 7. c	8. b 9. a 10. d

CHAPTER 14   Microevolution

14-I (p. 147)	1. F 2. T 3. F	4. F 5. T	6. F 7. T

14-II
(p. 149)

1. population
2. Hardy-Weinberg
3. allele frequencies
4. genetic equilibrium
5. mutations
6. genetic drift
7. Gene flow
8. Natural selection
9. Natural selection

10. Find (b) first, then (c) and, finally, (a).
  (a) $2pq = 2 \times (0.9) \times (0.1) = 2 \times (0.09) = 0.18$
  $= 18\%$, which is the percentage of heterozygotes
  (b) $p^2 = 0.81$
  $p = \sqrt{0.81} = 0.9 =$ the frequency of the dominant allele
  (c) $p + q = 1$
  $q = 1 - 0.9 = 0.1 =$ the frequency of the recessive allele

11. (a) homozygous dominant $= p^2 \times 200 = (0.8)^2 \times 200 = 0.64 \times 200$
  $= 128$ individuals
  (b) homozygous recessive $= q^2 \times 200 = (0.2)^2 \times 200 = (0.04) \times (200)$
  $= 8$ individuals
  (c) heterozygotes $= 2pq \times 200 = 2 \times 0.8 \times 0.2 \times 200 = 0.32 \times 200$
  $= 64$ individuals
  Check: $128 + 8 + 64 = 200$

14-III
(pp. 150–151)

1. natural selection
2. Stabilizing selection
3. Directional selection
4. Disruptive selection
5. Nautiloids
6. peppered moths
7. directional
8. directional
9. disruptive
10. malarial
11. stabilizing
12. polymorphism
13. sexual

14-IV
(pp. 152–153)

1. Speciation
2. species
3. Divergence
4. reproductive isolating
5. geographic barriers
6. polyploids
7. polyploidy
8. hybridization

Understanding and Interpreting Key Concepts

(pp. 153–154)

1. c
2. d
3. a
4. b
5. c
6. b
7. d
8. c
9. a
10. b
11. c
12. b
13. a

CHAPTER 15   Macroevolution

15-I
(p. 157)

1. macroevolution
2. fossils
3. Archean
4. Proterozoic
5. Paleozoic
6. Mesozoic
7. Cenozoic
8. changes
9. fossil
10. comparative morphology
11. regulatory genes
12. homologous
13. morphological divergence
14. analogous
15. morphological convergence
16. mutations
17. immunological comparisons
18. DNA hybridization

15-II
(p. 159)

1. 4.6
2. 3.8
3. Clay crystals
4. amino acids
5. water (steam)
6. amino acids
7. protein
8. spheres
9. cell membranes
10. Plate tectonics
11. mantle
12. Pangea
13. mass extinctions
14. adaptive radiations
15. adaptive radiation
16. adaptive zones
17. key innovation

15-III
(pp. 162–163)

1. prokaryotic cells (anaerobic bacteria)
2. oxygen, O_2
3. fermentation
4. 2.5
5. photosynthetic bacteria
6. eukaryotes
7. stromatolites
8. Proterozoic
9. organic compounds (chemicals)
10. adaptive zones
11. aerobic
12. Cambrian
13. trilobites
14. Ordovician
15. armor-plated
16. Lobe-finned
17. mass extinction
18. Carboniferous
19. Permian
20. Triassic
21. mass extinction
22. Dinosaurs (reptiles)
23. asteroid
24. Flowering plants
25. Cenozoic
26. mammals (birds, flowering plants)
27. birds (mammals, flowering plants)
28. flowering plants (birds, mammals)

15-IV   1. phylogeny   4. punctuation   7. inclusive
(p. 164)   2. lineage   5. generic (genus)   8. kingdoms
                     3. gradualism   6. specific (species)

Understanding and Interpreting Key Concepts

(pp. 164–166)   1. c   8. b   15. c
                     2. b   9. c   16. a
                     3. e   10. d   17. d
                     4. d   11. e   18. b
                     5. a   12. a   19. e
                     6. d   13. b   20. c
                     7. c   14. d

CHAPTER 16   Human Evolution: A Case Study

16-I   1. dentition   8. Hands   15. trees
(pp. 168–169)   2. dependency   9. daytime vision   16. monkeys (apes)
                     3. learning   10. diet   17. apes (monkeys)
                     4. primates   11. brain   18. hominoids
                     5. hominoids   12. behavioral   19. dryopiths
                     6. hominid   13. mammals   20. humans
                     7. Bipedalism   14. rodents

16-II   1. hominids   6. home   11. *Homo sapiens*
(pp. 170–171)   2. bipedalism   7. climate   12. *Homo erectus*
                     3. plasticity   8. *Homo erectus*   13. humans
                     4. australopiths   9. tools   14. Neandertals
                     5. rock-flaking   10. fire   15. cultural

Understanding and Interpreting Key Concepts

(pp. 171–172)   1. c   6. b   11. F
                     2. a   7. b   12. A
                     3. c   8. d   13. B
                     4. d   9. C   14. D
                     5. a   10. E

CHAPTER 17   Viruses, Monerans, and Protistans

17-I   1. DNA (RNA) or RNA 3. Viruses   6. animal viruses
(p. 177)   (DNA)   4. living cell (host cell) 7. plant viruses
                     2. protein coat   5. bacteriophages   8. viroids

17-II   1. chemosynthetic   5. Archaebacteria   9. nitrogen cycle
(p. 179)   2. heterotrophs   6. cyanobacteria   10. botulism
                     3. bacterial flagella 7. heterocysts   11. endospores
                     4. Gram-positive   8. nitrifying bacteria 12. human

17-III   1. Protista   5. cellular (plasmodial) 9. plankton
(p. 181)   2. bacterial   6. spores   10. dinoflagellates
                     3. prokaryotic   7. Euglenids   11. dinoflagellates
                        (independent)   8. chrysophytes   12. red tide
                     4. plasmodial (cellular)

17-IV   1. Trichomonad   7. *Plasmodium*   14. E
(pp. 182–183)   (*Trichomonas*) 8. sporozoan   15. G
                     2. *Giardia*   9. aquatic (watery)   16. D
                     3. cysts   10. contractile vacuoles 17. B
                     4. pseudopods   11. gullet   18. A
                        (pseudopodia)   12. vesicles (vacuoles) 19. F
                     5. Foraminiferans 13. H   20. C
                     6. amoebic dysentery

Understanding and Interpreting Key Concepts

(pp. 183–185)

1.	c	8.	b	15.	C
2.	a	9.	c	16.	A
3.	c	10.	d	17.	E
4.	b	11.	a	18.	D
5.	a	12.	I	19.	B
6.	c	13.	G	20.	F
7.	d	14.	H		

CHAPTER 18   Fungi and Plants
Part 1: Kingdom of Fungi

18-I
(p. 188)

1.	heterotrophs	7.	mycelium	12.	sac
2.	saprobes (saprobic)	8.	sexual	13.	club
3.	parasites (parasitic)	9.	asexual	14.	imperfect fungi
4.	mycelium	10.	Chytrids	15.	Lichens
5.	hyphae	11.	zygospore-forming	16.	Mycorrhizae
6.	spore				

Part 2: Kingdom of Plants

18-II
(pp. 189–190)

1.	Vascular	7.	two	12.	brown
2.	Xylem	8.	nonmotile	13.	algin
3.	phloem	9.	seeds	14.	photosynthetic
4.	sporophyte	10.	sporophyte	15.	cellulose
5.	gametophyte	11.	red	16.	starch
6.	sporophyte				

18-III
(pp. 191–192)

1.	cuticle	9.	gametophyte	14.	sporophyte
2.	reproductive	10.	lycophytes (horsetails) (ferns)	15.	sporangium
3.	sporangium	11.	horsetails (ferns) (lycophytes)	16.	meiosis
4.	archegonium	12.	ferns (lycophytes) (horsetails)	17.	spores
5.	antheridium	13.	zygote	18.	rhizoids
6.	inside			19.	gametophyte
7.	Mosses			20.	archegonium
8.	sporophyte			21.	antheridium

18-IV
(pp. 193–194)

1.	gymnosperms (angiosperms)	12.	megaspore	25.	monocots (dicots)
2.	angiosperms (gymnosperms)	13.	gametophyte	26.	dicots (monocots)
3.	Conifers	14.	male	27.	Angiosperms
4.	cone	15.	female	28.	flowers
5.	sporangia	16.	pollination	29.	pollinators
6.	two	17.	sperms	30.	sporophyte (embryo)
7.	two	18.	gametophytes	31.	female gametophyte
8.	microspores	19.	zygote	32.	male cones
9.	pollen	20.	embryo	33.	female cones
10.	male	21.	embryo	34.	meiosis
11.	spores	22.	gametophyte	35.	microspores
		23.	seed	36.	megaspores
		24.	angiosperms	37.	pollen tube

Understanding and Interpreting Key Concepts

(pp. 194–198)

1.	d	8.	c	14.	a
2.	b	9.	a	15.	d
3.	b	10.	a	16.	c
4.	e	11.	b	17.	c
5.	b	12.	d	18.	c
6.	c	13.	c	19.	c
7.	c				

Answers   *455*

CHAPTER 19 Animals

19-I
(pp. 201–202)

1. invertebrates
2. multicellular
3. heterotrophs
4. sexually
5. asexually
6. embryonic
7. motile
8. Radial
9. Bilateral
10. gut
11. sac-
12. tube-
13. coelom
14. peritoneum (mesoderm)
15. coelom
16. peritoneum (mesoderm)
17. segmented
18. Sponges
19. collar cells
20. larval
21. fragmentation
22. radial
23. nematocysts
24. planula
25. polyp (medusa)
26. medusa (polyp)
27. feeding polyp
28. reproductive polyp
29. female medusa
30. planula larva

19-II
(pp. 203–204)

1. flatworms
2. regenerating
3. leeches
4. definitive
5. blood fluke
6. scolex
7. proglottids
8. Nematodes (roundworms)
9. nematodes (roundworms)
10. false coelom
11. reproductive
12. Rotifers

19-III
(p. 205)

1. Deuterostomes
2. anus
3. mouth
4. protostomes
5. Mollusks
6. mollusks
7. gastropods
8. radula
9. gastropods
10. Bivalves
11. suspension feeding
12. respiration
13. jet propulsion
14. cephalopods

19-IV
(pp. 206–207)

1. Segmentation
2. Annelids
3. polychaetes
4. anus
5. coelom
6. blood
7. nerve cells
8. nephridia
9. swallow
10. blood
11. Polychaetes
12. oligochaetes
13. scavengers
14. setae
15. brain
16. pharynx
17. nerve cord
18. hearts
19. blood vessel
20. crop
21. gizzard

19-V
(p. 209)

1. cuticle
2. exoskeleton
3. development
4. metamorphosis
5. specialized (diverse)
6. appendages
7. tracheas
8. vision
9. chelicerates
10. Crustaceans
11. Millipedes (Centipedes)
12. centipedes (millipedes)
13. Insects
14. Malpighian tubules

19-VI
(p. 212)

1. deuterostomes
2. echinoderms
3. radial
4. bilateral
5. water-vascular
6. tube feet
7. Tunicates (Lancelets)
8. lancelets (tunicates)
9. agnathans
10. lack
11. cartilaginous
12. bony
13. lobe-finned
14. pharynx
15. amphibians
16. fertilization
17. amniotic
18. lungs
19. reptiles
20. bones
21. mammals
22. placental

Understanding and Interpreting Key Concepts

(pp. 213–215)

1. a
2. c
3. d
4. b
5. c
6. b
7. d
8. d
9. c
10. b
11. d
12. e
13. a
14. d
15. b
16. c
17. d
18. c
19. d
20. e

CHAPTER 20 Plant Tissues

20-I
(pp. 220–221)

1. shoot
2. root
3. Parenchyma
4. Collenchyma (Sclerenchyma)
5. sclerenchyma (collenchyma)
6. xylem
7. phloem
8. sieve-tube
9. Companion cells
10. epidermal
11. periderm

20-II (p. 222)	1. apical 2. Primary 3. Secondary 4. woody	5. lateral 6. monocots (dicots) 7. dicots (monocots) 8. cotyledon	9. monocot 10. cotyledons 11. dicot
20-III (pp. 223–224)	1. vascular 2. fibers 3. parenchyma 4. monocot 5. vascular bundles 6. dicots 7. node	8. terminal 9. lateral 10. leaves (flowers) 11. flowers (leaves) 12. deciduous 13. photosynthesis 14. veins	15. Stomata 16. gas 17. palisade mesophyll 18. spongy mesophyll 19. lower epidermis 20. vein 21. stoma(-ta)
20-IV (p. 225)	1. Taproot 2. fibrous 3. adventitious 4. cap 5. apical meristem 6. epidermis 7. cortex	8. vascular column 9. endodermis 10. pericycle 11. lateral roots 12. primary meristem (root) 13. primary meristem (root) 14. primary meristem (root)	15. root apical meristem 16. root cap 17. endodermis 18. pericycle 19. endodermis 20. pericycle
20-V (p. 227)	1. Herbaceous (nonwoody) 2. biennial 3. vascular 4. early wood	5. growing season 6. vascular cambium 7. bark	8. periderm 9. phloem 10. xylem

Understanding and Interpreting Key Concepts

(pp. 228–229)	1. b 2. a 3. b 4. c	5. a 6. d 7. d 8. a	9. c 10. c 11. b 12. b

CHAPTER 21   Plant Nutrition and Transport

21-I (p. 231)	1. hydrogen 2. mineral ions 3. macronutrients 4. micronutrients 5. Nitrogen-fixing 6. symbiotic 7. roots	8. nitrogen 9. B 10. A, I 11. B 12. B 13. B, H 14. A, D	15. B 16. B, F 17. A, G (D) 18. A, C 19. A, E 20. A 21. B
21-II (p. 233)	1. root hairs 2. mycorrhizae 3. symbiotic 4. Transpiration	5. cohesion 6. hydrogen bonds 7. stomata 8. guard cells	9. potassium 10. open 11. closes
21-III (p. 234)	1. accumulation 2. solutes 3. mineral ions 4. Starch 5. sucrose	6. translocation 7. aphids 8. pressure flow 9. source regions 10. sink regions	11. phloem tubes 12. water 13. fluid pressure 14. roots

Understanding and Interpreting Key Concepts

(pp. 235–236)	1. c 2. c 3. c 4. a 5. b	6. c 7. b 8. c 9. b 10. b	11. d 12. b 13. c 14. c

CHAPTER 22 Plant Reproduction and Development

22-I
(p. 239)

1. Sexual
2. Asexual
3. identical
4. sporophyte

5. flowers
6. meiosis
7. spores

8. sperms
9. eggs
10. asexually (vegetatively)

22-II
(p. 241)

1. C
2. F
3. A

4. E
5. B

6. G
7. D

22-III
(pp. 242–243)

1. anthers
2. stigmas
3. pollination
4. pollinators
5. pollen tube
6. Two
7. embryo sac
8. Double
9. zygote

10. triploid
11. endosperm
12. sporophyte
13. seedling (sporophyte)
14. embryo
15. seed
16. sperm
17. tube nucleus

18. anther (including pollen sacs)
19. megaspore mother cell
20. endosperm mother cell
21. egg
22. ovary wall
23. embryo sac
24. micropyle
25. ovule

22-IV
(p. 244)

1. seed
2. ovary
3. integuments
4. Cotyledons

5. Grains
6. fleshy
7. multiple

8. seed
9. dispersal
10. animals

22-V
(pp. 245–246)

1. germination
2. aerobic
3. germination
4. growth (development)
5. development (growth)
6. hormones

7. Auxins (Gibberellins)
8. gibberellins (auxins)
9. Abscisic acid
10. seed
11. Ethylene
12. apical dominance

13. Gibberellins
14. Florigen
15. Auxins
16. Abscisic acid
17. Ethylene
18. Cytokinins

22-VI
(p. 247)

1. B   2. D   3. C

22-VII
(p. 249)

1. photoperiodism
2. Phytochrome
3. germination
4. inactive
5. red
6. active

7. Pr is converted to Pfr
8. Daylength
9. Long
10. Short
11. Day
12. abscission

13. Senescence
14. dormancy
15. Short
16. long
17. gibberellins
18. abscisic acid

Understanding and Interpreting Key Concepts

(pp. 249–251)

1. c
2. c
3. d
4. b
5. b

6. c
7. c
8. d
9. c

10. a
11. d
12. d
13. d

CHAPTER 23 Animal Tissues, Organ Systems, and Homeostasis

23-I
(p. 256)

1. internal
2. nutrients
3. body
4. wastes
5. bacteria
6. nourish (feed)
7. tissues

8. organs
9. organ system
10. tissue
11. Somatic
(12–15 can be in any order)
12. epithelial
13. connective

14. muscle
15. nervous
16. Germ
17. Epithelial
18. exocrine
19. endocrine

23-II
(pp. 257–258)

1. ground substance
2. Dense
3. Loose

4. collagen (elastin)
5. elastin (collagen)
6. adipose

7. cartilage
8. bone
9. Blood

23-III
(pp. 258–259)

1. F   2. F   3. F   4. F   5. F   6. T

| 23-IV
(p. 260) | 1. cell (part)
2. metabolic
3. tissue
4. organism
5. stable | 6. cell
7. homeostasis
8. negative
9. positive
10. receptors | 11. integrator
12. effectors
13. F
14. F
15. F |

Understanding and Interpreting Key Concepts

| (pp. 261–262) | 1. c
2. d
3. b
4. e
5. b | 6. c
7. a
8. d
9. c | 10. a
11. c
12. c
13. d |

CHAPTER 24   Protection, Support, and Movement

| 24-I
(pp. 264–265) | 1. F
2. H
3. B | 4. G
5. A
6. E | 7. C
8. I
9. D |

| 24-II
(pp. 267–268) | 1. movement (protection, support)
2. protection (movement, support)
3. support (movement, protection)
4. mineral
5. blood cell
6. shapes
7. connective (hard)
8. collagen
9. ground
10. red | 11. blood cell
12. Yellow
13. red
14. embryo
15. Osteoblasts
16. osteocytes
17. turnover
18. osteoporosis
19. 206
20. axial (appendicular)
21. appendicular (axial)
22. intervertebral disks | 23. synovial
24. Cartilaginous
25. Fibrous
26. sternum
27. clavicle
28. scapula
29. radius
30. carpals
31. femur
32. tibia
33. tarsals
34. metatarsals |

| 24-III
(pp. 270–271) | 1. skeletal (cardiac, smooth)
2. cardiac (skeletal, smooth)
3. smooth (cardiac, skeletal)
4. elastic
5. myofibrils
6. actin (myosin)
7. myosin (actin) | 8. sarcomeres
9. sliding-filament
10. cross-bridges
11. ATP
12. actin
13. mitochondria
14. glycolysis
15. sarcoplasmic reticulum
16. actin
17. levers | 18. deltoid
19. pectoralis major
20. triceps
21. biceps
22. external oblique
23. rectus abdominis
24. sartorius
25. rectus femoris
26. tibialis anterior |

Understanding and Interpreting Key Concepts

| (pp. 271–273) | 1. e
2. c
3. b
4. a | 5. c
6. d
7. b
8. a | 9. a
10. b
11. c
12. a |

CHAPTER 25   Digestion and Human Nutrition

| 25-I
(pp. 275–276) | 1. digestive
2. circulatory
3. respiratory
4. urinary
5. movement
6. secretion | 7. digestion
8. absorption
9. peristalsis
10. Sphincters
11. D
12. B | 13. H
14. F
15. A
16. G
17. C
18. E |

| 25-II
(p. 277) | 1. mouth
2. 32
3. amylase
4. bicarbonate
5. Mucins | 6. pharynx
7. esophagus
8. epiglottis
9. gastric
10. secretions | 11. mucus
12. buffering
13. peptic
14. sphincter |

25-III (pp. 278–279)	1. T 2. T 3. F 4. F 5. T 6. salivary glands	7. oral cavity 8. liver 9. stomach 10. gallbladder 11. small intestine	12. large intestine 13. anus 14. pancreas 15. esophagus 16. pharynx
25-IV (p. 282)	1. kilocalorie 2. caloric 3. energy 4. Obesity 5. caloric 6. energy 7. carbohydrates 8. glucose	9. fleshy 10. energy 11. cushions (insulation) 12. anorexia nervosa 13. Bulimia 14. essential 15. saturated 16. cholesterol	17. circulatory 18. essential 19. incomplete 20. vitamins 21. food 22. vitamins 23. minerals 24. blood pressure
25-V (pp. 283–284)	1. molecules 2. fat 3. glycogen 4. glucose 5. glucose	6. amino acids 7. glucose 8. fats 9. glycerol 10. fatty	11. organic substances 12. toxic 13. hormone 14. amino acids 15. urea

Understanding and Interpreting Key Concepts

(pp. 284–286)	1. b 2. b 3. a 4. e	5. b 6. c 7. c 8. b	9. e 10. c 11. e

CHAPTER 26   Circulation

26-I (pp. 288–289)	1. blood vessels 2. closed 3. capillary beds 4. closed 5. lymph vascular 6. pH 7. 5	8. Plasma 9. red blood 10. hemoglobin 11. bone marrow 12. 120 13. scavenge (destroy) (remove)	14. damage 15. bacteria 16. bone marrow 17. platelets 18. clot
26-II (pp. 290–291)	1. pulmonary 2. right 3. carbon dioxide 4. left 5. systemic 6. oxygen 7. right 8. arteries 9. arterioles 10. venules 11. atria 12. ventricles	13. AV valves 14. semilunar valves 15. coronary 16. cardiac 17. AV valves 18. AV valves 19. semilunar valves 20. Ventricular 21. Communication 22. together 23. SA 24. AV	25. ventricles 26. aorta 27. left pulmonary veins 28. left semilunar valve 29. left ventricle 30. lower (inferior) vena cava 31. right AV valve 32. right pulmonary artery 33. upper (superior) vena cava
26-III (pp. 292–293)	1. blood pressure 2. drop (fall) 3. heart 4. artery 5. recoil 6. arterioles 7. pressure (volume) 8. capillary	9. endothelial 10. capillary bed 11. interstitial 12. venules 13. veins 14. Valves 15. contract	16. pressure 17. heart 18. pressure 19. veins (venules) 20. venules (veins) 21. F 22. F

26-IV (p. 295)	1. stroke 2. hypertension 3. heart attack 4. Atherosclerotic plaque 5. low 6. thrombus 7. embolus	8. arrhythmia 9. hemostasis 10. platelet plug 11. coagulation (clotting) 12. markers 13. antibodies 14. markers	15. A 16. B 17. O 18. agglutination (clumping) 19. Rh^+ 20. erythroblastosis fetalis
26-V (pp. 296–298)	1. tonsils 2. right lymphatic duct 3. thymus 4. thoracic duct 5. spleen 6. bone marrow 7. tissues 8. defenses	9. lymph 10. lymph vascular 11. capillaries 12. vessels 13. smooth 14. vessels 15. neck 16. lymphoid	17. lymphocytes 18. lymphocytes 19. marrow 20. node 21. spleen 22. blood 23. lymphocytes 24. thymus

Understanding and Interpreting Key Concepts

(pp. 298–299)	1. d 2. b 3. c 4. e	5. e 6. a 7. b 8. d	9. c 10. a 11. e 12. e

CHAPTER 27 Immunity

27-I (pp. 303–304)	1. pathogens 2. nonspecific defense 3. white blood 4. twenty 5. complement 6. phagocytes 7. attack	8. lysis 9. inflammatory 10. T (B) 11. B (T) 12. immune 13. Macrophages 14. helper T	15. B 16. killer T 17. suppressor T 18. memory 19. MHC 20. antigen
27-II (p. 306)	1. primary immune 2. antibody 3. antigen 4. plasma 5. virgin 6. surfaces 7. interleukin-1 8. lymphokines	9. dividing 10. antibodies 11. macrophages 12. complement 13. tissues 14. surfaces 15. killer T 16. natural killer (NK)	17. infected 18. exposed antigens 19. clonal selection 20. Memory 21. immunization 22. vaccine 23. passive
27-III (pp. 307–308)	1. allergy 2. autoimmune response 3. human immune deficiency 4. male homosexuals 5. vaccine	6. bodily fluids 7. H 8. D 9. E 10. J 11. B	12. G 13. C 14. I 15. F 16. A

Understanding and Interpreting Key Concepts

(pp. 308–309)	1. d 2. b 3. b 4. a	5. a 6. e 7. e	8. c 9. a 10. b

CHAPTER 28 Respiration

28-I (p. 312)	1. gill 2. gas exchange 3. internal 4. across	5. opposite 6. tracheas 7. spiracles	8. swim bladders 9. airways 10. blood vessels

28-II (pp. 313–314)	1. larynx 2. glottis 3. bronchi 4. bronchioles 5. lungs 6. alveoli 7. diaphragm 8. rib cage	9. increases 10. drops 11. pleural sac 12. intercostal (rib) muscles 13. diaphragm 14. pharynx 15. epiglottis	16. vocal cords 17. trachea 18. bronchus 19. bronchioles 20. thoracic (chest) cavity 21. abdominal cavity
28-III (pp. 316–317)	1. alveolus 2. interstitial 3. capillaries 4. opposite 5. hemoglobin 6. 70 7. metabolic 8. greater	9. lungs 10. bicarbonate 11. carbon dioxide 12. carbonic acid 13. bicarbonate 14. red blood 15. 250 16. carbon dioxide	17. bloodstream 18. alveoli 19. lower 20. Bronchitis 21. emphysema 22. Cigarette smoke 23. 80
28-IV (pp. 317–318)	1. air (blood) 2. blood (air) 3. alveoli 4. nervous 5. arterial	6. oxygen 7. carbon dioxide 8. respiration 9. oxygen 10. carbon dioxide	11. lungs (alveoli) 12. Hypoxia 13. hyperventilation 14. carbon monoxide (or carbon dioxide)

Understanding and Interpreting Key Concepts

(pp. 318–319)	1. c 2. c 3. c 4. d	5. a 6. a 7. a	8. a 9. a 10. c

CHAPTER 29   Solute-Water Balance

29-I (pp. 321–323)	1. F 2. F 3. T 4. F 5. T 6. gastrointestinal tract (digestive tract) 7. metabolism 8. excretion 9. "insensible" 10. solutes 11. ammonia 12. urea	13. uric acid 14. kidneys 15. urine 16. volume 17. solute 18. cortex (medulla) 19. medulla (cortex) 20. nephrons 21. water 22. ureter 23. urinary bladder 24. urethra 25. glomerulus	26. Bowman's capsule 27. proximal 28. Henle 29. distal 30. capillaries 31. capillaries (glomerulus) 32. proximal tubule 33. Bowman's capsule 34. distal tubule 35. capillaries (threading around nephron tubules) 36. collecting duct 37. loop of Henle
29-II (p. 325)	1. urine 2. blood pressure 3. Bowman's capsule 4. 45 5. solutes 6. nephron 7. water 8. hypothalamus	9. ADH (antidiuretic hormone) 10. permeable 11. urine 12. inhibited (decreased) 13. sodium 14. renin	15. aldosterone 16. reabsorption 17. water 18. hypertension 19. nephron 20. ions 21. nephron
29-III (pp. 326–327)	1. acid-base 2. pH 3. kidneys 4. hydrogen 5. bicarbonate 6. hydrogen 7. Diabetes	8. autoimmune 9. urine 10. kidney dialysis 11. hypertonic 12. lose 13. drink 14. solutes	15. gain (absorb) 16. hypotonic 17. water 18. urine 19. metabolic 20. concentrated

Understanding and Interpreting Key Concepts

(pp. 327–328)	1. d	5. d	8. b
	2. c	6. d	9. b
	3. e	7. d	10. e
	4. c		

CHAPTER 30 Neural Control and the Senses

30-I
(p. 330)

1. neuron	5. spinal cord	8. Motor
2. three	6. interneurons	9. dendrites
3. Sensory	7. motor	10. axon
4. brain		

30-II
(pp. 332–333)

1. Na^+ (K^+)	10. Na^+	19. node
2. K^+ (Na^+)	11. K^+	20. action potentials
3. action potential	12. K^+	21. chemical synapse
4. repolarized (refractory)	13. resting potential	22. endings
5. resting	14. propagates	23. transmitter
6. trigger	15. refractory	24. contraction
7. threshold	16. Na^+	25. synaptic integration
8. Na^+	17. K^+	26. toxin
9. all-or-nothing	18. myelin	

30-III
(pp. 334–335)

1. synaptic integration	8. action potentials	15. sympathetic
2. neurons	9. ACh	16. parasympathetic
3. local	10. withdrawal	17. sympathetic
4. nerve	11. central	(parasympathetic)
5. reflex arc	12. peripheral	18. parasympathetic)
6. receptors	13. somatic	(sympathetic)
7. motor	14. autonomic	19. Biofeedback

30-IV
(p. 336)

1. spinal cord	9. cerebellum	16. forebrain (cortex)
2. gray matter	10. cerebellum	17. olfactory
3. white matter	11. cerebellum	18. cerebrum
4. brain	12. pons	19. thalamus
5. hindbrain	13. midbrain	20. hypothalamus
6. respiration	14. tectum	21. cerebral cortex
7. blood circulation	15. mammals	22. cerebral cortex
8. medulla		

30-V
(pp. 338–339)

1. D	13. Short-term	23. hallucinogens
2. C	14. long-term	(psychedelics)
3. E	15. reticular activating	24. marijuana smoking
4. A	16. Drugs	25. limbic
5. B	17. hypothalamus	26. hypothalamus
6. cerebrum	18. Stimulants	27. cerebral cortex
7. cerebral cortex	19. Depressants	28. hypothalamus
8. white	20. hypnotics	29. cerebellum
9. Motor	21. analgesics	30. thalamus
10. sensory	22. psychedelics	31. hypothalamus
11. association (cerebral)	(hallucinogens)	32. pons
12. Memory		33. medulla oblongata

1. Receptors
2. stimulus
3. Chemoreceptors
4. mechanoreceptors
5. photoreceptors
6. thermoreceptors
7. brain
8. strong stimulation
9. stimuli
10. Taste
11. taste buds
12. pheromones
13. Somatic
14. referred
15. hair cells
16. sound waves
17. eardrum
18. bones
19. eardrum
20. inner ear (oval window)
21. action potentials
22. brain
23. light
24. photoreceptors
25. brain
26. lens
27. eyes
28. mollusks
29. compound
30. retina
31. cornea
32. rod
33. cone
34. rhodopsin
35. optic
36. hammer (one of the middle ear bones)
37. cochlea
38. auditory nerve
39. eardrum (tympanic membrane)
40. vitreous humor (body)
41. cornea
42. iris
43. lens
44. aqueous humor
45. retina
46. fovea
47. optic nerve
48. sclera

Understanding and Interpreting Key Concepts

(pp. 343–346)

1. d
2. c
3. b
4. c
5. a
6. c
7. a
8. b
9. d
10. e
11. d
12. d
13. e
14. c
15. c
16. e
17. d
18. a
19. c
20. b

CHAPTER 31   Endocrine Control

1. secretions
2. hormones
3. boundaries
4. neural
5. hormones (transmitters)
6. signaling
7. target
8. hormones
9. transmitter
10. signaling
11. pheromones

1. hypothalamus (pituitary)
2. pituitary (hypothalamus)
3. pituitary
4. antidiuretic (ADH)
5. oxytocin
6. Releasing
7. anterior
8. anterior
9. endocrine (target)
10. hormones
11. Prolactin
12. somatotropin
13. target
14. pituitary dwarfism
15. gigantism
16. acromegaly

1. adrenal
2. adrenal cortex
3. hormones
4. glucose
5. medulla
6. epinephrine (norepinephrine)
7. norepinephrine (epinephrine)
8. carbohydrate (glucose)
9. fight
10. flight
11. thyroid
12. Hypothyroidism
13. mental
14. dwarfism
15. iodine (TSH)
16. Goiter
17. calcium
18. Rickets
19. pancreatic islets
20. glucagon
21. insulin
22. somatostatin
23. insulin
24. target (receptor) cells
25. pineal
26. melatonin
27. puberty
28. pineal
29. pituitary
30. parathyroid
31. thyroid
32. adrenal
33. kidney
34. pancreatic islets
35. ovary
36. testis

1. G
2. E
3. J
4. K
5. D
6. A
7. F
8. C
9. H
10. B
11. I
12. L
13. receptors
14. different
15. Steroid
16. hormone-receptor
17. genes
18. alter
19. Testicular feminization
20. Nonsteroid
21. second messengers
22. cAMP
23. amplify

Understanding and Interpreting Key Concepts

(pp. 355–356)
1. d
2. c
3. c
4. d
5. b
6. a
7. c
8. b
9. a
10. c
11. a

CHAPTER 32 Reproduction and Development

32-I
(p. 359)
1. budding
2. fission
3. genetically identical
4. gametes
5. cycles
6. gamete
7. energy
8. external
9. internal
10. yolk
11. gamete
12. fertilization
13. zygote
14. blastula
15. cleavage
16. Gastrulation
17. germ
18. organ
19. differentiation
20. morphogenesis
21. F
22. F
23. T

32-II
(pp. 361–362)
1. gonads
2. accessory
3. gametes
4. sperm development/ production
5. seminiferous
6. epididymis
7. vasa deferentia
8. ejaculatory
9. urethra
10. semen
11. prostate
12. bulbourethral
13. testosterone
14. luteinizing
15. follicle-stimulating
16. hypothalamus
17. hypothalamus
18. testes
19. negative feedback
20. hypothalamus
21. seminal vesicle
22. urinary bladder
23. prostate gland
24. bulbourethral glands
25. epididymis
26. seminiferous tubule

32-III
(pp. 363–365)
1. ovaries
2. estrogen (progesterone)
3. progesterone (estrogen)
4. 400
5. fertilization
6. Oviducts
7. endometrium
8. cervix
9. vagina
10. menstrual
11. follicle
12. ovulation
13. corpus luteum
14. estrogen
15. ovulation
16. endometrium
17. endometrium
18. cycle
19. Endometriosis
20. penis erection
21. coitus
22. orgasm
23. oviduct
24. ovary
25. uterus
26. urinary bladder
27. urethra
28. clitoris
29. labium minor
30. labium major
31. endometrium
32. vagina

32-IV
(p. 367)
1. ovulation
2. oviducts
3. sperm
4. ovum
5. diploid
6. blastocyst
7. embryonic disk
8. embryo
9. extraembryonic
10. amnion
11. chorion
12. endometrium
13. extraembryonic
14. chorionic villi
15. embryonic
16. gastrulation
17. fetus
18. organs
19. lanugo
20. birth
21. uterine
22. dilates (expands)
23. amniotic
24. umbilical cord
25. lactation
26. Prolactin
27. oxytocin
28. brain
29. malformations
30. mental retardation (fetal alcohol syndrome)

32-V
(pp. 369–370)
1. adult
2. aging
3. division
4. fertilization
5. zygote
6. abstention
7. vasectomy
8. tubal ligation
9. sexually transmitted diseases

Understanding and Interpreting Key Concepts

(pp. 370–371)
1. e
2. b
3. c
4. d
5. e
6. a
7. b
8. c
9. e
10. d
11. a
12. d
13. a
14. b
15. c

CHAPTER 33   Population Ecology

33-I (p. 377)	1. C 2. D 3. E 4. A 5. B 6. dynamics	7. size 8. density 9. dispersion 10. age structure 11. *r* 12. exponential	13. biotic 14. limiting 15. resistance 16. carrying capacity 17. Logistic
33-II (pp. 378–379)	1. Density-dependent 2. Density-independent 3. density-independent 4. density-dependent	5. Survivorship 6. I 7. II 8. III	9. III 10. I 11. II 12. table
33-III (p. 379–380)	1. 5.3 2. 6 3. natural controls	4. habitats 5. climatic 6. carrying capacities	7. limiting factors 8. Pollutants 9. exponential
33-IV (p. 381)	1. demographic transition 2. preindustrial 3. transitional 4. industrial	5. postindustrial 6. industrial 7. transition 8. family	9. China 10. older 11. growth 12. carrying capacity

Understanding and Interpreting Key Concepts

(pp. 382–383)	1. c 2. d 3. a 4. b	5. d 6. d 7. d 8. b	9. b 10. a 11. d 12. b

CHAPTER 34   Community Interactions

34-I (pp. 385–386)	1. niche 2. community 3. habitat 4. species	5. populations (species) 6. dispersion 7. neutral 8. commensalism	9. mutualism 10. interspecific 11. predation (parasitism) 12. parasitism (predation)
34-II (p. 387)	1. mutualism 2. interspecific	3. exploitative 4. interference	5. competitive exclusion 6. growth
34-III (pp. 388–389)	1. Predators 2. Parasites 3. carrying capacity 4. reproductive 5. adaptive capacity 6. cycling 7. coevolution 8. moment-of-truth	9. warning 10. mimicry 11. camouflage 12. parasites 13. Parasitoids 14. social 15. B 16. C	17. F 18. A 19. A 20. D 21. D, G 22. G, H 23. E 24. A, I
34-IV (pp. 390–391)	1. stable 2. partition 3. resource partitioning 4. Natural selection 5. dissimilar	6. resource partitioning 7. competition 8. reduced 9. ecological	10. water hyacinth 11. honeybees 12. extinction 13. natural
34-V (pp. 392–393)	1. succession 2. climax 3. Primary 4. secondary 5. local	6. cyclic 7. fire 8. increase 9. fewer 10. more	11. lower 12. equator 13. rainfall (sunlight) 14. sunlight (rainfall) 15. speciation

Understanding and Interpreting Key Concepts

(pp. 393–395)	1. b 2. b 3. c 4. a	5. b 6. c 7. b 8. b	9. c 10. a 11. d 12. a

CHAPTER 35   Ecosystems

35-I
(pp. 397–398)

1. autotrophs
2. energy (sunlight)
3. nutrients
4. herbivores
5. carnivores
6. parasites
7. omnivores
8. decomposers
9. detritivores
10. ecosystem
11. energy
12. materials (nutrients)
13. open
14. energy
15. nutrient
16. energy
17. Nutrients (materials)
18. trophic level
19. food chain
20. food webs

35-II
(p. 399)

1. primary
2. Gross
3. Net
4. energy
5. grazing
6. herbivores
7. detrital
8. pyramid
9. producers
10. producers
11. energy pyramid

5-III
(p. 401)

1. nutrient
2. thirteen
3. phosphorus
4. Biogeochemical
5. hydrologic
6. atmospheric
7. sedimentary
8. hydrologic
9. Watersheds
10. nutrients
11. increases
12. reservoirs (gases)
13. reservoirs
14. organic
15. rapid
16. slow
17. greenhouse

35-IV
(pp. 402–403)

1. Bacteria
2. nitrogen fixation
3. Ammonification
4. ammonia (NH_3)
5. ammonium (NH_4^+)
6. Nitrification
7. nitrite (NO_2^-)
8. nitrate (NO_3^-)
9. denitrification
10. fertilizers
11. biological magnification
12. Ecosystem

Understanding and Interpreting Key Concepts

(pp. 403–404)

1. c
2. c
3. b
4. b
5. a
6. c
7. a
8. d
9. d
10. b
11. b

CHAPTER 36   The Biosphere

36-I
(pp. 407–408)

1. biosphere
2. hydrosphere
3. atmosphere
4. climate
5. atmosphere
6. ocean
7. semiarid
8. trees
9. subalpine
10. subalpine
11. rain shadow
12. seasons

36-II
(pp. 409–410)

1. vegetation
2. species (plants)
3. biogeographic realms
4. Biomes
5. Deserts
6. dry shrublands
7. dry woodlands
8. Grasslands
9. Shortgrass
10. Tallgrass
11. Tropical
12. monsoon
13. Humus
14. Loam
15. Topsoil

36-III
(pp. 411–412)

1. Evergreen broadleaf
2. rainfall
3. humidity
4. rapid
5. deciduous broadleaf
6. monsoon
7. temperate deciduous
8. Evergreen coniferous
9. Conifers
10. boreal (taiga)
11. Montane coniferous
12. temperate
13. pine barrens
14. Arctic tundra
15. permafrost
16. Alpine

36-IV
(p. 413)

1. littoral
2. limnetic
3. profundal
4. overturns
5. thermocline
6. Oligotrophic
7. Eutrophic

36-V
(pp. 414–415)

1. estuary
2. intertidal
3. upper littoral
4. mid-littoral
5. lower littoral
6. Sandy (Muddy)
7. muddy (sandy)
8. detrital
9. pelagic
10. neritic
11. oceanic
12. benthic
13. hydrothermal vents
14. Upwelling

Understanding and Interpreting Key Concepts

(pp. 415–417)
1.	a	6.	a	11.	e
2.	c	7.	b	12.	c
3.	d	8.	b	13.	d
4.	b	9.	b	14.	b
5.	d.	10.	c	15.	c

CHAPTER 37 Human Impact on the Biosphere

37-I
(pp. 419–420)
1.	pollutants	8.	PANs	14.	Acid rain
2.	Thermal inversions	9.	deposition	15.	ecosystems
3.	Industrial smog	10.	sulfur	16.	ultraviolet
4.	Photochemical smog	11.	nitrogen	17.	Skin
5.	Nitric	12.	dry acid (wet acid)	18.	Chlorofluorocarbons
6.	nitrogen dioxide	13.	wet acid (dry acid)		(CFCs)
7.	nitrogen dioxide				

37-II
(p. 422)
1.	salination (salt buildup)	7.	downstream	12.	Nutrients
2.	waterlogging	8.	throwaway	13.	slash-and-burn
3.	Primary	9.	marginally	14.	extinction
4.	Secondary	10.	Deforestation	15.	Desertification
5.	Tertiary	11.	rapid decomposition	16.	overgrazing
6.	upstream				

37-III
(p. 424)
1.	energy	6.	greenhouse	11.	Fusion
2.	Net	7.	low	12.	nuclear winter
3.	Fossil	8.	high	13.	anticipate
4.	kerogen	9.	Meltdowns	14.	ecosystems
5.	coal	10.	Breeder		

Understanding and Interpreting Key Concepts

(pp. 425–426)
1.	a	6.	a	11.	d
2.	b	7.	b	12.	a
3.	a	8.	a	13.	c
4.	b	9.	c	14.	d
5.	a	10.	c		

CHAPTER 38 Animal Behavior

38-I
(pp. 429–430)
1.	behavior	5.	genetic	8.	Instinct
2.	responses	6.	organizes	9.	shape
3.	Heredity	7.	activates	10.	red
4.	Learning				

38-II
(p. 431)
1.	Learning	6.	associative	11.	Imprinting
2.	genetically	7.	insight	12.	compass (navigational)
3.	Associative	8.	Imprinting	13.	navigational (compass)
4.	trial-and-error	9.	human	14.	imprint
5.	Insight	10.	imprint	15.	stars

38-III
(p. 433)
1. D	2. A	3. E	4. B	5. F	6. C

38-IV
(pp. 434–435)
1.	B	6.	communicate	11.	visual
2.	A	7.	social	12.	bioluminescent
3.	D	8.	communicate	13.	acoustical
4.	C	9.	communication signal	14.	tactile
5.	Social	10.	chemical		

38-V
(p. 436)
1.	costs (benefits)	5.	victim	9.	protection
2.	benefits (costs)	6.	dilution	10.	reproduction (mating)
3.	reproductive success	7.	selfish	11.	protection
4.	Predation	8.	self-sacrifice		

38-VI
(pp. 437–438)

1. altruistic
2. natural selection
3. kin selection
4. genes
5. suicide
6. queen
7. adaptive
8. reproductive
9. cultural
10. biological
11. behavioral
12. genetic
13. emotions
14. proteins
15. behave
16. reproduction (survival)
17. survival (reproduction)
18. moral
19. transmission

Understanding and Interpreting Key Concepts

(pp. 438–440)

1. e
2. b
3. b
4. b
5. d
6. c
7. a
8. a
9. c
10. b
11. d
12. e
13. c
14. c
15. e

Crossword Number One

¹E	C	O	²S	Y	³S	T	E	⁴M		⁵E		⁶S	U	P	E	⁷R	N	A	⁸T	U	R	A	⁹L		¹⁰O	R	¹¹D	E	¹²R
N		R			R			O		X		H				T			W				¹³A		U		E		S
¹⁴E	R	G		¹⁵C	L	A	S	S		P		¹⁶G	R	O	U	¹⁷P		¹⁸P	O	P	U	L	A	T	I	O	N		S
R		A			I		²¹E	Y	E			O				R		F				O			R		P
²²G	E	N	U	²³S		T		S		²⁴R	A	N	D	O	M	I	Z	²⁵A	T	I	O	N		M		²⁶G	O	B	I
Y			T			I		E		S		N		T			A			R
	²⁷M	²⁸E	T	A	²⁹B	O	L	I	S	M		³⁰C	O	M	M	³¹U	N	I	³²T	Y		³³N	O	V	A
³⁴A		G		³⁵S	O			³⁶T		N		I		N		E		T
³⁷T	A	R		³⁸I	N	D	E	³⁹P	E	N	D	E	N	T		⁴⁰P	R	O	T	I	S	T	A		⁴²P	L	A	N	I
	⁴³E	A	S	E		O		T		S		H		L		B		T		C		L		O
⁴⁴S		T			⁴⁵P	L	A	N	T	A	E		E		J		Y		⁴⁶B	U	D	G	E	⁴⁷T		N
U			⁴⁸T	I	S	⁴⁹S	U	E		L		S		E		P		A
B			A			⁵⁰S		⁵¹I	N	D	U	C	T	I	O	N		⁵⁴T	H	E	O	R	⁵⁵Y		⁵⁶
⁵⁷J	A	F	F	A		⁵⁸M		⁵⁹T		⁶⁰D	E	S	N	T		W		R		A
E		U		⁶¹P	A	R	T	I	C	L	E		⁶²V	A	R	I	A	B	L	E		⁶³D	A	R	W	I	N
⁶⁵C	O	N	T	R	O	⁶⁶L		A		F		E		V		I		O		G
T		G		⁶⁷S	I	G	N	I	F	I	C	A	N	T		⁶⁸D	E	C	O	M	⁶⁹P	O	S	E	R
I		I			C		N		S		E		T		⁷²P		S		R		T		⁷³A
V		⁷⁴D	E	N	G		⁷⁵F	E	R	R	I		⁷⁶S		R		⁷⁷R	E	P	R	O	D	U	C	⁷⁸T	I	O	N
E		⁷⁹T		L		E		E	O		P	O		H		V		N		A		I
⁸⁰P	H	Y	L	U	M		⁸¹R	E	N	I	N		⁸²E	N	D	O	R	S	E		⁸³O	R	G	A	N	I	S	M
⁸⁶L	I	E		L			T		C	U		R		I		G		A
I		⁸⁷S	Y	S	T	E	M	⁸⁸A	T	I	C		⁸⁹F	I	N	C	H		⁹⁰D	E	D	U	C	T	I	O	N		L
F		I		R		I	A			E		E		O		E		I
⁹¹E	O	S		⁹²N	A	T	U	R	A	L		⁹³O	B	S	E	R	V	A	T	I	O	N		⁹⁴M	O	N	E	R	A

Crossword Number Two

1 CHLOROPLAST
5 CHLOROPHYLL
9 CYTOPLASM
13 ELECTRON
15 CARBOHYDRATE
17 PHOSPHATE
19 ZIT
20 COA
21 GAS
22 STARCH
24 GRADIENT
25 LUX
26 NADH
28 MTOC
33 SHY
34 MITOCHONDRION
36 DOG
37 KREBS
38 ACTIVE
40 ATP
41 PYRUVATE
43 ELEMENT
47 ORGANIC
48 CONDENSATION
50 PSI
55 ADENOSINE
56 EVAPORATION
58 CAROB
59 NIL
61 OZONE
62 NANOMETER
64 CHEMOSYNTHESIS
66 HYDROPHILIC
67 PHOTORESPIRATION
68 POLAR
70 AMINO
72 LACTATE
75 ZINC
80 PHOSPHORYLATION
81 ISOTONIC
84 NUT
86 ACID
88 HYPERTONIC
89 BOND
90 PROTON
92 OXIDATION
93 BIOSYNTHETIC
96 STROMA
97 DUO
98 BIOLUMINESCENCE
99 ADP
101 TURGOR
102 CHEMIOSMOTIC
103 CAROTENOID
104 NUCLEOLUS

Crossword Number Three

```
A   T R A N S C R I P T A S E     C L E A V A G E     A B N O R M A L
U       L   O       R         W H       U     M             A   O
T E M P L A T E   C O D O M I N A N T     C A R C I N O G E N     C
O       U   E   H   T       N   I     B   N     N       E     U
S E T T L E D   I   O N C O G E N E     R   I     O O G E N E S I S
O       A   E   I   S       L       A   N           E
M A G U S   F     T R I S O M Y   G   N U C L E O S O M E   R F L   P
A       E   F   O       N       E   U   E       I       R
L I N K A G E   N O I S E     U R A C I L     E   M R N A       O
            R   E     C   M   L   E       N   U   C       M
P   E X P R E S S I O N   X Y Y   F E R T I L I Z A T I O N     O
L   U   I   N     V   I   T   T   O       Y   A   M       T
E   G E N E T I C   A M N I O C E N T E S I S   M E T A P H A S E
I   E   U   I     T   K   I   E   P   E   I   L       R
O W N   P L A S M I D   R   I   P E D I G R E E S   O P E R O N
T   I   T       O   N   O   E   R   R     N   T       C
R E C E S S I V E   P U N N E T T   S P E R M   C   G E N O M E
O         O   R         S       G   A   V           N
P   A C H O N D R O P L A S I A   G A L A C T O S E M I A     T
Y   E     O   O       S       M   T   O       A       R
  O P E R A T O R   L I M P   S A M P L I N G   D I P L O I D   O
    O   E     Y   E     P   N   L   O       R   A       M
C O D O N   R E C O M B I N A N T   I   N O N S I S T E R   V   E
A   N     M   L   E   O   I   I   F   E   P     S   A   R
N   A D E N I N E   R E S T R I C T I O N   S I S T E R   T R U E
C   L   N   F   A   I       O   C     I     L   I
E     M E T A S T A S I S   C D N A   B A S E   D O M I N A N T
R   T   M   T     E     I   I O   T     A     G   T   R
  B R E E D I N G   B E N I G N   I   T R A N S L O C A T I O N
  I   N   O         C   A     O   N   H     S   O   A
C O N T I N U O U S   C H O R I O N I C   T H Y M I N E   N
```

Answers   471

Crossword Number Four

1 Across: BIPEDALISM **4:** EQUILIBRIUM **7:** PLATE **9:** SOIL **10:** PITH **13:** HYBRID **17:** HOMINID **19:** LAPEL **20:** LILY **21:** GRADUALISM **23:** WHO **26:** SAPIENS **27:** LXI **28:** MUTATION **30:** GRAIN **32:** SAL **33:** DIVERGENCE **34:** DIFFERENTIAL **37:** PUNCTUATION **38:** ALLELE **39:** GELEE **41:** YET **43:** TEAL **44:** MACROEVOLUTION **47:** STABILIZING **51:** SPECIATION **52:** SHE **54:** LI **56:** TECTONICS **58:** KEY **59:** SARI **60:** HOMO **61:** DIRECTIONAL **63:** FREQUENCY **64:** ENOLA **66:** ADAPTIVE **68:** MESOZOIC **69:** OAT **71:** NEUTRAL **73:** EXTINCTION **77:** PRIMATE **78:** DISRUPTIVE **80:** HOMOLOGOUS **81:** CONVERGENCE **82:** MASS **83:** POPULATION

Crossword Number Five

1 Across: EYESPOT **5:** MESODERM **7:** CHEMOSYNTHESIS **12:** GUT **13:** CORD **14:** EUBACTERIA **17:** RUM **18:** LYSOGENIC **20:** FLAME **22:** PERITONEUM **23:** MANTLE **25:** CARAT **26:** OUGHT **30:** ACORN **31:** TABOO **32:** DAM **33:** PARASITES **34:** COLON **36:** ARCHAEBACTERIA **40:** AMNIOTIC **41:** CUTICLE **42:** I **43:** METHANOGENS **44:** VIRUS **48:** SPORE **49:** DRONE **50:** RANT **53:** MAPLE **55:** SLITS **56:** NOTOCHORD **58:** SYMBIOSIS **60:** GILLS **62:** MICROBE **64:** METAMORPHOSIS **65:** GYMNOSPERMS **66:** PHARYNX **67:** MILLIPEDE

Crossword Number Six

COHESION · CAP · GUARD · A P A W · SUM
CA · A · SO · E · I B · HIPPO · H Y
S · MICROSPORE · R · CC · S S O · I O · O C
POR · O K · M A O C · T C D · O O
AN · EMIGRE · SCHISM · T I · OVARY · TAR
ROOT · N O F · N B · S P L · I C
IC · S T · POLLINATION · SEED · ISIS · H I
ANODE · E H O T U · I R · I T O Z A
N T N · RAY W · THIGMOTROPISM · M O U Z A
E N T · R O · N O E M R A
MEGASPORE · GRAIN · L · DORMANCY · E B
A C D · R N · E S · I I · E U
M E E · TRANSLOCATION · SYSTEM · D
E N E · V P V F N · CM T E · S D
CUTICLE · A I I U K · E E A
O O E N · C T R L · P · PHLOEM · VIP · P
TAP · D · HERBACEOUS · O L · A · R
Y H O E O T N L · S E · E
L Y A · STRIP I · F D · FLOWER · C M S
E T X P · I O R L E · P U B U
D · EPIDERMIS · NAUSEA · NODE · I L R R
O A R M I I · T A Y Y
N · XYLEM · PHOTOTROPISM · HORMONE

Crossword Number Seven

AXON · LIMBIC · ADIPOSE · R · ARTERIOLE E
D · V · L I G N · N E P · Y
REFLEX · O · RAGES · N · CAPILLARY · E
E T O C L U E O E E E
NERVOUS · D U U · LIP M · P N · KIDNEY S
A E O L T I H E T D · S K
L · INFLAMMATION · RETROVIRUS · T E
N A T N O R R C · I R L
SAC · FERTILIZATION · I · AUTOIMMUNE
O O T N U · EL · M S T
MAP · ANTIGEN · INTEGUMENTARY · U O
A U N E D R N N
I L G · MECHANORECEPTOR · HEART
I M L O O V I C F I E C
CLONAL · THERMORECEPTOR · IONA · OSLO
N N E M B R S U I L T C P M
ALDOSTERONE · ORION T E O I P
D R T N S E P T E R G R R L
ERYTHROCYTE I B H A · MACROPHAGE
R M T R A R L T A U T M
MHC · SALIVARY · ANGRY U · INTESTINE
I A C L U N O O O N
SYMPATHETIC · ESTROGEN · RAS · NET

Crossword Number Eight

```
 D E M O G R A P H I C     C     P R E D A T O R       B I O S P H E R E
 I       E                 O     R         N       B               H       X
 V       L         O C     D E P E N D E N T       N I Ñ O     T R O P H I C
 E       T         C         V   Y     E U   F   O M     E         F O O D   U
 R     D I S P E R S I O N   L         S A   A   T     N     C     R       S
 S       O         A   E     L   P A R A S I T I C     C   I   H   O       I
 I     W E B       N   L     U       T       T   O     A   T   E   T       O
 T I N         E   C   C     M I M I C R Y     P U R I M       H E N
 Y         W     T O O T H   H         F N   A       I     E
 H A B I T A T   I         I   I N T E R S P E C I F I C   R       D
 I         U     O         E   C     E       I   I   A   M   E
 F I X A T I O N   P E R M A F R O S T     T   C   L E A R N
 D         R     O         A   T   T           Y   A       L       S
 E C O S Y S T E M   P A R T I T I O N I N G     T   I       I
 C         E     U         C   O   D       S     I   L       T
 O     P O T E N T I A L   H   N   A D A P T A T I O N   I       Y
 M         R     E         A   Y   L       I       N   M
 P R O D U C T I V I T Y     N   P I O N E E R     C H A I N
 O         N     L         I   Z E       A   C           T
 S Y M B I O S I S   O V E R T U R N     T H E R M O C L I N E
 E         T     A         R   U       A       N
 R A I N S H A D O W   C O M M U N I T Y   H Y D R O L O G I C
```